Neuroethological Studies of Cognitive and Perceptual Processes

Neuroethological Studies of Cognitive and Perceptual Processes

EDITED BY

Cynthia F. Moss
University of Maryland at College Park

and

Sara J. Shettleworth
University of Toronto

Routledge
Taylor & Francis Group
New York London

First published 1996 by Westview Press

Published 2019 by Routledge
52 Vanderbilt Avenue, New York, NY 10017
2 Park Square, Milton Park, Abingdon, Oxon OX14 4RN

First issued in hardback 2019

Routledge is an imprint of the Taylor & Francis Group, an informa business

A CIP catalog record for this book is available from the Library of Congress.

ISBN 13: 978-0-367-31700-3 (hbk)
ISBN 13: 978-0-8133-2655-9 (pbk)

Contents

PART THREE
Specializations for Spatial Perception and Orientation

1

Introduction: Neuroethology, Perception, and Cognition

Sara J. Shettleworth

Abstract

Neuroethology is the branch of behavioral neuroscience concerned with analyzing the neural circuitry of naturally occurring behaviors. Perhaps uniquely among the specialized subfields that have grown out of ethology, neuroethology retains a commitment to an integrated approach to behavior, combining Tinbergen's (1963) four questions, about causation, function, evolution, and development. Inherent in the neuroethological research program is a tension between studying ethologically meaningful behaviors and using artificial situations to dissect them. The chapters in this book illustrate a range of ways to resolve this tension. Some of them provide evidence that the causal factors important in the field can remain surprisingly relevant in very artificial contexts. A second potential tension is that generated by studying the impressive specializations of particular species while searching for scientific generalities. Several chapters illustrate how choosing species with particular specializations provides a means to address general issues about perception and cognition.

Introduction

Neuroethology is firmly grounded in Tinbergen's (1963) "four whys" of ethology: the "whys" of immediate causation, development in the individual, evolution, and current function. While trying to answer the causal "why" question in the most complete and detailed sense possible, by tracing the neural circuitry of ethologically meaningful behaviors, neuroethologists also have in mind the other three whys. As the chapters in this book illustrate, this means that neuroethological research takes place in a context of information about the natural history, evolution, and development of the species under study.

Much neuroethological research analyzes the causation of particular behaviors in particular animals. How do bats catch insects in the dark? How do birds learn to sing? How do frogs hear potential mates and fish see potential rivals? Questions like these are distinctively different from those in the rest of behavioral neuroscience which commonly concern general issues about structures and neural function: What does the hippocampus do? What synaptic changes underlie learning? Many such questions, like the one about hippocampal function, are most relevant to mammals, and they are most often studied in convenient laboratory species like rats, mice, and monkeys. In contrast, a hallmark of neuroethology is its eclectic choice of species. Bats, frogs, owls, and bees, all are candidates for a neuroethological analysis. Solving a functionally important problem like catching prey or finding mates in a striking and unusual way sometimes seems to be one of the chief qualifications for the study of particular species.

This book contains a sample of current research on the neuroethology of cognitive and perceptual processes. A more general book on neuroethology might also have included research on motor systems and motivation. Research on a wide range of species and information-processing problems is represented here, including learning about flowers in bees (Gould, Chapter 2), classical conditioning in invertebrates (Colwill, Chapter 3), song learning, neighbor recognition, imprinting, food-storing and migrating in birds (Chapters by Cynx, Weary, Bolhuis, and Healy), processing of calls by frogs (A. Simmons and Buxbaum, Chapter 9), and processing of visual signals by fish (Fernald, Chapter 10). One whole section of the book (Chapters 11-13) describes mechanisms for auditory localization of prey in two now-classic cases, bats and owls. The chapters vary in their relative emphasis on *neuro* and *ethology*. Some chapters that are almost entirely about behavioral work emphasize how a thorough understanding of phenomena at that level is a necessary underpinning to sound analyses of neural mechanisms. For instance, Colwill (Chapter 3) prescribes some previously under-emphasized criteria for choosing appropriate preparations in neural analyses of associative learning. Dooling et al. (Chapter 6) describe behavioral methods for detecting species-specific processing of vocal signals, methods that could be used to select stimuli for later neural analyses of possible specialized perceptual mechanisms. A wide range of methods from neuroscience is also represented, including lesions, electrophysiology, hormonal manipulations, neuroanatomy and, in Healy's chapter, the comparative evolutionary approach.

Bats, bees, and songbirds inhabit a very different world from ours (Boden, 1988). Each of these worlds poses its own challenges to the person attempting to understand it. Nevertheless, the research unraveling the diverse animals' specialized cognitive and perceptual abilities that is represented in this book does have common themes. Some of these arise out of two sorts of tensions inherent in the neuroethological research program. One is that between keeping the focus of research on behavior as it takes place in its natural context and using artificial situations to dissect it. The other tension is that between fascination with the remarkable specializations of individual species and science's equally compelling search for generalities. Ways in which these apparent conflicts are productively resolved can be seen throughout the book.

The Field and the Lab

The most obvious way to avoid the potential conflict between dissecting behavior and keeping it "natural" is to do one's experiments in the field. Gould's chapter on flower learning by honey bees exemplifies this approach. Using von Frisch's time-honored technique, Gould entices foraging bees to artificial flowers and then manipulates the flowers' characteristics to discover what the bees are responding to. It is but a small step from here to teaching a hungry forager how to find food in the laboratory, as described in the chapters by Moss and Simmons and by Haresign et al. on audition in bats. It seems reasonable to assume that these animals are using their abilities to the fullest whether they are sitting on a platform locating phantom targets for a reward of mealworms or flying around outside catching real insects.

Students of bird song face a more difficult task than people wishing to analyze the sensory and neural control of foraging, because singing takes place in a more limited context than foraging. Someone like Weary (Chapter 7), wishing to understand how a male songbird's territorial defense reflects his ability to recognize his neighbors will have as background the results of experiments in which birds have been induced to respond to recorded songs in the field. However, as Weary points out, recognition cannot be separated from response in the field. If a subject responds to two songs in the same way, this may not mean he does not recognize them as coming from two different individuals. Moreover, because the females' responses to song are often difficult to observe, whole realms of questions about song recognition are impossible to answer in the field. And, unlike the case with the fish described in Fernald's chapter, the questions that can be asked about captive birds are limited by the unwillingness of many species to breed in captivity. (Notice that the willingness of budgerigars to carry out their whole life cycle in the laboratory is one feature making them attractive subjects for the studies described by Dooling et al. in Chapter 6.)

Until recently, most behavioral work on song learning was based on analyzing the songs of adults that had been exposed to different early experiences (see review in Chapter 4). As in field studies of song recognition, with this method performance (here, singing a given song) is equated with learning. However, the three chapters in this book on bird song describe experiments in which evidence of learning and discrimination has been obtained independently of singing and territorial behavior by training hungry birds to work for food with songs as discriminative stimuli. But when the animal's primary motivation as well as its environment have been changed, how can we be sure of what we are measuring? Cynx (Chapter 4), Dooling et al. (Chapter 6), and Weary (Chapter 7) all meet this question head on, and all answer, in effect, that "the proof of the pudding is in the eating." In other words, performing a food-rewarded auditory discrimination can validly be used to study the song system if the operant behavior is sensitive to the same variables that affect singing or response to song. And food-rewarded operant behavior is surprisingly sensitive to such variables.

Cynx (Chapter 4) directly addresses this issue in his experiments on song discrimination learning in zebra finches. For instance, he shows that the speed with which these birds learn a discrimination between two zebra finch songs depends on the subjects' gender and the season. The latter is a surprising result because zebra finches had been thought to be opportunistic breeders, not influenced by seasonal cues such as day length. As Cynx describes, his findings have stimulated a re-examination of the factors that control readiness to breed in this species. Further, the fact that zebra finches' discrimination between zebra finch songs is influenced by hormone levels, whereas their discrimination between canary songs is not, suggests that the neural system for acquisition and production of species-specific song is somehow involved in song discrimination learning in adult zebra finches. The fact that a subject's own song is especially salient in food-rewarded operant song discriminations also supports this conclusion.

Accepting that food-rewarded responding in a laboratory is controlled by the same variables that influence behavior in functionally meaningful situations in the wild opens up for the researcher a whole kit of well-established tools for analyzing cognitive and perceptual processes. The cognitive mechanisms used in mate choice or territorial defense become amenable to the same methods applied to "general" cognitive processes in humans and other species. Dooling et al. (Chapter 6) provide one of the more sophisticated examples. Using a clever behavioral test, they have gathered data suitable for multidimensional scaling to reveal how birds categorize the vocalizations and faces of their own and other species. Work of this kind will likely be used more and more to address issues in behavioral ecology concerning how the forms of animal signals are perceived by their receivers and how they may have evolved (cf. Guilford & Dawkins, 1991).

Specializations and Generalities

Nearly every chapter in this book contains examples of animals' biases to learn best about or respond best to species-relevant events. Gould (Chapter 2) describes honey bees' bias to learn about the best cues to nectar. Dooling et al. (Chapter 6) describe how budgerigars, canaries, and zebra finches each discriminate better among calls of their own species than among songs of the other two species. A perceptual specialization for processing species-specific sounds is also evident in the frogs discussed by A. M Simmons and Buxbaum (Chapter 9), though its neural and developmental basis may be different. Moss and Simmons (Chapter 11), Haresign et al. (Chapter 12), and Moiseff et al. (Chapter 13) describe the auditory specializations of bats and owls for nocturnal foraging. As Moiseff et al. discuss, these are necessarily somewhat different because echolocating bats themselves are generating the sounds they use for localization whereas owls must listen for sounds made by the prey.

In his chapter, Gould refers to both biases and specializations. By a bias, Gould means a tendency to selectively respond to or learn about some features of flowers. Any perceptual system has biases, in that it does not respond equally to all stimuli. But, as Gould points out, learning theorists in psychology have traditionally assumed that learning is without bias. Whatever the animal could perceive it could learn about, more or less indifferently. Gould reviews developments within the last 20 years that have led to modification of this belief and acceptance by many psychologists that learning, like perception, is fine-tuned to the requirements of each species' ecology (for further discussion see Rozin & Schull, 1988). Attempts to document this fine-tuning, however, are fraught with difficulties. For the implication is that the observed biases are adaptive specializations (Rozin & Kalat, 1971). Supporting this conclusion is as difficult as supporting any hypothesis about adaptation in evolutionary biology (see Williams, 1966). It may appear self-evident that the fit between the details of some behavioral mechanism and the animal's natural history is too intricate to have arisen by chance. Ideally such arguments for adaptation can be taken further with comparative studies involving large numbers of species, as Healy's chapter illustrates. However, in behavioral work at least, comparisons can also be done within a species to test assertions about the specialization of a given system to do a particular job.

It is obvious, for example, that the visual system is specialized for processing light and the auditory system, for sound. The eyes simply don't respond to sound nor the ears to light. However, it is less obvious that there are specialized systems for learning and cognitive processing. Suggestions that such specializations exist have occasioned much debate among those concerned with both animal and human cognition. Just as in perception, however, learning and cognition can pose a number of different functionally incompatible requirements, and these would be expected to lead to the evolution of separate, adaptively specialized, memory systems within the same animal (Sherry & Schacter, 1987; Shettleworth, 1993). In their development of this idea, Sherry and Schacter (1987) cite as an example the different requirements of song learning and food-storing in some passerine species. In song learning in many species, passive exposure to song at a restricted period early in life results in learning which is retained for many months, if not the bird's lifetime. In food-storing, at least in some species, the locations of a constantly changing inventory of food caches must be remembered for a few days or weeks and then forgotten. Similar functional incompatibilities can be identified among candidates for separable memory systems in people. Healy (Chapter 5) discusses anatomical evidence for separable song and spatial memory systems in birds and how evolution has fine-tuned them for the requirements of each species.

To these specialized memory systems in birds can be added the recognition system involved in imprinting, as discussed by Bolhuis (Chapter 8). He describes a particularly well worked out example of a species-specific bias or predisposition. Young chicks will become filially attached to a wide variety of artificial stimuli in the laboratory. However, they show filial behavior most readily to a stuffed hen of the Burmese junglefowl, the ancestral form of modern domestic chickens. This does not mean, though, that they learn most quickly about junglefowl. Rather, a predisposition to respond to junglefowl hens combines with a learned preference for a familiar imprinting object. As Bolhuis shows, this species-specific predisposition itself depends in subtle ways on age and experience.

Once it is accepted that cognitive and perceptual processes are the products of evolution (and ethologists have always accepted this; Fernald's chapter can be consulted for a fine example), it becomes obvious that the same forces producing specializations and differences among species are also responsible for generalities. This point is made forcefully by Weary (Chapter 7) in his discussion of the types of stimuli birds use to recognize their own species' songs. He suggests that both similarities and differences can be understood in terms of what sounds are transmitted best in the bird's natural habitat. Taking this line of argument one step further, Simmons and Buxbaum (Chapter 9) propose that general issues can be better understood by selecting "special" species. In the case they describe, studies of how frogs perceive pitch are instructive because these animals have available to them only one of the theoretically possible mechanisms that any animal could use. And of course what makes these frogs suitable for studies of pitch perception in the first place is their demonstrated ability to use the pitch of calls in intraspecific aggressive and sexual contexts. That is to say, general issues in neuroscience and cognition can be addressed by neuroethological studies of species that confront given cognitive problems in particularly demanding ways. "Specializations," in this approach, are general processes writ large. The neuroethologist's zoo is the ideal laboratory for studying them.

Ethology, Neuroethology, and Behavioral Neuroscience

Just like this introductory chapter, most introductory texts on animal behavior or ethology begin with Tinbergen's four questions, the questions of immediate causation, function, evolution, and development. It is usually pointed out that, in Tinbergen's (1963) view, a complete understanding of behavior includes answers to all four of them. However, while such an integrated approach to animal behavior may still be promoted in textbooks, the 1980's and 90's have seen a fragmentation of the field at the level of professional organizations, journals, and general communication that would horrify Tinbergen even as the findings of some of these new subfields might fascinate him. (See Dawkins, 1989, for a discussion of this point with respect to behavioral ecology and ethology.) Besides neuroethology, we now have behavioral ecology, sociobiology, comparative psychology, developmental psychobiology, cognitive ecology (Real, 1993), and cognitive ethology (Ristau, 1991), not to mention the comparative aspects of cognitive neuroscience, among other named specialized subfields. However, many of the chapters in this book show very clearly that neuroethologists — perhaps uniquely among behavioral neuroscientists — retain a commitment to an integrated approach to behavior, to combining evolutionary, functional, and developmental information to illuminate neural mechanisms of perception and cognition.

Acknowledgment

Camera ready copy for all the chapters was expertly and patiently created by Charles Tuchinda. The editors and authors are very grateful for his help. Some of the chapters are based on material presented at the 14th Symposium on Quantative Analyses of Behavior held at Harvard University in June, 1991. Those who participated thank Michael Commons and the late R. J. Herrnstein for arranging that occasion.

C.F.M. and S.J.S.

References

Boden, M. A. (1988) Artificial intelligence and biological intelligence. In H. J. Jerison & I. Jerison (Eds.), *Intelligence and evolutionary biology*, (pp. 45-71) Berlin: Springer-Verlag.

Dawkins, M. (1989) The future of ethology: How many legs are we standing on? *Persp. Ethol. 8*: 47-54.

Guilford, T., & Dawkins, M. S. (1991) Receiver psychology and the evolution of animal signals. *Anim. Behav. 42*: 1-14.

Harvey, P. H., & Pagel, M. D. (1991) *The comparative method in evolutionary biology.* Oxford: Oxford University Press.

Real, L. (1993) Toward a cognitive ecology. *Trends Ecol. Evol. 8*: 413-417.

Ristau, C. (Ed.) (1991) *Cognitive ethology: The minds of other animals.* Hillsdale, N.J.: Erlbaum.

Rozin, P., & Kalat, J. W. (1971) Specific hungers and poison avoidance as adaptive specializations of learning. *Psych. Rev. 78*: 459-486.

Rozin, P. & Schull, J. (1988) The adaptive-evolutionary point of view in experimental psychology. In R. C. Atkinson, R. J. Herrnstein, G. Lindzey, and R. D. Luce (Eds.) *Stevens's Handbook of Experimental Psychology, Second Edition,* Vol. 1 New York: Wiley, pp. 503-546.

Sherry, D. F., & Schacter, D. L. (1987) The evolution of multiple memory systems. *Psych. Rev. 94*: 439-454.

Shettleworth, S. J. (1993) Varieties of learning and memory in animals. *J Exp Psych: Anim. Beh. Proc. 19*: 5-14.

Tinbergen, N. (1963) On aims and methods of ethology. *Zeits für Tierpsychol 20*: 410-433.

Williams, G. C. (1966) *Adaptation and natural selection.* Princeton: Princeton University Press.

PART ONE

Specializations in Learning

2

Specializations in Honey Bee Learning

James L. Gould

Abstract

Honey bees display a number of cognitive specializations that suggest the existence of multiple memory systems. One handles flower-pattern learning, which is accomplished on a forager's approach to a flower and is stored pictorially at relatively low resolution. A second specialized system deals with learning landmarks, which are used to triangulate the location of floral targets; this learning seems to occur during the departure phase of foraging and is stored pictorially at a higher resolution than flower-pattern memory. Another specialization of honey bee memory serves to index floral information temporally: foragers learn how to harvest individual flowers, and store that information (along with flower recognition) according to time of day. These specializations, along with others also described, seem well adapted to the needs of a flower-constant nectar-foraging insect.

Introduction

The central organizing principle of modern biology is evolution by natural selection. This is not to say that evolution cannot occur by other means — chance and genetic drift, for instance, are clearly important in many contexts. Rather, to the extent that order and predictability can be educed from natural patterns, they are usually the consequence of natural selection. Thus we are not surprised to find, for instance, that honey bees are well adapted to their niche as perennial social foragers of nectar and pollen. Selection has doubtless operated to favor variants that are more efficient at the hundreds of tasks involved — more efficient than competing species and competing honey bee colonies.

Optimization (in the face of chance and genetic drift, and within the limits of physiological possibility and existing or potential variation in the population) is easy to see in the context of anatomical specialization: honey bee workers come provided with mouthparts suitable for taking nectar from flowers, and pollen baskets (and the associated presses) well adapted to collecting and transporting pollen. Such structures are not found in species with other sorts of food sources, or even in honey bee castes that do not collect food (drones and queens). Niche-specific physiological and biochemical specializations are equally easy to identify (reviewed by Gould and Gould, 1988). Behavioral specializations appropriate for gathering nectar and pollen are also well known, such as the elaborate cleaning sequence that systematically moves the pollen to the pollen presses.

Logically, there is no reason to expect that specializations in learning appropriate for the tasks faced by social bees will not have been strongly favored by natural selection. The apparent existence of learning specializations in animals has been evident from the work of field ethologists for decades (e.g. Lorenz, 1935; Tinbergen, 1932, 1938). The concept received a more formal and explicit formulation by Rozin and Kalat (1971) in the restricted context of well-controlled laboratory studies of food-avoidance learning; they pointed out that the requirements of food-avoidance learning involved a "functional incompatibility" with the more general sort of learning thought to be evident in laboratory studies. Gould (1982) proposed the term "learning programs" to describe the widespread cases of context-specific learning observed in the field. Gould and Marler (1984, 1987) sought to synthesize studies and discipline-specific jargon from ethological and psychological studies into a single, selection-based formulation emphasizing innately guided learning; they pointed out that both field and laboratory work indicates that there must be multiple learning programs in animals, and specifically in rats and pigeons. At the same time, they argued against the wholesale abandonment of general-process learning theory advocated by some (see papers in Marler and Terrace, 1984).

Sherry and Schacter (1987), in a comparison of song learning and cache-location memory in birds, further elaborated this argument; they referred to Gould and Marler's learning programs as "memory systems," and reviewed the evidence for separate procedural versus declarative learning systems in humans. They pointed out that more conventional general-purpose learning can deal with many or most contingencies; learning programs will evolve only when selection favors some degree of specialization in acquisition, retention, or retrieval. Barkow et al. (1992) carried this application of learning specializations in our species further, referring to the human brain as a "multimodular mind." Pinker and Bloom (1992) have applied this perspective to human language learning (see also Gould and Marler, 1987). Tooby and Cosmides (1992) have described the slow but (they believe) steady acceptance of an evolutionary perspective on human learning and behavior.

The idea that learning may be "customized" to suit a species' needs continues to encounter some hostility (e.g. Terrace, 1984; Bitterman and Couvillon, 1991). In part this is the legacy of the old Behavioristic assumption that learning is a unitary phenomenon which depends on a single general process (reviewed by Gould and Marler, 1984), and perhaps the (mistaken) fear that recognition of widespread learning specializations somehow entails the rejection of general-process learning. Even when the reality of species-specific learning is granted, the discussion is frequently couched in terms of "constraints on learning" (e.g. Hinde and Stevenson-Hinde, 1973). In evolutionary discussions, a constraint is generally a physiological limitation which prevents an individual from expanding or shifting its niche in a particular way; the limitation may be an artifact, or a necessary consequence of some other more critical adaptation. Learning "constraints," on the other hand, appear to mean any sort of specialization or bias (I distinguish between "specialization" and "bias" below) evident in learning which deviates from the traditional Behavioristic view that an animal can be conditioned to respond equally well to any cue it can sense, and taught to perform any novel behavior of which it is physically capable (Watson, 1925; Pavlov, 1927; Skinner, 1938). Thus, while a pollen press is an obvious specialization, and could not reasonably be called a constraint, a learning routine specialized for remembering salient details about flowers is said by many students of behavior to be "constrained."

This view makes some sense if one believes that learning began as an unconstrained system for picking out correlations, and evolution has acted to focus it by limiting the cues and contexts available. As a survey of many vertebrate examples of specialized learning makes clear, however, the data (both intuitively and by Cladistic criteria) clearly indicate that much (perhaps most) learning has evolved as specialized programs dedicated to solving specific problems (Gould and Marler, 1984; Gould, 1986a); the "constraints" are, in fact, conceptual limits imposed by experimental results over the last three decades on the theoretical framework of Behavioristic thinking; they have nothing to do with the idea of evolutionary constraint.

The object of this chapter is to look at studies of learning specializations in honey bees, and to try to relate them to the niche-specific tasks they serve.

Odor and Color Learning

Early in this century, von Frisch (reviewed, 1967) demonstrated that forager honey bees learn the odor and color of the feeders (and, presumably, the flowers) they visit. More quantitative work (reviewed by Gould, 1984; Menzel, 1985; Gould and Towne, 1988) has revealed a number of asymmetries in this learning. For example, odor and color are learned at different rates and to different degrees of precision (Figure 2.1). Shape (as we will see) and time are learned still more slowly. One could argue, of course, that these observations result not from learning specializations but rather from perceptual biases: perhaps honey bees are simply more sensitive to odors than they are to visual stimuli. This seems unlikely in view of the marked differences in the three classes of visual learning already mentioned (and described in more detail below); moreover, bees (rev. von Frisch, 1967) like wasps (Tinbergen, 1932) seem unable (or unwilling) to learn odor cues in association with nest location.

Within any one modality there can be additional learning specializations. With regard to odor learning, for example, floral odors are learned

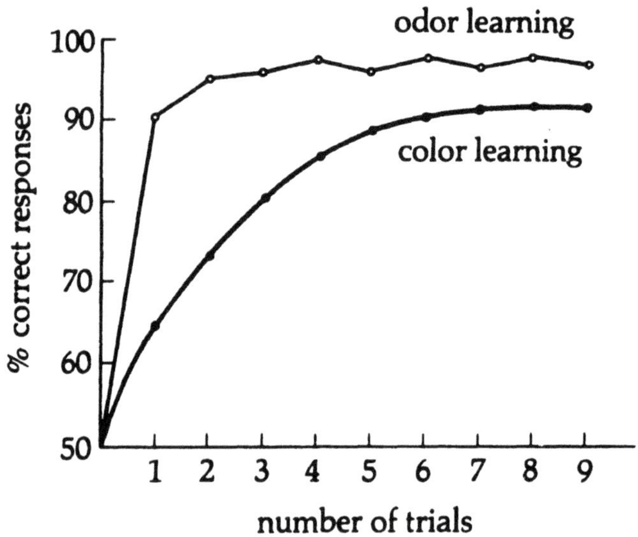

FIGURE 2.1 Foragers learn odor quickly and reliably, but color takes longer and never elicits choices as reliable as those seen with odor training. (The training involved a rewarded target, the S+, and an unreinforced alternative, the S-; by varying the stimulus intensity of the two targets until naïve bees choose them equally often, it is possible to balance the inherent attractiveness of the alternatives; this is why both curves begin at approximately a 50%-correct-choice rate before any training.) Gould and Gould, 1988, adapted from Menzel et al. 1974.

faster than non-floral scents, though all are eventually learned roughly equally well (Menzel, 1985). Again, the argument could be made that bees are simply better at *detecting* floral odors, so that the results reflect a sensory bias; in fact, however, the nonpheromonal odors that bees seem most sensitive to — oleic acids and related compounds (used to identify dead bees) — are the ones learned most slowly (Menzel, 1985). Moreover, there are differences in odor-specific learning rates between at least two European races of honey bees which correspond to differences in local flora (Koltermann, 1973; Lindauer, 1976). The readiness of naïve honey bees to learn floral odors more rapidly than the many other (and likely irrelevant) scents they might encounter during foraging probably leads to faster and more reliable memorization of this important flower-identification cue than the sort of unbiased US/CS correlation process described in many textbooks.

Specializations are equally evident in color learning. For instance, violet is learned faster than most other colors, while green is learned more slowly, though all are eventually learned to about the same degree of precision (Figure 2.2). Again, simple sensory biases seem unlikely to play a role: the spectral sensitivity of honey bees (von Helversen, 1972) is in no way correlated with initial color-learning rates of bees, and there is no evident asymmetry in color learning in the context of nest-site recognition (von Frisch, 1967). The advantage lies in a natural world marked by a paucity of green flowers, but an abundance of green vegetation located distractingly close to the blossoms bees need to remember.

Flower-pattern Learning and Matching

Honey bees learn to recognize the shapes of flowers, and store this information pictorially with a resolution of about 8-10° (Gould, 1985; 1986b; 1988a; Figure 2.3); real-time visual resolution, by contrast, is on the order of 1-1.5° (Seidl and Kaiser, 1982). The shape is learned primarily upon approach (Opfinger, 1931; Gould, 1988; Figure 2.4), as is color (Menzel and Erber, 1978); bees can remember at least 16 elements and four colors (Gould, 1988a). The possible ecological relevance of this arrangement will be discussed in comparison with landmark learning below.

One evident specialization in shape learning is that bees learn complex patterns faster than simple ones (Figure 2.5), at least up to the limit of their resolution (Schnetter, 1972). This preference is based on the spatial frequency of the pattern rather than any particular aspect of its symmetry. It goes almost without saying that flowers, in general, are

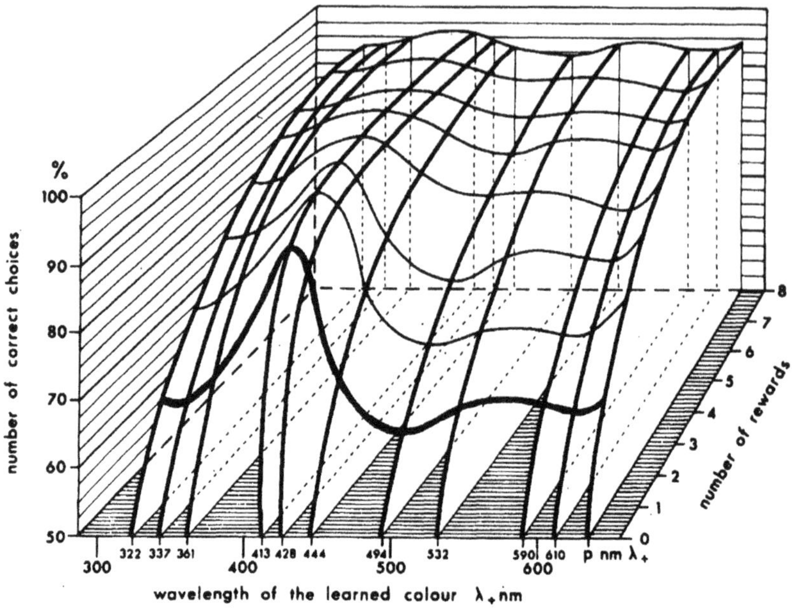

FIGURE 2.2 This 3-dimensional representation shows learning curves for 11 colors. The choice ratio after one trial is highlighted. Violet targets (413nm) are learned much faster than green ones (500nm), though after enough trials this difference disappears. Gould and Gould, 1988, adapted from Menzel et al. 1974.

more complex (that is, have a far higher spatial frequency) than most non-floral objects — in particular, flowers are generally more complex than the nearby leaves. Again, a bee whose attention and learning is focussed by innate programming toward the portion of its sensory world that is likely to be a flower is more apt to memorize the appropriate target quickly and reliably; a blank-slate approach would be more prone to error and slower in picking out correlations.

The apparent ability of bees to store pictorial representations leads naturally to the question of how these memories are matched against real-time experience to judge whether a particular target is of the species on which the forager is currently specializing. The capacity to recognize a mirror-image (left-right) reversal as similar or identical to a training pattern is widespread among animals. Some workers argue that this ability arises as an artifact of bilateral nervous systems — that is, it is a inherent ambiguity rather than an adaptive ability (eg, Corballis and Beale, 1970). Tests with honey bees (Gould, 1988b) indicate that in this species at least there is no essential mirror-image ambiguity — bees can choose the training pattern reliably over a left-right reversal — but,

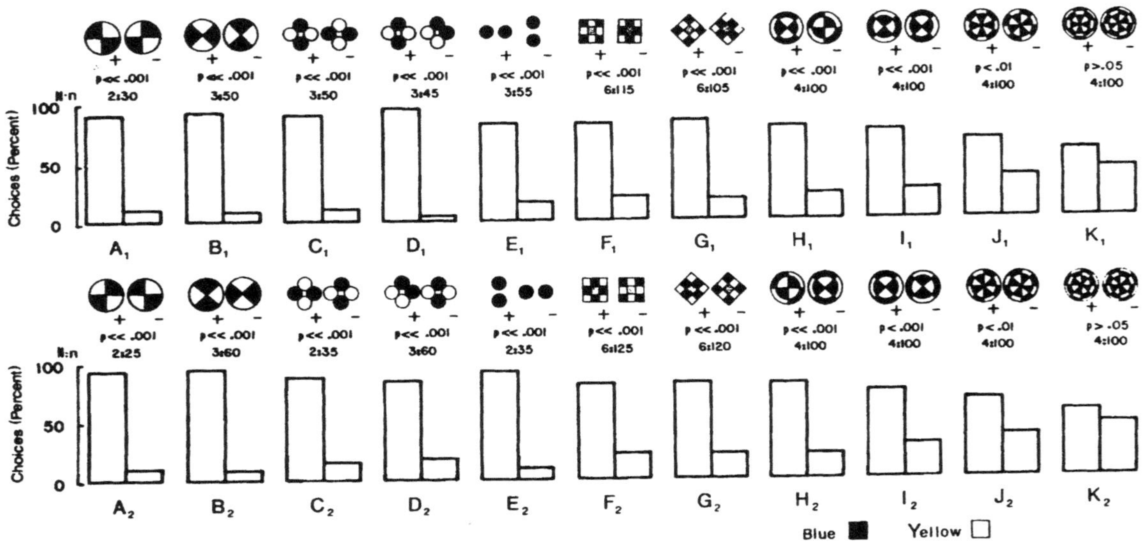

FIGURE 2.3 Resolution of pattern learning by foragers. During training bees were offered two patterns, one (indicated by "+") with a sucrose solution available in small recesses at the center of the pattern; the other (marked "-") offered only water. The left-right placement of the two patterns was reversed after each visit. To control against sensory biases, each experiment was repeated with naïve bees using the previously unreinforced pattern as the one offering a reward, and the previously reinforced pattern as the one stocked with water; this takes the place of balancing the stimulus intensity of the two targets. As in all tests of bee learning, new patterns and display surfaces were used during testing; neither sugar solution nor water was offered. Two to six bees were used in each part, and each was tested in three to five rounds involving five landings each; the numbers (in the form N:n) over each graph indicate the number of bees and number of choices, respectively. (The same basic methods were used in all the experiments reported in this chapter.) As the patterns became more complex (those on the right side of the figure), choice accuracy declined. In J and K the foragers became very hesitant, and hovered in front of each pattern in alternation before choosing. Their highly regular hovering point permitted calculation of memory resolution, yielding a value on the order of 8-10°.

when offered a choice between the reversal and a novel pattern, bees readily opt for the reversal. Hence, there appears to be no inescapable confusion, while at the same time there seems to be a positive ability to recognize reversals as similar. (This discrimination may not be unique to bees: most tests have not offered simultaneous alternatives, but rather have measured response rates to a single pattern, whether the training figure, its reversal, or a novel shape.) The ability to recognize a mirror-image reversal is of self-evident value when a forager approaches a flower from, say, the left side, having previously viewed it only from the right.

Some species, such as pigeons, are also capable of rapid (perhaps automatic) recognition of rotated objects. Others (like humans) are less adept, having apparently to rotate the target mentally, and so showing a choice delay proportional to the degree of disparity between the training and testing targets. Other species cannot match rotated figures at all (eg, Shepard and Metzler, 1971; Hollard and Delius, 1983). Honey bees apparently lack the ability to recognize vertically oriented patterns that have been rotated more than 45° — a deficiency unlikely to cost them anything in the real world, given the natural symmetry of flowers. But for horizontally oriented blossoms, an inability to recognize a rotated pattern might force a forager to circle each target to look for a match. Tests using horizontal targets placed in a device that allows foragers to view artificial flowers from only 90° of the 360° normally available, followed by tests involving 180° rotations, indicate that honey bees can recognize horizontal rotations (Figure 2.6). The context-specific nature of this ability — horizontal versus vertical patterns — has not yet been reported (or perhaps tested for) in any other species.

Landmark Learning

Honey bees can use prominent unambiguous landmarks for navigation, both to substitute for the sun under overcast (Dyer and Gould, 1981) and to take precedence over available celestial cues under many circumstances (von Frisch and Lindauer, 1954). There is even some evidence that bees can use landmarks in a maplike way to navigate over familiar terrain after being displaced (Gould, 1986c; Gould and Gould, 1994). Bees also use landmarks to localize inconspicuous food sources (reviewed by Gould, 1984); given the low resolution of honey bee vision, almost any food source will be inconspicuous until a forager is relatively near. Landmark memory is pictorial and has a horizontal resolution of 2.5-3° (Gould, 1987a; Figure 2.7); this value agrees well with the qualitative measure obtained by von Frisch and Lindauer (1954). Landmark learning appears to take place predominantly on departure (Opfinger, 1931; Gould, 1988c; Figure 2.8).

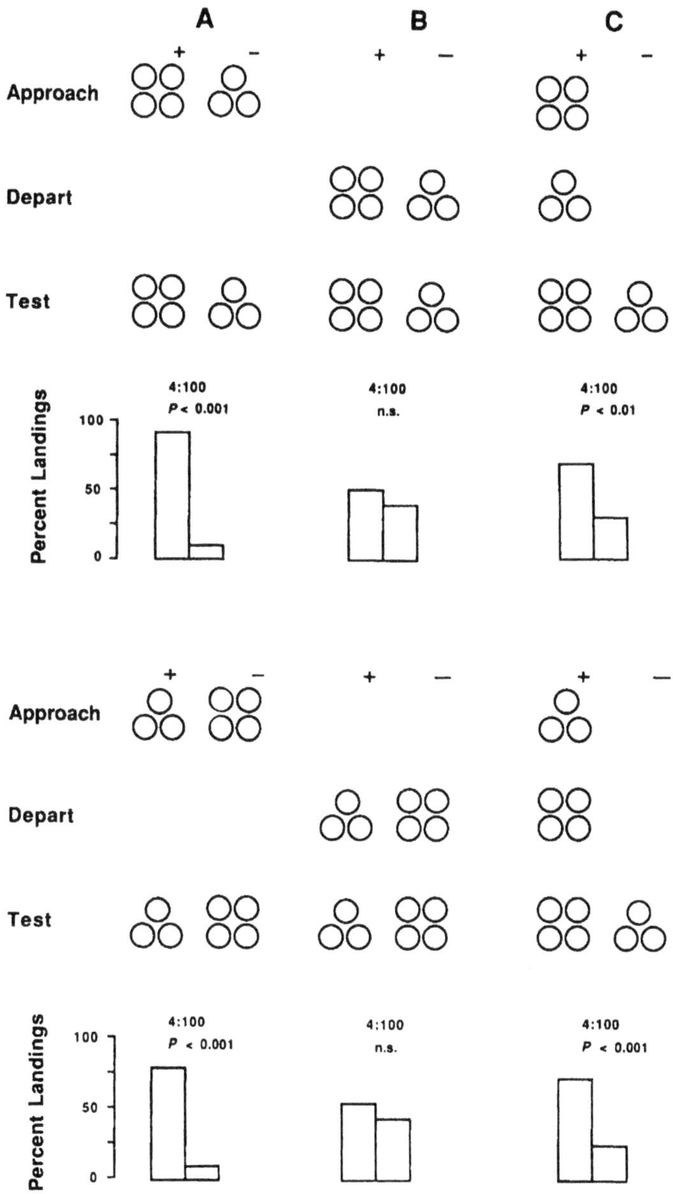

FIGURE 2.4 Timing of pattern learning by foragers. In A, both the reinforced and unreinforced patterns were visible during approach, while in B they could only be seen during departure. In C the visible pattern was different during the two phases of foraging. The data indicate that the pattern seen during approach was the one chosen most reliably during testing.

The possible logic of the differences in timing and memory resolution between flower-pattern learning and landmark learning is open to conjecture. With regard to timing, if two "photographs" are to be taken, one panoramic and at relatively high resolution while the other is directed at a particular target and stored at lower resolution — two photographs that will be used at different points during the next approach — it may be a matter of simple neurophysiological necessity to acquire the information at two different times rather than simultaneously.

Bees must store many separate images in the number of neurons available if they are to be able to recognize landmarks throughout their home range, and be able to identify the many different types of flowers they may forage from over the course of the day. The difference in storage resolution might represent different optima in the balance between reliability and speed of matching (which would benefit from higher resolution) versus neural economy. The optimum resolution may well be different for the two tasks: foragers may be able to live with a lower resolution for flower memory because they can approach targets with

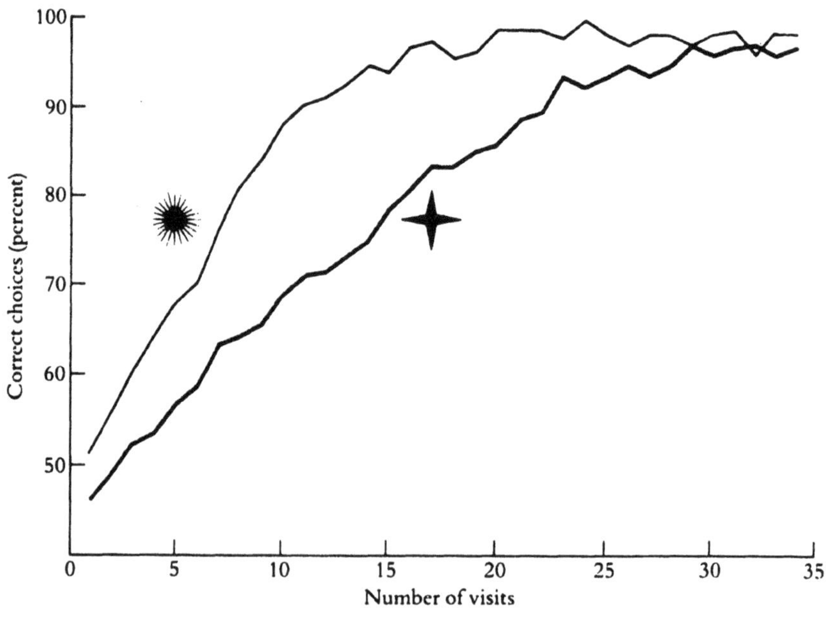

FIGURE 2.5 When trained on a 4-pointed shape (with a 23-pointed Figure as the S-, balanced in color and size to yield a 50:50 spontaneous-choice ratio), foragers learn far more slowly than when the 23-pointed shape is the S+. Even the faster learning is slow compared to color and odor learning.

appropriate colors as closely as they like to make their identification, whereas landmarks must be uniquely recognizable from the location of the food itself if bees are to be able to use them for triangulation.

Bees also show clear preferences in their use of landmarks, choosing correct angular position over shape and size, for example (Cartwright and Collett, 1983), and shape over color (Gould, in preparation). Bees are also quite discriminating about which landmarks to use for triangulation when several are available (Gould, 1987a). Just as human navigators learn to do, foragers put more emphasis on landmarks with an

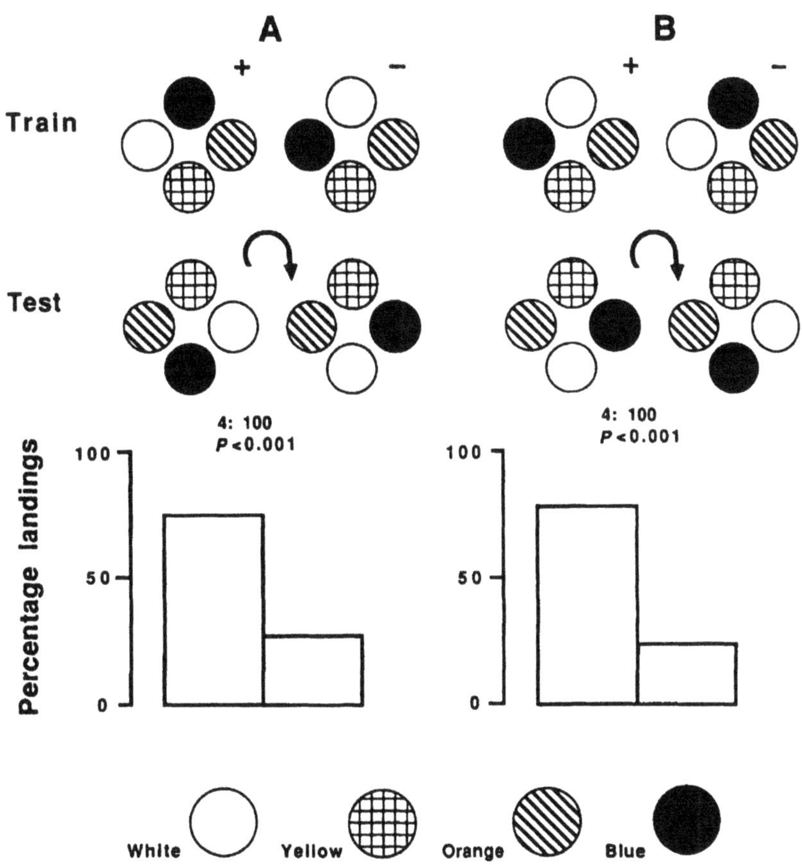

FIGURE 2.6 Recognition of horizontal rotations by foragers. In these preliminary tests, foragers were trained on one pair of patterns visible from only one quadrant. They were then tested on the same set to yield control values (right) or a pair rotated 180° (left). The rotated patterns were chosen almost as reliably as the unrotated targets.

optimal degree of angular separation (90°); this observation lays to rest a puzzling controversy over the way landmark memory is stored. (Figure 2.9). Anderson (1977) had reported that when some of the landmarks around a food source are removed, returning bees search too close to the remaining cues, which suggested to him that landmark memory is not pictorial. In fact, bees, in their attempt to triangulate their location on the basis of landmarks with the most useful angular separation, must have confused some of the remaining landmarks with ones that had been there previously (Gould, 1987a).

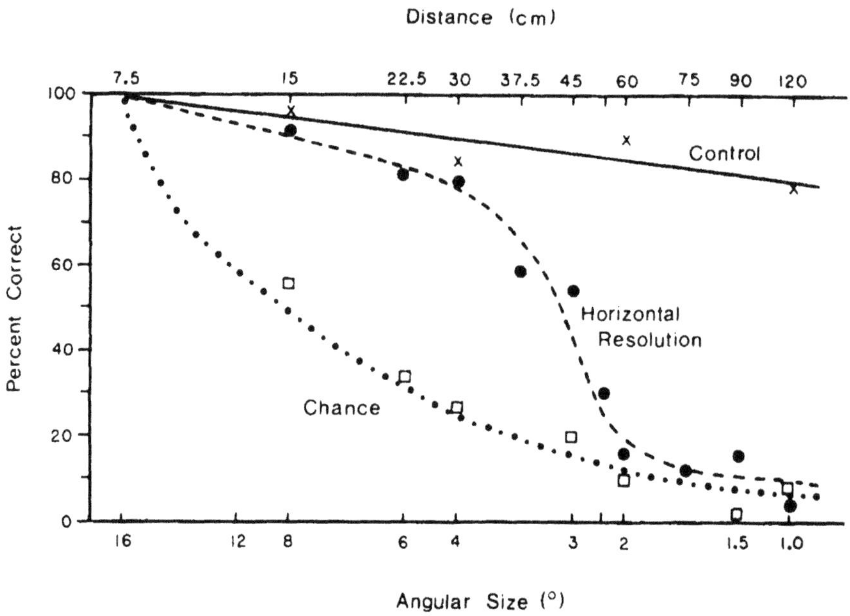

FIGURE 2.7 Resolution of landmark memory in foragers. The upper curve (Control) is derived from bees trained to landmarks at the distances indicated (7.5-120cm) which were increased in size as the separation increased so as to keep a constant visual angle (sample size — 6:150 at each distance); this line represents the best foragers may be expected to do at the choice task when apparent cue size does not vary. The bottom curve (Chance) represents the chance of a forager finding the correct hole accidentally under the conditions and with the criteria used in these tests (2:50 at each distance) The middle curve (Horizontal Resolution) shows the choice accuracy of foragers tested with increasingly small landmarks (6:150); three different training/testing protocols were used, each of which yielded similar results. Choice accuracy appears to fall to 50% (relative to the control and chance values) at about 2.5-3°.

Choice Hierarchies

Although foragers are able to learn color, odor, shape, and the relative location of nearby landmarks, they do not weight these different cues equally when making a choice between alternative targets (reviewed by Gould and Gould, 1988). For example, if one trains bees to, say, a blue triangular peppermint-scented feeder, and then offers the returning foragers a choice of two targets arranged so that the cues have been dissociated, the bees show a clear preference for the correct odor over all other cues (Figure 2.10). Such tests also indicate that color is second in importance, followed by shape and then landmarks in the Italian subspecies (*Apis mellifera ligustica*), or landmarks and then shape in the German race (*A. m. carnica*) (Koltermann, 1973; Lindauer, 1976).

It is ironic that the choice hierarchy is quite different from the order in which cues are encountered: bees normally see and use landmarks before they can make out the flowers, detect color before they can perceive shape, and (unless the scent is very strong and they happen to be

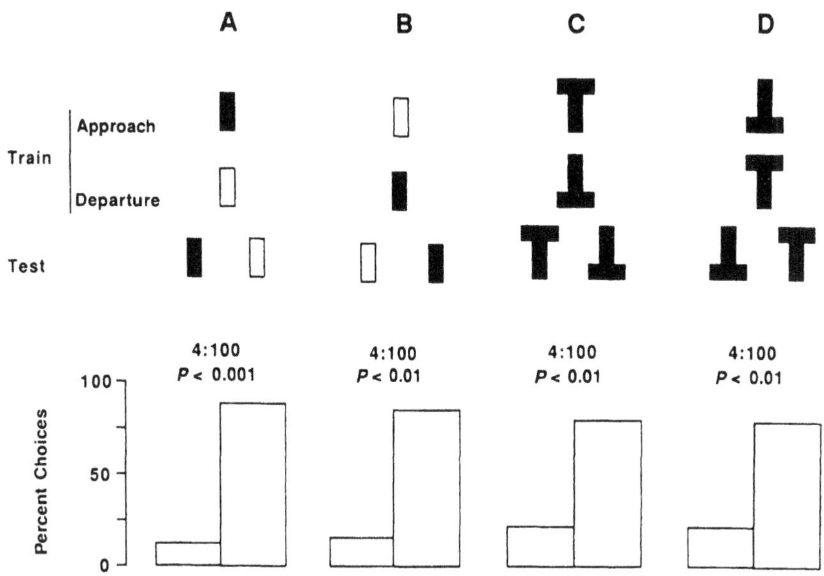

FIGURE 2.8 Timing of landmark learning by foragers. During training, foragers observed one set of 44-66mm-tall landmarks during approach and another upon departure; testing consisted of offering both pairs simultaneously and recording forager choices. The landmarks were positioned 150mm from the feeding hole, which was one of 225 holes, each 3mm in diameter, on the training and testing platforms. One set of tests (A and B) established the timing of landmark-color learning; the other (C and D) examined the timing of landmark-shape learning.

approaching from downwind) encounter odor last of all. What the odor —>color —>shape hierarchy does seem to correspond to is the usual reliability of these cues under natural conditions: odor is absolutely diagnostic of species identity, while there can be substantial variation in both actual and (depending on lighting and background) apparent color. Shape varies most of all, depending greatly on the angle of approach. Since this hierarchy is innate, it is reasonable to assume that it exists because its use confers a substantial advantage on foragers.

Another point worth making is that there are certain readily perceivable cues (including polarization direction, and speed or direction of rotation) that cannot be remembered in the context of flower learning even if they are the only stimuli that have predictive value (Menzel, 1985). In nature, of course, such cues have no use whatsoever in this context, and bees predisposed to ignore them could be expected to learn faster and with greater reliability than more open-minded foragers.

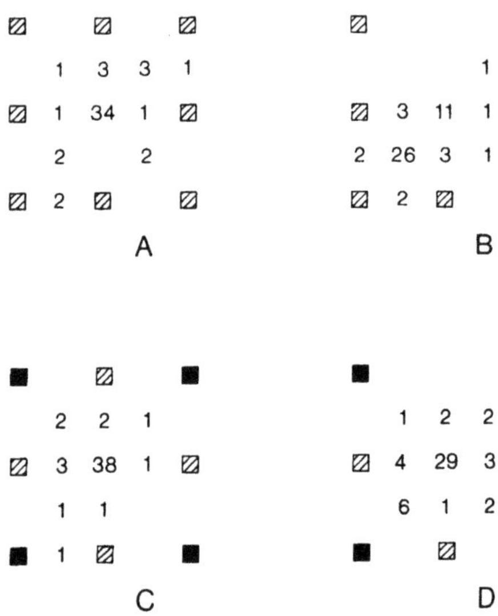

FIGURE 2.9 Foragers were trained to find food in one of the 225 holes on the same device used in the work described in Figure 1.8. When the correct site was surrounded by 8 identical landmarks, the foragers were fairly accurate, as indicated by the number of landings recorded during testing without food present (upper left) When some landmarks were removed before testing (upper right), the pattern of landings shifted toward the remaining markers. This occurs because the foragers confuse the remaining landmarks with some of the more widely spaced markers that have been removed. This is illustrated by the repetition shown at bottom using two different colors of markers. In this case, when some landmarks are removed for testing there is no possibility of confusion.

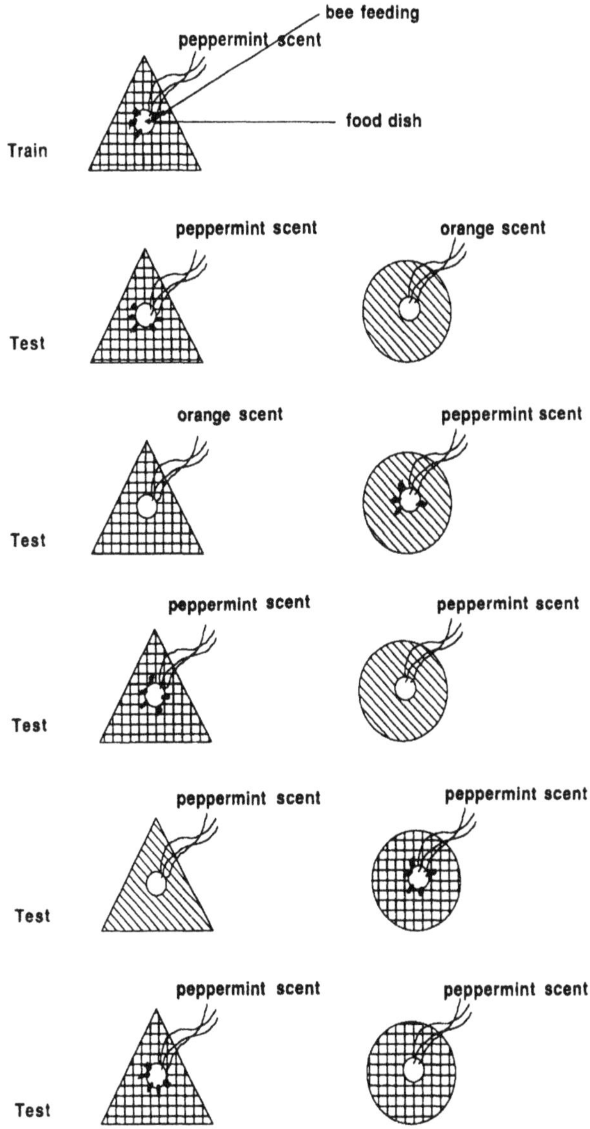

FIGURE 2.10 Foragers were trained on a triangular blue peppermint-scented target; when tested against a circular yellow orange-scented alternative, the bees correctly chose the training target. When, however, the odor was dissociated from color and shape (so that the circular yellow target had peppermint scent while the blue triangle was supplied with orange scent), the bees chose on the basis of odor. When the odor was correct on both targets but shape and color were dissociated, color was the decisive cue. The final step in this qualitative summary of such hierarchy experiments is the demonstration that only when both color and odor are correct does shape become decisive.

Timing and Flower-handling Learning

The ability of honey bees to remember the time of day that a particular food source is available has been known for nearly 80 years (reviewed in von Frisch, 1967). More recently Bogdany (1978) reported that the cues associated with flower recognition are stored as a time-linked set: changing one cue led to the forager's relearning them all, even if only one had changed. Information used at one time of day was not apparently readily available at others, and only one set of cues could be remembered at one time of day — though several sets could be learned if offered at distinctly different times.

A time-linked, one-flower-at-a-time storage system is well suited to a flower-constant nectar forager: flowers generally offer nectar over a limited species-specific portion of the day (a strategy that helps focus foraging — and therefore pollination — efforts), and a bee that can keep and use a time-linked list of nectar-producing species ought to be more efficient at harvesting food for its colony. Working this out by trial and error would obviously be time consuming and wasteful.

One aspect of foraging that has received relatively little attention is the physical act of harvesting. Different species of plants require different approach flights, landing techniques, and nectar-extraction behavior. Learning curves obtained from bumble bees in the field (Heinrich, 1979) indicate a clear role for experience, but since bumble bees usually forage two or three species simultaneously the nature of the curve might well be different for honey bees. Indeed, the curve differs between specialist and generalist species of bumble bees: foragers of a species that specializes on monkshood learns to harvest that blossom far faster than foragers from a generalist species (Laverty and Plowright, 1988). It was Darwin (1876) who first suggested that the need to learn how to work a particular species of blossom efficiently might help account for flower constancy. The object of the experiment described here was to confirm that foragers can learn to use different techniques on morphologically similar flowers, and then to see if the techniques are stored according to time of day.

Control tests showed that bees land spontaneously most often on the lowest petal of a vertically orientated six-petalled target. The training technique involved using hidden solenoids to flick foragers off any but the correct petal. After confirming that foragers could be taught to alight on either the lower-left or lower-right petal, the bees were trained to land on one petal from 0930-1100 hours, and the other from 1100-1230 (Gould, 1987b). Even though the targets used over the course of training and testing were identical in color, odor, position, and shape, foragers returning on subsequent days tended to land on the petal that had been appropriate at that time on the previous day (Figure 2.11).

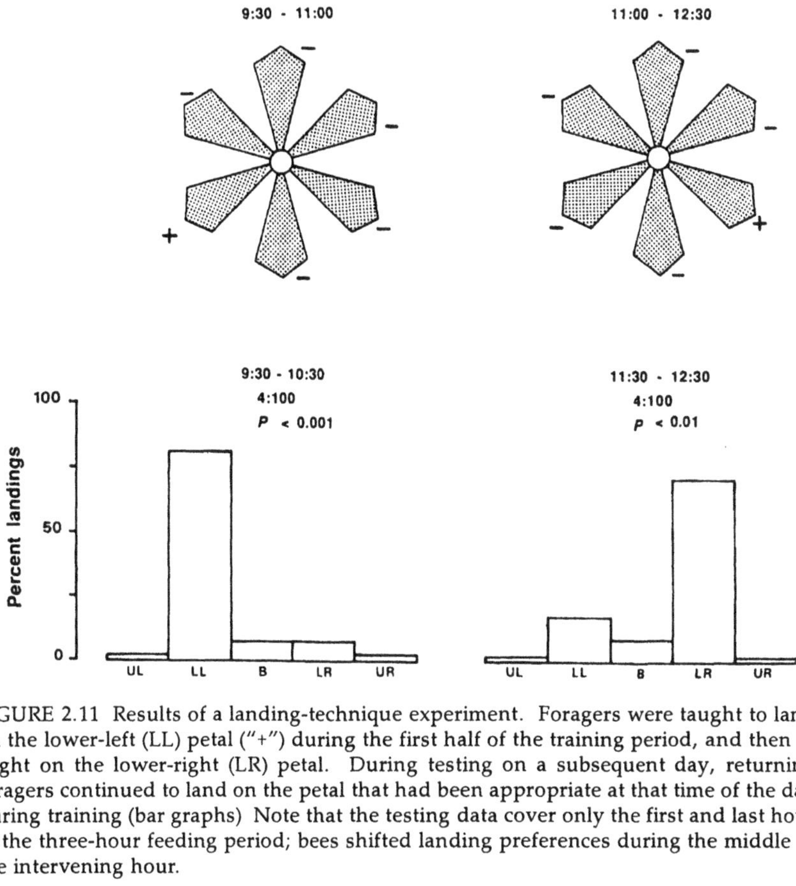

FIGURE 2.11 Results of a landing-technique experiment. Foragers were taught to land on the lower-left (LL) petal ("+") during the first half of the training period, and then to alight on the lower-right (LR) petal. During testing on a subsequent day, returning foragers continued to land on the petal that had been appropriate at that time of the day during training (bar graphs) Note that the testing data cover only the first and last hour of the three-hour feeding period; bees shifted landing preferences during the middle of the intervening hour.

Conclusion

As suggested at the outset, natural selection should favor specialization of learning when the context, the appropriate cues to commit to memory, and the way to organize the learned information for later use can be specified in advance. The life history of most animals provides at least a few such circumstances — cases in which the practicability of the situation is combined with a need to learn something (such as the appearance of one's offspring, or the characteristics of a threat or a food item). Animals facing such situations without specializations that can be built into the nervous system will, in general, learn more slowly and less reliably, and so be less fit.

There seems little chance that honey bees have inherited their suite of learning programs independently of selection, since species for which flower constancy is maladaptive lack the temporal and operant specializations prominent in honey bees (e.g. Lewis, 1986; Dukas and Real, 1991). For instance, nonsocial species, in which each forager must harvest a diverse range of foods in order to balance its diet, learn very little about flowers and forget quickly what they do learn; "better" learning could lead to serious nutritional deficiencies, particularly for solitary bees gathering pollen and nectar to provision offspring. The diet of social bees, in contrast, is diverse without such individual precautions because the foragers specialize independently and pool their harvest. Bumble bees, which, though social, have far smaller colonies than honey bees, are suitably intermediate in their ability to learn and remember. The honey bees' readiness to learn, store, and value the appropriate stimuli, studiously ignoring others that (outside of experimental paradigms) are irrelevant, and to index them in the most sensible way given the nature of the task before them, is probably the consequence of millions of years of evolutionary "fine tuning."

References

Anderson, A. M. (1977) A model for landmark learning in the honey bee. *J. Comp. Physiol. 114*: 335-55.

Bogdany, F. J. (1978) Linking of learning signals in honey bee orientation. *Behav. Ecol. and Sociobiol. 3*: 323-336.

Barkow, J. H., Cosmides, L., and Tooby, J. (Eds.) (1992) *The Adapted Mind*. Oxford: Oxford University Press.

Cartwright, B. A., and Collett, T. S. (1983) Landmark learning in bees. *J. Comp. Physiol. 151*: 521-543.

Corballis, M. C., and Beale, I. L. (1970) On telling left from right. *Psychol. Rev. 77*: 451-464.

Darwin, C. (1876) *On the Effects of Cross and Self Fertilisation in the Vegetable Kingdom*. London: John Murray.

Dukas, R., and Real, L. A. (1991) Learning foraging tasks by bees: a comparison between social and solitary species. *Anim. Behav. 42*: 269-76.

Dyer, F. C., and Gould, J. L. (1981) Honey bee orientation: a backup system for cloudy days. *Science 214*: 1041-1042.

Frisch, K. v. (1967) *The Dance Language and Orientation of Bees*. Cambridge, Massachusetts: Harvard University Press.

Frisch, K. v., and Lindauer, M. (1954) Himmel und Erde in Konkurrenz bei der Orientierung der Bienen. *Naturwissenschaften 41*: 245-253.

Gould, J. L. (1982) *Ethology: The Mechanisms and Evolution of Behavior*. New York: W. W. Norton.

Gould, J. L. (1984) The natural history of honey bee learning. In P. Marler and H. S. Terrace ed. *The Biology of Learning*. Berlin: Springer-Verlag, pp. 149-180.

Gould, J. L. (1985) How do bees remember flower shape? *Science 227*: 1492-1494.
Gould, J. L. (1986a) The biology of learning. *Ann. Rev. of Psychol. 37*: 163-192.
Gould, J. L. (1986b) Pattern learning by honey bees. *Anim. Behav. 34*: 990-997.
Gould, J. L. (1986c) The landmark map of honey bees: do insects have cognitive maps? *Science 232*: 861-863.
Gould, J. L. (1987a) Landmark learning by honey bees. *Anim. Behav. 35*: 26-34.
Gould, J. L. (1987b) Honey bees store learned flower-landing behavior according to time of day. *Anim. Behav. 35*: 1579-1581.
Gould, J. L. (1988a) Resolution of pattern learning by honey bees. *J. Insect Behav. 1*: 225-233.
Gould, J. L. (1988b) A mirror-image "ambiguity" in honey bee pattern matching. *Anim. Behav. 36*: 487-492.
Gould, J. L. (1988c) Timing of landmark learning by honey bees. *J. Insect Behav. 1*: 373-378.
Gould, J. L., and Gould, C. G. (1994) *The Animal Mind.* New York: W. H. Freeman.
Gould, J. L., and Gould, C. G. (1988) *The Honey Bee.* New York: W. H. Freeman.
Gould, J. L., and Marler, P. (1984) Ethology and the natural history of learning. In P. Marler and H. S. Terrace ed. *The Biology of Learning.* Berlin: Springer-Verlag, pp. 47-74.
Gould, J. L., and Marler, P. (1987) Learning by instinct. *Sci. Am. 255* (1): 74-85.
Gould, J. L., and Towne, W. F. (1988) Honey bee learning. *Adva. Insect Physiol. 20*: 55-86.
Heinrich, B. (1979) *Bumblebee Economics.* Cambridge, Massachusetts: Harvard University Press.
Helversen, O.v. (1972) *J. Comp. Physiol. 80*: 439-472.
Hinde, R. A., and Stevenson-Hinde, J. (Eds.) (1973) *Constraints on Learning.* New York: Academic Press.
Hollard, V. D., and Delius, J. D. (1983) Rotational invariance in visual pattern recognition by pigeons and humans. *Science 218*: 804-806.
Koltermann, R. (1973) Rassen-bzw. artspezifische Duftbewertung bei der Honigbiene und Ekologische Adaptation. *J. Comp. Physiol. 85*: 327-60.
Laverty, T. M., and Plowright, R. C. (1988) Flower handling by bumblebees: a comparison of specialists and generalists. *Anim. Behav. 36*: 733-40.
Lewis, A. C. (1986) Memory constraints and flower choice in *Pieris rapae. Anim. Behav. 232*: 863-65.
Lindauer, M. (1976) Recent advances in the orientation and learning of honey bees. In *Proc. XV Inter. Cong. Entomol.* Washington, D. C.: International Entomological Association, pp. 450-60.
Lorenz, K. (1935) Der Kumpan in der Umwelt des Vogels. *J. für Ornithologie 80*: English translation: Companions as factors in the bird's environment. In K. Lorenz (1970), *Studies in Animal and Human Behavior (vol. 1).* Cambridge, Mass.: Harvard University Press, pp.101-258.
Marler, P., and Terrace, H. S. (Eds.) (1984) *The Biology of Learning.* Berlin: Springer-Verlag.

Menzel, R. (1985) Learning in honeybees in a behavioral and ecological context. In B. Hölldobler and M. Lindauer (Eds.) *Experimental Behavioral Ecology and Sociobiology*. Stuttgart: Gustav Fischer Verlag, pp. 55-74.
Menzel, R., and Erber, J. (1978) Learning and memory in bees. *Sci. Am. 239*: (1): 102-110.
Menzel, R., Erber, J. and Mashur, T. (1974) Learning and memory in the honeybee. In L. Barton Browne ed. *Experimental Analysis of Insect Behavior*. New York: Springer-Verlag, pp. 195-217.
Opfinger, E. (1931) Über die Orientierung der Bienen an der Futterquelle. *Zeits. für Vergl. Physiol. 15*: 431-487.
Pavlov, I. (1927) *The Conditioned Reflex*. New York: W.W. Norton.
Pinker, S., and Bloom, P. (1992) Natural language and natural selection. In H. Barkow, L. Cosmides, and J. Tooby, (Eds.) *The Adapted Mind*. Oxford: Oxford University Press, pp. 451-93.
Rozin, P., and Kalat, J. W. (1971) Specific hungers and poison avoidance as adaptive specializations in learning. *Psychological Review 78*: 459-86.
Schnetter, B. (1972) Experiments on pattern discrimination in honey bees. In R. Wehner ed. *Information Processing in the Visual System of Arthropods*. Berlin: Springer-Verlag, pp. 195-200.
Seidl, R., and Kaiser, W. (1982) Visual field size, binocular domain, and the ommatidial array of the compound eyes in worker bees. *J. Comp. Physiol. 143*: 17-26.
Shepard, R. N., and Metsler, J. (1971) Mental rotation of three-dimensional objects. *Science 171*: 701-703.
Sherry, D. F., and Schacter, D.L. (1987) The evolution of multiple memory systems. *Psychol. Rev. 94*: 439-54.
Skinner, B. F. (1938) *The Behavior of Organisms*. New York: Appleton-Century-Crofts.
Terrace, H. S. (1984) Animal learning, ethology, and biological constraints. In P. Marler and H. S. Terrace ed. *The Biology of Learning*. Berlin: Springer-Verlag, pp. 15-46.
Tinbergen, N. (1932) Über die Orientierung des Bienenwolfes. *Ziets. vergl. Physiol. 16*; 305-34; English translation: The homing of Philanthus triangulum. In N. Tinbergen (1972), *The Animal in Its World (vol. 1)*. Cambridge, Mass.: Harvard University Press, pp.103-27.
Tinbergen, N., and Kruyt, W. (1938) Über die Orientierung des Bienenwolfes. III: Die Bevorzugung bestimmiter Wegmarken. *Zeitschrift für vergleichende Physiologie 25*: 292-334; English translation: Landmark preference by homing Philanthus. In N. Tinbergen (1972), *The Animal in Its World (vol. 1)*. Cambridge, Mass.: Harvard University Press, pp.146-96.
Tooby, J., and Cosmides, L. (1992) The psychological foundations of culture. In J. H. Barkow, L. Cosmides, and J. Tooby, (Eds.) *The Adapted Mind*. Oxford: Oxford University Press, pp. 19-136.
Watson, J. B. (1925) *Behaviorism*. New York: W. W. Norton.

3

Detecting Associations in Pavlovian Conditioning and Instrumental Learning in Vertebrates and in Invertebrates

Ruth M. Colwill

Abstract

Many vertebrates and invertebrates display behavioral changes following exposure to Pavlovian and instrumental conditioning procedures. These behavioral changes are generally thought to represent the output of an associative learning process. Thus, Pavlovian conditioning leads to the formation of stimulus-outcome associations and instrumental learning leads to the development of response-outcome associations. In this chapter, I discuss a relatively unappreciated criticism of the conventional view that stimulus-outcome and response-outcome learning may be inferred simply on the basis of standard operational criteria. Then, I describe two techniques that are popular in the modern conditioning literature for detecting associations between a stimulus and its consequent outcome and between a response and its consequent outcome. One of these techniques depends on the identity of the outcome and the other technique depends on the value of the outcome. Finally, I review the possible limitations regarding the use of these methodologies for detecting associative learning.

Introduction

One of the most valuable cognitive assets to have evolved is the ability to predict the occurrence of future events as a result of past experience. Through learning, an individual is thus able to modify its subsequent behavior in such a way as to maximize beneficial events and to minimize aversive ones. Various functional advantages of the capacity to develop expectations have been suggested, particularly in the areas of territorial defense and courtship (see Bolles and Beecher, 1988; Dewsbury, 1990; Hollis, 1982; Kesner and Olton, 1990). For example, Hollis (1984) has shown that male blue gouramis (*Trichogaster trichopterus*) mount a more aggressive and successful defense of their territories against anticipated than unanticipated male conspecific intruders. Furthermore, following exposure to a signal that has previously announced the appearance of a female conspecific, male blue gouramis engage predominantly in courtship appeasement behaviors when encountering a receptive female. Without that prior warning, however, males greet receptive females with aggressive displays (Hollis et al. 1989).

Efforts to discover how an animal acquires, stores and uses predictive information have been considerable in the discipline of psychology. With few exceptions (e.g. Gallistel, 1990, 1992), it is generally assumed that an associative mechanism underlies the capacity of most species to learn; that is to represent information about relations between events (Pavlovian conditioning) and the relations between instrumental acts and their consequences (instrumental learning). These two learning procedures are outlined in Figure 3.1. The top panel of Figure 3.1 illustrates a typical Pavlovian conditioning procedure: One event, the conditioned stimulus or CS, immediately precedes presentation of another event, the unconditioned stimulus or US. Pigeons that are exposed to repeated presentations of the illumination of the response key for 5 sec. followed by access to food for 5 sec will eventually peck reliably at the illuminated response key. The bottom panel of Figure 3.1 illustrates a typical instrumental conditioning procedure: Performance of a target behavior (the response or R) leads to the delivery of a reward (the outcome or O). Rats that receive food rewards for pressing a lever will show a substantial increase in the rate at which they subsequently press the lever.

Contemporary learning theorists have sharpened their characterization of the conditions under which stimulus-outcome and response-outcome associations are thought to develop (Dickinson, 1980; Mackintosh, 1983; Rescorla, 1988). However, despite their rejection of a simple temporal contiguity principle, modern learning theorists share with their predecessors the view that an association between two events enables one event to excite an internal representation of the other event (Pavlov, 1927). Thus, it has become common to describe the effect of CS-US pairings as enabling the CS to activate a representation of the US, and the effect of R-O pairings as enabling the response to activate a representation of O. The spirit of this associationist approach is embodied in the efforts of neurobiologists searching for the neural basis of informa-

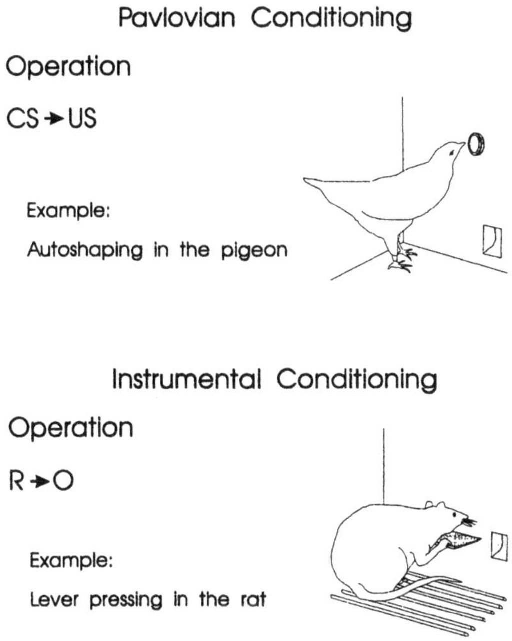

FIGURE 3.1 Top panel illustrates a basic Pavlovian conditioning procedure in which the illumination of a keylight (CS) is followed by the presentation of food (US). Pigeons exposed to such CS-US pairings will come to peck at or autoshape to the illuminated key. Bottom panel illustrates a basic instrumental conditioning procedure in which a lever press produces a food reward (O). Rats exposed to such an R-O contingency will increase their rate of lever pressing. These behavioral changes are thought to be mediated by the formation of stimulus-outcome and response-outcome associations, respectively.

tion storage and is reflected in the selection of learning algorithms used in neural network models (Byrne and Berry, 1989; Hebb, 1949).

In this chapter, I outline the kind of evidence that is needed to assert that associative learning has taken place. Traditionally, the burden of demonstrating associative learning lay only in showing a difference between the response to a stimulus (CS+) paired with the unconditioned stimulus (US) and one that was not paired (CS-). This criterion has been used almost exclusively to guide the development of behavioral preparations suitable for a neural analysis of associative learning processes. Unfortunately, this simple prescription for demonstrating associative learning is not without interpretative problems. After discussing an important but relatively unappreciated ambiguity of the differential conditioning procedure that may limit its usefulness, I describe two alternative approaches that provide superior evidence for associative learning. One method relies on observations that the behavioral consequences of conditioning operations are contingent on the identity of the outcome; the other method evaluates that dependency with respect to the value of the outcome.

Conventional Approaches to Identifying Associative Learning

Historically, associative learning has been identified through the use of particular experimental operations. Guided by the belief that the formation of an association between two events requires their contiguous presentation, learning theorists debated the appropriate comparison conditions for assessing the effects of paired presentations (see Gormezano et al. 1983; Mackintosh, 1974). Several strategies were developed to eliminate purely nonassociative explanations for behavioral changes resulting from a paired treatment (see Figure 3.2). The consensus to emerge from this debate was that it was satisfactory to compare a condition in which two events are paired with one in which they are unpaired. Because this design equates exposure to both events, it was assumed that any nonassociative effects arising from habituation or sensitization to the events per se would be equivalent across the paired and unpaired conditions. Consequently, any behavioral differences between these conditions could be attributed to the variation in the temporal relation of the two events. In fact, it has been quite common to assert that Pavlovian conditioning has occurred if the stimulus signaling an outcome evokes behavior that is different from behavior evoked by a stimulus not paired with the outcome (Mackintosh, 1983; Rescorla, 1967).

This recommendation for detecting associative learning has influenced the design and interpretation of the work of many individuals studying learning in simple systems. Many invertebrate species have been judged to have shown Pavlovian conditioning because differential behavior has been observed to paired and unpaired stimuli. Some of these studies involve appetitive conditioning procedures in which a stimulus is paired with an attractive or positive US (e.g. food). For example, using the blow fly, *Phormia regina*, Ricker et al. (1986) paired one of two salt solutions (1M KCl and 1M NaCl) with a sucrose US. They found that proboscis extensions occurred more often to the CS+ than to the CS-. Similar results were reported by Nelson (1971). Fukushi (1979) demonstrated discriminative conditioning in the housefly, *Musca domestica*, using two different odors (acetic acid and *n*-butanol) as conditioned stimuli (CSs). Proboscis extension was greater to the CS paired with a 0.5M sucrose solution than to the unpaired CS. Bitterman et al. (1983) found differential conditioning of proboscis extension in honeybees, *Apis mellifera*. Both

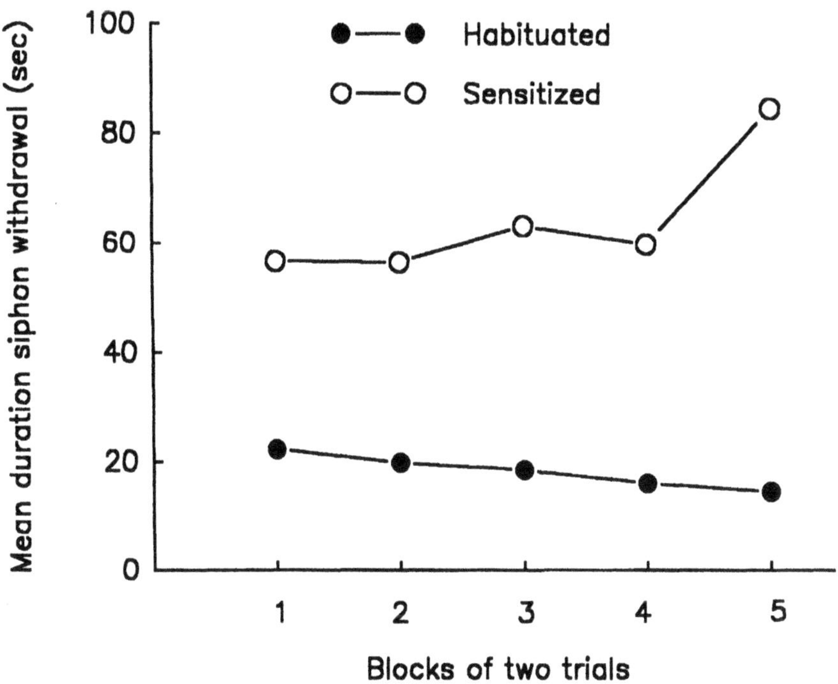

FIGURE 3.2 Behavioral effects of a putative CS (siphon stimulation) and a putative US (tail shock) on a defensive reflex (siphon withdrawal) in *Aplysia californica*. Mean duration of siphon withdrawal is shown in response to the administration of a brief squirt of water to the siphon every 8 min for 10 trials. Data are plotted in blocks of two trials. Habituated subjects showed a decrease in responding to the repeatedly presented siphon stimulus. Sensitized subjects received a single aversive tail shock 15 min. prior to the siphon stimulus presentations. These subjects showed an immediate and sustained increase in the duration of siphon withdrawal evoked by the siphon stimulus. These findings highlight the need for control procedures so that the behavioral effect of CS-US pairings may be attributed to the pairing rather than to experience with the events per se.

worker bees and drones showed more responses to an odor (e.g. carnation) that was paired with a 20%-40% sucrose solution than to the unpaired odor (e.g. geraniol).

Discriminative appetitive conditioning has also been demonstrated in the marine mollusc, *Aplysia californica*, illustrated schematically in Figure 3.3 (Colwill et al. 1994). On CS+ trials, a plastic open-tipped probe containing one of two chemosensory stimuli (either banana- or lemon-flavored seawater) was applied to the lips and mouth area for 10 sec. A piece of food held in forceps was applied to the lips coincident with the removal of the CS+ and was kept in contact with the lips until it was grasped by the radula. On CS- trials, the other chemosensory CS was presented for 10 sec but no food was given. Following a series of conditioning sessions, a test session was administered by an experimenter blind to the conditioning history of the subjects. Each CS was presented twice for 15 seconds but no food USs were given. Observations of each subject's behavior during CS presentations were recorded at 5 second intervals. Figure 3.4 shows the mean percentage of observations recorded as bites; a bite was defined as retraction of the lips and extension of the radula. The mean percentage of bites was significantly higher to the CS+ than to the CS-, Wilcoxon T (15) = 12, $p < .01$.

Several invertebrates have also shown differential performance to a CS+ and CS- in situations that employ an aversive US (e.g. electric shock). Using an electric shock US, Block and McConnell (1967) demonstrated conditioned body contractions in planarians, *Dugesia dorotocephala*. The frequency of the conditioned response increased considerably to the CS+ (e.g. light) but remained at a constant low level to the CS- (e.g. weak vibration) over the course of discrimination training. Sahley and her colleagues have obtained discriminative conditioning in the leech, *Haemopis marmorata* (Karrer and Sahley, 1988) and in the terrestrial mollusc, *Limax maximus*, (Sahley et al. 1981). Karrer and Sahley (1988) exposed leeches to two different foods (raw chicken and raw liver), one of which (CS+) was always paired with a quinidine sulfate solution applied

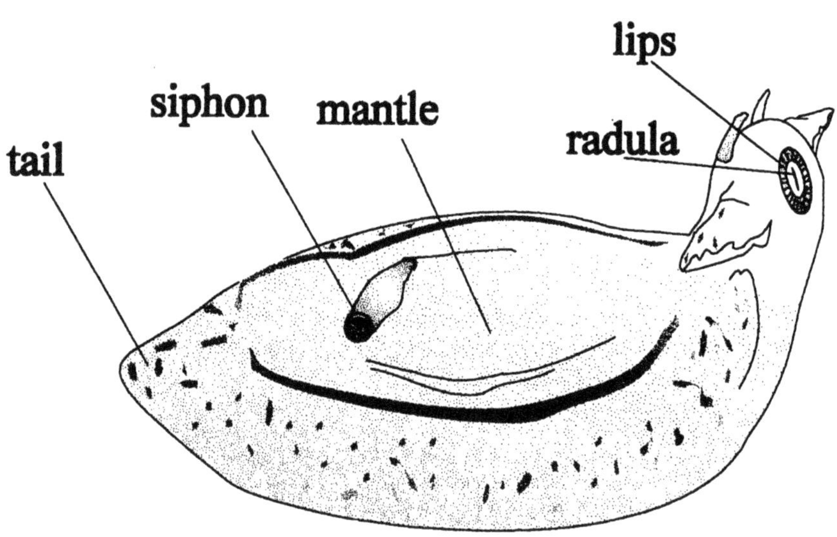

FIGURE 3.3 Schematic diagram of *Aplysia californica* following surgical removal of the parapodia. Labelled organs indicate sites of CS and US administrations in the appetitive and aversive Pavlovian conditioning procedures described in text.

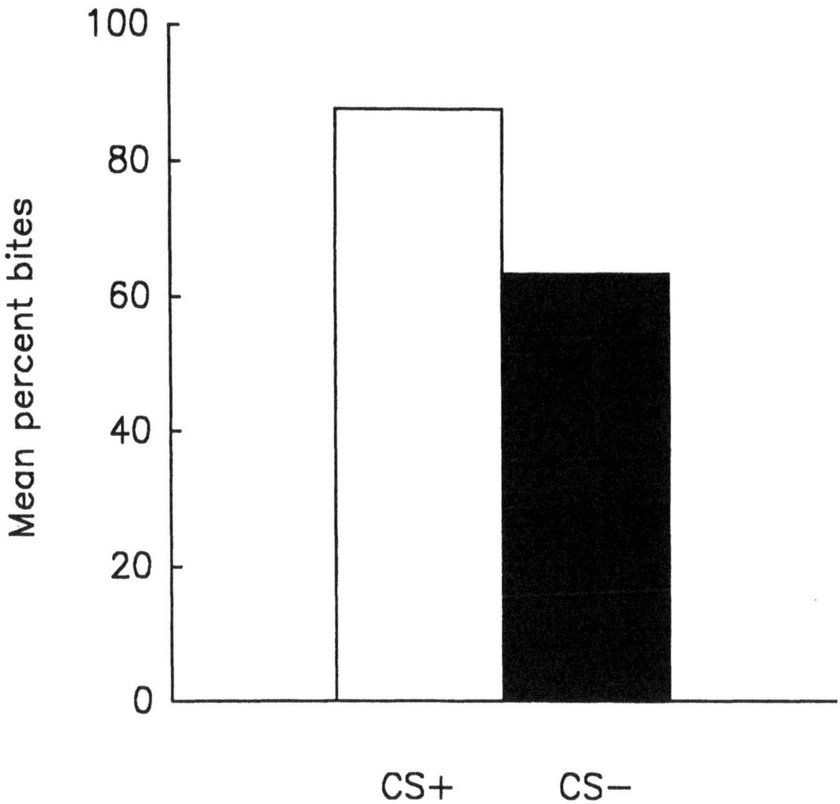

FIGURE 3.4 Results of the discriminative appetitive conditioning experiment with *Aplysia* in which one chemosensory cue was paired with food (CS+) and a different chemosensory cue was not paired with food (CS-). The mean percentage of observations recorded as bites on the nonreinforced test trials is shown separately to the CS+ (white bar) and to the CS- (black bar).

to the prostomium region. On test trials, subjects were given a choice between raw beef and either the CS+ or the CS- foods. Whereas the preference for the CS- relative to raw beef did not change over the course of conditioning trials, subjects showed a marked decline in their choice of the CS+. Henderson and Strong (1972) also obtained conditioning in the leech, *Macrobdella ditetra*. Subjects received either paired presentations of a light CS with a shock US or unpaired CS and US presentations. Anteroposterior contractions to the CS increased substantially in the paired group but not in the unpaired group.

Differential aversive conditioning has also been obtained in the nudibranch mollusc, *Hermissenda crassicornis*, (Crow and Alkon, 1978), and in *Aplysia californica* (Carew et al. 1981; Carew et al. 1983; Walters et al. 1979, 1981). *Hermissenda* exposed to pairings of a spot of light with rotation showed a CS-elicited shortening of the foot (Lederhendler et al. 1986) and were less likely to enter the illuminated area (Crow and Alkon, 1978) than animals given unpaired presentations of these events. *Aplysia* given pairings of a weak tactile stimulus applied to the siphon with shock to the tail evidenced significantly longer siphon withdrawals to the CS than subjects receiving unpaired presentations (Carew et al. 1981). A discriminative conditioning paradigm has also been developed in *Aplysia* in which mild stimulation is applied to different parts of the body (e.g. mantle and siphon). In this case, siphon withdrawal was enhanced to the CS paired with a tail shock US relative to the CS- (Carew et al. 1983).

Analysis of these preparations has yielded significant insights into the neural basis of plasticity. In the last decade, there has been tremendous success in identifying the neural mechanisms that mediate the behavioral effects of the differential conditioning paradigm (see Byrne et al. 1991; Hawkins et al. 1986b; Macphail, 1993). For example, a cellular mechanism called activity-dependent neuromodulation has been discovered in *Aplysia* (Hawkins et al. 1983; Walters and Byrne, 1983) that selectively enhances the ability of a sensory neuron activated by the CS+ to trigger the response system producing the conditioned response (CR). The development of biochemical analogues of the differential conditioning paradigm has allowed further investigation of the mechanism that underlies the superior synaptic transmission induced by activity-dependent neuromodulation (see Byrne and Crow, 1991). For instance, Ocurr et al. (1985) used a high concentration of potassium (K^+) applied to an isolated cluster of pleural sensory neurons to mimic the effect of spike activity produced by a CS, and an application of serotonin (5-HT) to mimic the effects of a US. Presentations of the CS and US were either paired or unpaired. Ocurr et al. (1985) discovered elevated levels of cyclic AMP, an intracellular second messenger with broad effects on membrane currents, transmitter release and the phosphorylation of specific proteins, in the paired sensory neurons relative to the unpaired sensory neurons (see also Abrams et al. 1984, 1985).

Ironically, demonstrations that some invertebrate species are, by this criterion of differential responding to CS+ and CS-, endowed with the capacity for associative learning has prompted some serious scrutiny of the assumptions underlying this recommendation. Several authors have cautioned that the paired presentation of two events may have effects other than that of establishing an association between them (Farley and Alkon, 1985; Mitchell et al. 1975, 1976, 1977; Pfautz et al. 1978; Rescorla, 1984a; Rescorla and Holland, 1976). That concern stems in part from observations that the presentation of a salient event in the period immediately following a stimulus can disrupt habituation to that stimulus (Wagner, 1978). Thus, the presentation of an outcome or US immediately or shortly after the CS+ may reduce the amount of habituation to the CS+ that would otherwise occur. Such interference is absent on trials on which the CS- is presented. Accordingly, a stimulus paired with an outcome may yield effects different from those of an unpaired control stimulus not because of differential opportunity to form an association but because of differential habituation to the signals. Consequently, it may not always be possible to infer associative learning simply from differential responding to the CS+ and CS-.

An unpublished experiment from my laboratory using *Aplysia* may aid in understanding this critique of the use of a differential conditioning paradigm for demonstrating associative learning. One group of subjects received 10 presentations of tactile stimulation of the siphon (CS+) paired with an aversive shock US applied to the mantle. Another group received the same events but the siphon stimulation (CS-) was never temporally contiguous with the shock US. In a test conducted 24 hours later, the CS+ elicited more defensive responding as indexed by the duration of siphon withdrawal than the CS- (136 sec. and 47.8 sec., respectively, Mann-Whitney $\underline{U}$ (10,10) = 19, $p < .05$). This difference meets the classical requirement for an associative change: The differential temporal arrangements of the events resulted in differential performance.

However, a nonassociative account of that differential performance is also possible. The defensive response evoked by the CS in the two groups during testing may reflect no more than the effect of sensitization induced by the US. Even if that response is not part of the original reaction to the stimuli, it may nevertheless emerge as a result of sensitization. The fact that the response is greater to the CS+ may only be due to reduced habituation to that stimulus caused by the presentation of the US. In other words, exposure to the US would lead to longer siphon withdrawals in both the paired and unpaired animals (an effect of sensitization); exposure to the CS would lead to a decrease in the duration of siphon withdrawal (an effect of habituation) but this decrease would be smaller for the paired animals because administration of the US immediately after the CS+ would interfere with the habituation process. In this way, nonassociative factors may be entirely responsible for the difference in responding to the CS+ and CS-.

Three additional comments concerning this nonassociative account of the difference in behavior evoked by a CS+ and a CS- should be noted. First, the argument is applicable not only to experiments using a between-subjects design but also to those using a within-subjects design to demonstrate discriminative conditioning. The US presentation may sensitize responses to both the CS+ and the CS- but because of its greater contiguity to the CS+ it will have a potentially more potent effect in disrupting habituation to the CS+. Second, this argument can, in principle, be applied to the results of any simple conditioning study, regardless of whether the subjects are invertebrates such as *Hermissenda, Limax, drosophila* or bees, or vertebrates such as rats, rabbits or pigeons. Third, this argument is considerably weakened if the form of the CR is qualitatively different from that of the response produced by sensitization.

This interpretative difficulty is not intended to minimize the significance of demonstrations of differential conditioning in simple systems. At the very least, the nonassociative explanation of differential responding to CS+ and CS- offers a simple mechanism for prolonging attention to events of importance. Sensitivity to the temporal arrangement of a CS and US through nonassociative processes may represent an intermediate step in the evolution of a mechanism for actually associating those events. To determine, however, if an associative mechanism does in fact mediate differential responding will require the application of more sophisticated techniques described in the next section.

The limitation on the inference that may be drawn about associative learning from a successful differential conditioning experiment also extends to experiments using the random control procedure (Rescorla, 1967). The random control procedure arranges for two events to be presented in such a way that there is an overall zero correlation between them. In other words, the random control procedure ensures that one event is completely uninformative about the potential occurrences or absences of the other event. This control procedure is assumed to provide a neutral (nonassociative) baseline for assessing the effects of paired and unpaired presentations of a CS and a US. Thus, it has been common to attribute a difference in responding to a paired CS and a random CS to the development of an excitatory association between the CS+ and the US; similarly, it has been common to attribute any differences between an unpaired CS and a random CS to the development of an inhibitory association between the CS- and the US.

However, proper application of the random control procedure involves equating exposure to the CS and matching the frequency of CS-US pairings to the paired condition. Thus, in order to generate a random relation between the CS and the US, it is necessary to distribute additional USs during the periods in which the CS is absent. Consequently, the random treatment offers more opportunities for habituation to that US. To the degree that an habituated US interferes less with the processing of another event, there will be less disruption of habituation to the random CS on those occasions when the random CS is followed by the US. As a result, a difference may emerge to a CS+ and a random CS not because of a difference in the opportunity of those CSs to develop an association with their USs but because of a difference in the ability of those USs to interfere with habituation to their CSs. Thus, operational criteria per se are inadequate to determine the source of the behavioral differences generated by paired, random and unpaired treatments.

The idea that outcomes may act in ways other than by being associated with the events that immediately precede them is strikingly evident in the historical treatment of instrumental learning (Guthrie, 1952; Hull, 1943; Thorndike, 1911). Classical stimulus-response (S-R) theory conceived of outcomes as catalysts whose function was to strengthen the connection between behavior and the stimulus context in which that behavior occurred. Although instrumental learning may not be understood entirely in these terms, several authors have found it useful to invoke a catalytic process to explain some consequences of response-outcome and stimulus-outcome contiguities (Colwill and Rescorla, 1986; Rescorla, 1982b). In these cases, a contiguous outcome promotes learning about two other events rather than itself participating in the association. Because the contiguous occurrence of two events may have consequences for responding through means other than the establishment of an association between them, operational criteria per se are insufficient for inferring the presence of an association.

Contemporary Approaches to Identifying Associative Learning

In this section, two techniques are examined which, combined with standard operational criteria, can be used for identifying associations. Both assessments must be made in situations in which subjects are exposed to at least two outcomes (e.g. two different foods) each of which is uniquely signaled either by Pavlovian stimuli or by instrumental responses. In one case, patterns of behavior are inspected for their dependence on the identity of the outcomes; in the other case, the behavioral consequences of conditioning are examined for their dependence on the value of the outcomes. To the degree that performance depends on the identity or value of the particular outcome that followed the CS or the instrumental response, it cannot simply be that the outcome is preserving a sensitized response.

A. Dependence on Outcome Identity

Three procedures exploit the dependence on outcome identity to detect the presence of associations. One method inspects the conditioned response for features that are uniquely correlated with the outcome; one method examines the subsequent ability of the signal to influence preferentially instrumental responses established with the same outcome; and one method examines the effect of noncontingent outcome deliveries in selectively impairing learning about stimuli and responses paired with that outcome.

1. The Form of the CR

Pavlov (1928) was among the first to note that the form of the conditioned response may be tuned to specific aspects of the US. The magnitude and composition of the salivary reaction was influenced by whether a dry food or acid served as the US. Analogous observations were later made by Jenkins and Moore (1973) using the pigeon autoshaping preparation. They exposed pigeons that were both hungry and thirsty to presentations of two very different illuminated discs; one of these stimuli was followed by food and the other by water. Both stimuli came to produce pecking, but the topography of the pecks was quite different to the two stimuli. The CS paired with food elicited a brief, forceful peck in which the beak opened just before key contact; the CS paired with water produced a softer, shallower peck in which the beak was held slightly open throughout a sweep across the key. These conditioned responses mimicked the manner in which the pigeons ate their grain and drank their water (Skelton et al. 1980; Spetch et al. 1981). More recently, LaMon and Zeigler (1984) have shown that the size of the grain determines how wide the beak will be held open during pecks to the CS.

LaMon and Zeigler's (1984) observations are especially pertinent to the issue of whether these data can also be understood in terms of differential habituation. Some critics might argue that different USs interfere with habituation of different aspects of the stimulus. The results of Jenkins and Moore's (1973) study may show only that food disrupts habituation of those stimulus features that support forceful pecking, and water selectively reduces habituation of the stimulus features that sustain soft pecks. However, this argument stretches credibility to the limit in order to account for the relation between seed size and the magnitude of beak openings.

Measurement of the dependence of the behavior produced by the CS on the identity of the US has indicated an associative basis for the behavioral changes evidenced by *Aplysia* in a Pavlovian conditioning paradigm. Different sites of noxious stimulation have been found to elicit qualitatively different movements of the siphon. Specifically, shock USs delivered to anterior sites (e.g. head or mantle) produce constriction of the siphon whereas shock USs administered to posterior sites (e.g. tail) produce flaring of the siphon. Hawkins et al. (1989) found that when a mild tactile CS applied to the siphon was used to signal the occurrence of such USs, the response that emerged to the CS resembled the response to the US. Similar results have been obtained by Walters (1989).

The fact that the CR elicited by a CS may incorporate components of the unconditioned response (UR) evoked by the US is consistent with the view that CSs activate representations of the outcomes whose arrival they predict. Of course, for the purpose of identifying stimulus-outcome associations, the CR need not resemble the UR very closely at all. Rather, the logic of this approach requires only that some component of the CR be unique to that particular US.

This method has also yielded suggestive evidence for associative encoding of the outcome in instrumental learning situations. Hull (1977) examined the way in which the details of lever press responses varied as a function of whether a food pellet or water served as the reinforcer. Most notably, contact time with the lever manipulandum was significantly longer when responses earned food pellets than when they earned water. Although not without alternative interpretation (Davey and Cleland, 1982), these data suggest that anticipation of the outcome was mediated by its association with the response.

2. *Transfer to Instrumental Responses*

Simple observation of the effect of US identity on the CR provides a relatively direct measure of stimulus-outcome associations. An alternative approach has been to examine the impact of CSs as a function of US identity in a variety of different testing situations. One example of such a test that shows a US-specific effect of contiguity assesses the ability of CSs to modulate instrumental performance. Some studies have measured the impact of Pavlovian CSs through their ability to acquire control over instrumental responses whereas other studies have examined their initial effect on the performance of previously established instrumental responses. In either event, CS-US associations are inferred if the impact of the CS is found to be conditional upon the identity of the outcome used to train the instrumental response.

Several investigators have transferred Pavlovian CSs to well-established instrumental responses and looked for specific interactions as a function of outcome identity (Baxter and Zamble, 1982; Colwill and Motzkin, 1994; Colwill and Rescorla, 1988; Kruse et al. 1983). An experiment by Colwill and Rescorla (1988) shows how the transfer technique can be used to detect stimulus-outcome associations in both Pavlovian and instrumental conditioning. Rats were allowed to earn one outcome by responding in the presence of a stimulus (Sd). During another stimulus (CS+), performance of that instrumental response was not permitted and a different outcome was simply delivered. Thus, for some subjects, displacements of a recessed panel earned food pellets during presentations of a light Sd; during presentations of a noise CS+, access to the response manipulandum was prevented and sucrose liquid was simply delivered. Particular stimulus, response and outcome combinations were counterbalanced across subjects. Subsequently, two new responses (lever pressing and chain pulling) were trained, one with the outcome that had occurred during the Sd, and one with the outcome that had occurred during the CS+. Finally, these two transfer responses were made available with occasional presentations of the Sd and CS+. The basic design of this experiment is summarized in Figure 3.5.

The results of the transfer test with the lever press and chain pull manipulanda and intermittent presentations of the Sd and CS+ are shown in Figure 3.6. The impact of both stimuli was conditional upon the identity of the outcome used to train the transfer responses: Both stimuli augmented responses with which they shared an outcome. These results establish that associations develop between different kinds of signals and the outcomes that occur in their presence. Incidentally, they also demonstrate that instrumental responses become associated with their outcomes; in the absence of response-outcome associations, the stimuli would not have been able to affect differentially responses trained with the same outcome.

Similar conclusions were reached by Trapold and his colleagues (Trapold, 1970; Trapold and Overmier, 1972) in a series of elegant studies examining the effect of stimulus-outcome associations on the course

Basic Design of Transfer Experiment

Training	Transfer	Test
Sd: Rn-O1	R1-O1	Sd: R1 v R2
CS-O2	R2-O2	CS: R1 v R2

FIGURE 3.5 Basic design of a transfer experiment using rats. Sd and CS denote an instrumental discriminative stimulus and a Pavlovian conditioned stimulus, respectively. Rn denotes nose poke responses; R1 and R2 are lever press and chain pull responses. O1 and O2 are food rewards, 45-mg Formula A Noyes food pellets and 8% sucrose liquid.

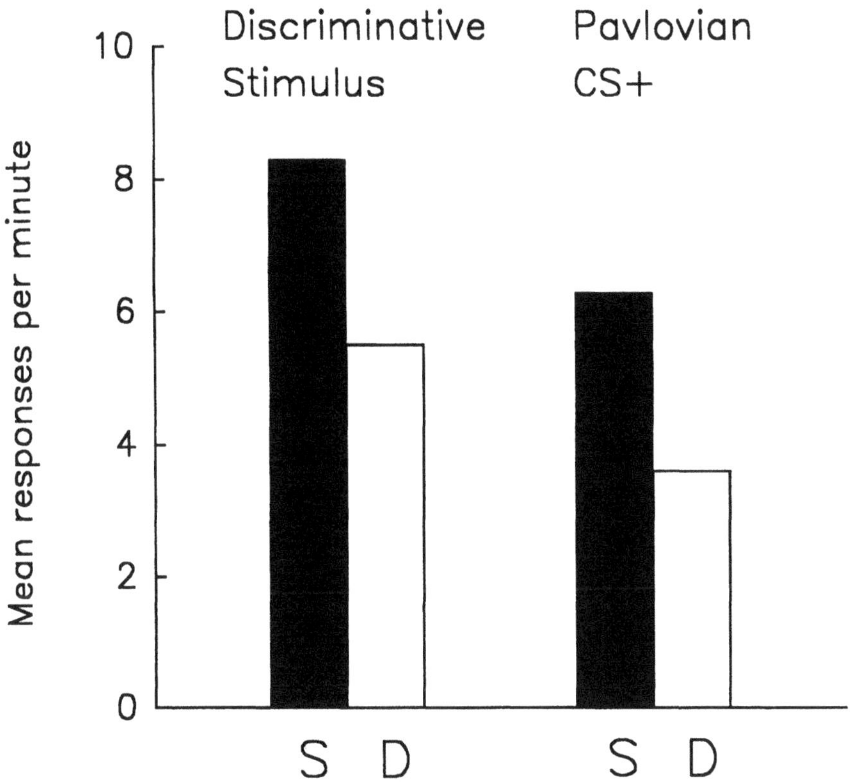

FIGURE 3.6 Results of the transfer experiment. The two bars on the left show responding during an instrumental discriminative stimulus and the two bars on the right show responding during a Pavlovian CS+. Responding is plotted separately when the training outcome was the Same (black bar) for both the stimulus and the response, and when that outcome was Different (white bar) for the stimulus and the response. The Sd enhanced performance of the response trained with the same outcome, Wilcoxon $\underline{T}$ (15) = 18, $p < .05$; similarly, the CS+ also augmented performance of the response trained with the same outcome, $\underline{T}$ (16) = 29, $p < .05$. (Based on data from Colwill and Rescorla, 1988).

of new learning. Trapold (1970) gave rats training designed to establish two stimuli as Pavlovian signals for different outcomes. One stimulus was paired with food and the other stimulus was paired with sucrose. Then, in the second phase of the experiment, the rats were required to learn an instrumental task in which the two stimuli signaled which one of two lever press responses would lead to reward. For one group, correct responses during a stimulus were followed by the outcome previously paired with that stimulus; for the other group, the outcomes were

Basic Design of Trapold (1970) Experiment

Pavlovian Training	Instrumental Training	
S1-O1	S1: R1-O1, S2: R2-O2	Consistent
S2-O2	S1: R1-O2, S2: R2-O1	Inconsistent

FIGURE 3.7 Basic design of Trapold (1970) experiment. S1 and S2 denote tone and clicker stimuli; R1 and R2 denote levers located to the left and right of the food magazine; O1 and O2 denote 45-mg food pellets and 12% sucrose liquid. The S-O combinations in instrumental training were either the same as (Consistent) or different from (Inconsistent) those in the previous Pavlovian training.

switched. Figure 3.7 illustrates the design of this experiment. The rats in the condition in which the stimulus-outcome relationships remained consistent across the two phases showed superior acquisition of the discrimination relative to the rats in the inconsistent condition. This difference suggests that each stimulus was associated with its specific outcome during the Pavlovian conditioning phase.

The transfer technique can be used not only to determine what animals learn about the identity of the outcomes that follow Pavlovian CSs and instrumental responses but to inspect what they know about the absences of such outcomes (Colwill, 1991, 1994; Kruse et al. 1983). For example, Colwill (1991) analyzed stimuli trained as signals that instrumental responses will not be followed by their outcomes. Such stimuli were shown to provide information about the identity of those omitted outcomes. Rats were allowed to earn food pellets for one response (e.g. nose poking) except during presentations of a stimulus (e.g. a light) and sucrose liquid for another response (e.g. handle pulling) except during presentations of a different stimulus (e.g. a noise). Then, two new responses (lever press and chain pull) were established; one of these responses earned food pellets and the other earned sucrose liquid. In a subsequent transfer test, there were occasional presentations of the light and noise stimuli with the lever press and chain pull manipulanda available. Colwill (1991) found that each stimulus suppressed to a greater degree the response trained with the outcome whose omission had previously been signaled by that stimulus. This finding indicates that the subjects had encoded the identity of the omitted outcome as an associate of the discriminative stimulus (see also Bonardi, 1989).

3. *Noncontingent Outcome Deliveries*

A third technique that has been used to detect associations exploits the observation that learning about the relation between two events is determined by the contingency between those events. It is now well-established that the formation of a stimulus-outcome association in Pavlovian conditioning may be impeded by noncontingent deliveries of the outcome in the absence of the stimulus (Durlach, 1983; Durlach and Shane, 1993; Rescorla, 1968; Stanhope, 1990). Analysis of the effect of degrading the contingency between the CS and the US suggests that formation of a CS-US association is blocked by the association that develops between the context and the US (Durlach, 1983, 1989; Rescorla, 1984b). Given this view, the degree to which a detailed representation of the outcome is associated with the context may be assessed by the severity with which CS-US learning is disrupted by the presentation of identical outcomes during the intertrial interval (ITI).

Several experiments conducted with pigeons and rats provide evidence that the identity of the ITI outcome influences formation of the stimulus-outcome association. An unpublished pigeon autoshaping experiment from my laboratory shows that pairings of a keylight CS with a food US are less effective if identical food USs are presented during the ITI. The basic design of this experiment is shown in Figure 3.8. In each conditioning session, pigeons were given 8 presentations of each of two 10-sec CSs. One CS was the display of an inverted triangle on the response key and the other CS was the display of an X on the same key. Each of these CSs was uniquely paired with one of two food USs, corn or wheat. The probability of receiving food coincident with the offset of the CS was .25. One of these USs was also scheduled to occur during the ITI with a probability of .25. In this way, two CSs were each paired

Basic Design of Outcome-Specific Contingency Experiment

Training
CS1-O1, CS2-O2, O1
CS1-O1, CS2-O2, O2

FIGURE 3.8 Basic design of an outcome-specific contingency experiment using pigeons. CS1 and CS2 are projections of an inverted white triangle and a white X on the response key. O1 and O2 are different grains, corn and wheat.

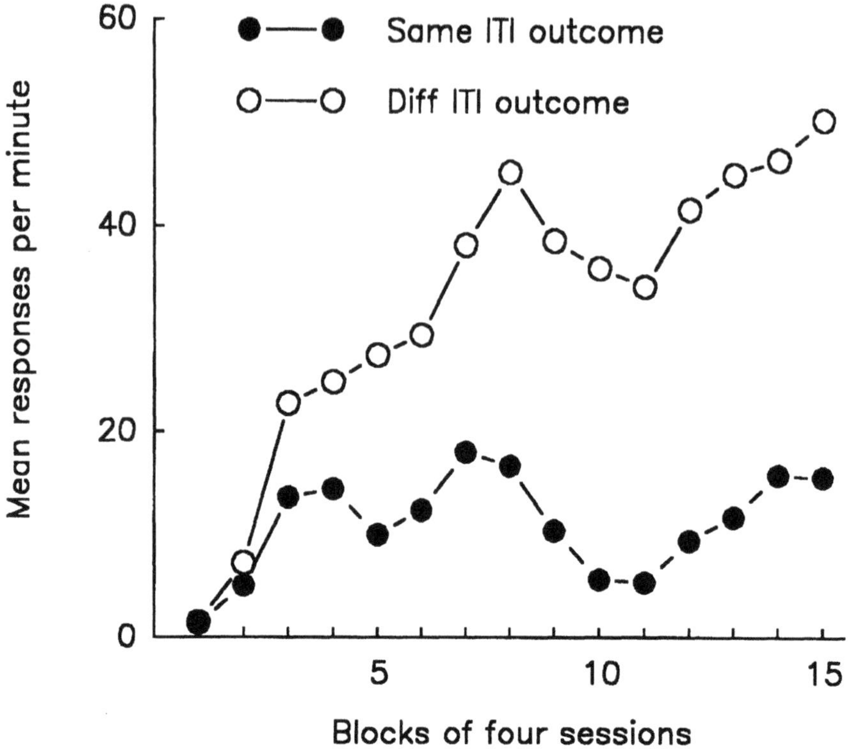

FIGURE 3.9 Results of the outcome-specific contingency experiment. The mean number of keypecks per min is shown separately for the CS paired with the same outcome as that delivered during the ITI (filled circles) and for the CS paired with a different outcome from that delivered during the ITI (open circles). Each point represents the mean response rate for blocks of four sessions.

equally often with different outcomes but the relation of one of those CS-US pairs was made random by presenting its outcome during the ITI.

The results of this experiment are shown in Figure 3.9. Pecking is shown separately to the CS paired with the US that was the same as that delivered during the ITI and to the CS paired with a different outcome. It is clear that differential responding emerged to the keylights over the course of training. Whereas pecking developed to the CS signaling the different outcome, acquisition to the CS paired with the same outcome was impaired. These results confirm those reported by Stanhope (1992) using water and grain as USs in a pigeon autoshaping preparation, and are consistent with those reported by Colwill and Motzkin (1994) using pellets and sucrose USs in a rat appetitive conditioning procedure. Together, they provide evidence consistent with the view that a detailed representation of the outcome is associated with the stimulus that predicts its occurrence.

Manipulations of the contingency between the response and its outcome have also generated evidence for outcome encoding in instrumental learning. Colwill and Rescorla (1986) reported that noncontingent outcome presentations selectively disrupt performance of a response previously trained with that outcome relative to a response trained with a different outcome. Rats were initially trained to make one response (e.g. chain pulling) for a sucrose liquid and a different response (e.g. lever pressing) for a food pellet. Then, in subsequent sessions, additional presentations of one of the two outcomes were delivered randomly. Performance of the response earning the same outcome as that delivered noncontingently was significantly reduced relative to the response earning a different outcome.

Each of these observations of the outcome-specific effects of contingency has been accompanied by additional findings that reinforce an associative explanation of the performance decrement. For example, Colwill and Rescorla (1986) found that the selective depression of the same response persisted into extinction when all outcomes were withheld, and that the disruptive effect of noncontingent outcomes was alleviated by signaling those outcomes with a Pavlovian CS. These results encourage the view that noncontingent outcome presentations weakened the association between the response and that outcome rather than affecting behavior through nonassociative processes.

Outcome-specific contingency effects have yet to be demonstrated in simple invertebrate preparations. The use of this strategy for revealing associations with the outcome in invertebrates, however, will be governed largely by the degree to which a context blocking hypothesis can be substantiated for the disruptive effects of inserting additional USs into the ITI. Initial studies with *Aplysia* and *Hermissenda* have shown both mollusks to be sensitive to the contingency between a CS and a US (Farley, 1987; Hawkins et al. 1986a). That the disruptive effect of inserting USs into the ITI is due to the development of a context-US association that blocks CS-US learning is suggested by the fact that signaling these ITI USs with another stimulus alleviates the deficit (Hawkins et al. 1986a).

Consistent with a context-blocking analysis of the effects of contingency are studies demonstrating discriminative conditioning of contextual stimuli (Colwill et al. 1988; Rogers et al. in press). For example, Colwill et al. (1988) exposed *Aplysia* to two different contexts; one context was a smooth, round, white plastic bowl filled with lemon-flavored artificial seawater, and the other context was a dark gray rectangular chamber with a ridged surface and an aerator in one corner. Subjects received several aversive shocks to the mantle in one of the contexts but not in the other. In a subsequent test session, the duration of siphon withdrawal in response to a mild tactile stimulus applied to the siphon was measured in the shocked and nonshocked contexts by an experimenter blind to the specific conditioning histories of the individual subjects. Longer siphon withdrawals were obtained in the context paired with the aversive shock USs relative to the unpaired context. That this difference reflects acquisition of a context-US association is supported by additional evidence showing that subsequent pairings of the tactile stimulus and the shock US were less effective in the paired context relative to the unpaired context.

B. Dependence on Outcome Value

The major alternative to the identity-based detection method is one that involves altering the value of one of the elements of an association and inspecting the impact on behavior to the other target element. To the degree that responding to the target element tracks the new value of the altered element, one may infer that the two elements were associated.

This technique has been used extensively to analyze the structure of Pavlovian conditioning (Colwill and Motzkin, 1994; DeCola and Fanselow, 1995; Holland, 1990; Holland and Straub, 1979; Rescorla, 1982a; Stanhope, 1989, 1992). In an experiment with rats, Colwill and Motzkin (1994) used two different stimuli, a 10-sec tone and a 10-sec light, to signal two different outcomes, a food pellet and a liquid sucrose. The basic design is outlined in Figure 3.10. One of the CSs predicted one outcome, and the other CS predicted the other outcome. As a result of this treatment, both CSs came to elicit a variety of behaviors directed towards the maga-

Basic Design of US Devaluation Experiment

Conditioning	Devaluation	Test
CS1-O1	O1+, O2-	CS1 v CS2
CS2-O2	O2+, O1-	

FIGURE 3.10 Basic design of Colwill and Motzkin (1994) US devaluation experiment with rats. CS1 and CS2 are Pavlovian CSs, tone and light; O1 and O2 are food USs, 45-mg Formula A Noyes pellets and 8% sucrose liquid; + and - indicate the administration or not of LiCl.

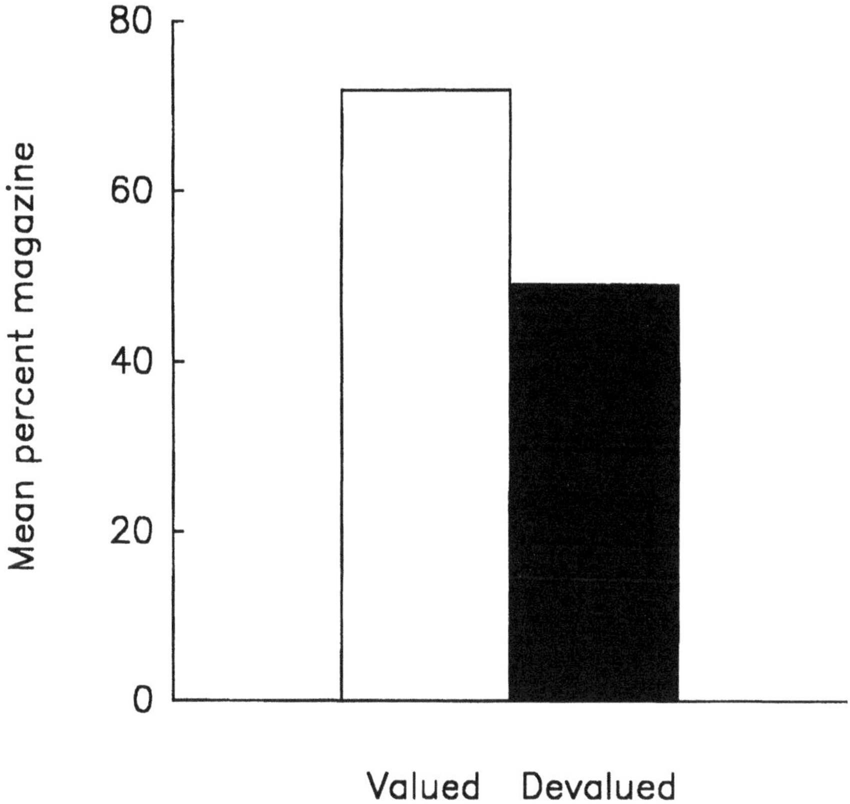

FIGURE 3.11 An observer blind to the conditioning treatments observed each subject once during the first 5 sec. CS period and once during the last 5 sec. period on each of the 8 nonreinforced test trials of each CS. The mean percentage of magazine-directed behaviors during the last 5 sec. is shown as a function of whether the CS had been trained with an outcome that was subsequently devalued (black bar) or not (white bar). Significantly fewer magazine-directed behaviors were observed during the CS whose outcome had been devalued, Wilcoxon $\underline{T}$ (15) = 15.5, $p < .05$). (Based on data from Colwill and Motzkin, 1994).

zine where the outcomes were delivered. These magazine-directed behaviors included approach to the magazine, orienting to the magazine and insertion of the snout or head into the magazine. Then, the value of one of the outcomes was reduced through pairings with a toxin. Finally, responding to the CSs was observed in an extinction test. The results of that test are shown in Figure 3.11. The CS whose outcome had been reduced in value evoked significantly less magazine-directed behavior than the CS whose outcome had not been altered. The specificity of this reduction in responding eliminates many nonassociative explanations for the result and suggests that specific associations developed between signals and their outcomes (Rescorla and Cunningham, 1978).

Similar conclusions have also been reached by investigators studying Pavlovian second-order conditioning (Rescorla, 1980). Of considerable interest is a report by Sahley and her colleagues (Sahley et al. 1984) of results in the terrestrial mollusc, *Limax maximus*, that parallel those obtained with vertebrates. They found that responding to a second-order stimulus was sensitive to a change in the value of its outcome. Initially, subjects received pairings of a carrot odor CS and a quinidine US. Then, subjects were given simultaneous pairings of a potato odor CS with the carrot odor CS. These treatments were sufficient to establish a conditioned response to the second-order stimulus (the potato odor) when tested alone. That response, however, was abolished if prior to testing the second-order stimulus, subjects were given nonreinforced presentations of the first-order stimulus (carrot odor). Demonstrations of this sort are routinely interpreted as evidence of an association between the second-order and the first-order stimuli.

Use of the outcome revaluation technique has also provided important insights into the structure of instrumental learning. Several experiments suggest that rats learn about the consequences of their instrumental acts (Adams and Dickinson, 1981; Colwill, 1994; Colwill and Rescorla, 1985). This point was made clearly in a series of experiments by Colwill and Rescorla (1985). The basic design of one experiment is outlined in Figure 3.12. Rats were trained to make two different responses, lever press and chain pull. One of the responses was followed by food pellets and the other by liquid sucrose. Then, the value of one of the outcomes was reduced by conditioning a taste aversion to that outcome. Finally, subjects were given a test in which lever and chain were both available but no outcomes were delivered. The results of that test are displayed in Figure 3.13. Responding established with the cur-

Basic Design of Instrumental Outcome Devaluation Experiment

Training	Devaluation	Test
R1-O1	O1+, O2-	R1 v R2
R2-O2	O2+, O1-	

FIGURE 3.12 Basic design of Colwill and Rescorla (1985) instrumental devaluation experiment with rats. R1 and R2 are instrumental responses, lever press and chain pull; O1 and O2 are food rewards, 45-mg Formula A Noyes pellets and 8% sucrose liquid; + and - indicate the administration or not of LiCl.

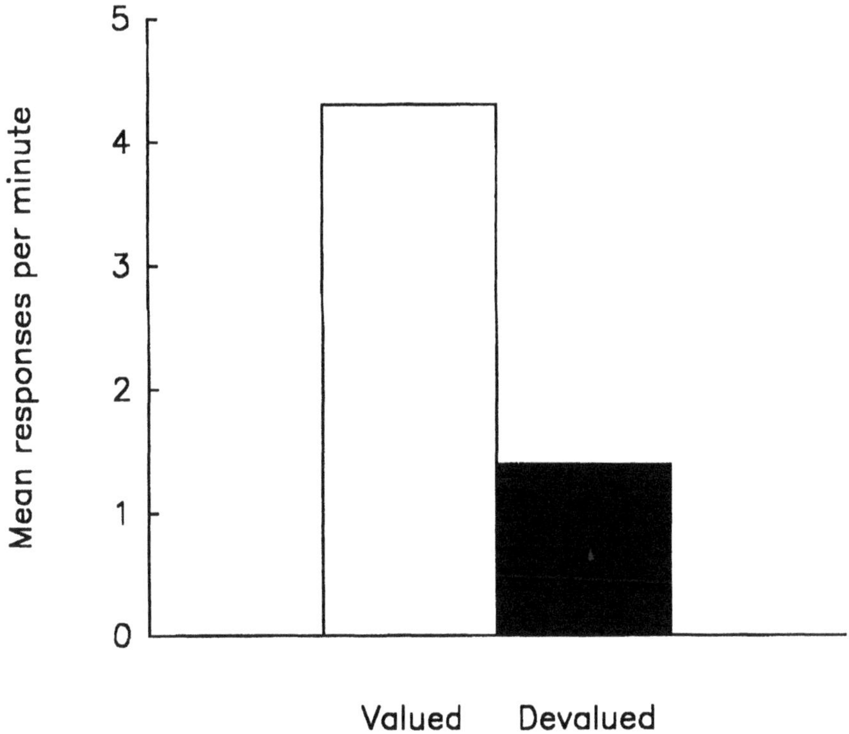

FIGURE 3.13 Mean responses per min during an extinction test are shown when the outcome used to train the response had been devalued (black bar) and when it had not (white bar). Responding was significantly depressed following devaluation of its outcome, Wilcoxon $\underline{T}$ (31) = 45, $p < .01$. (Based on data from Colwill and Rescorla, 1985).

rently devalued outcome was substantially reduced relative to responding trained with the outcome whose value was unchanged.

Dickinson (1989) and Shipley and Colwill (in press) have also obtained evidence for associative encoding of the outcome when the value of the outcome is increased by manipulating drive or motivation. For example, Shipley and Colwill (in press) trained hungry rats to make one response (e.g. lever press) for food pellets, and a different response (e.g. chain pull) for liquid sucrose. In the second phase, the rats were given a choice between the two responses in the absence of the outcomes. During this test, the animals were either hungry or thirsty. The purpose of inducing thirst was to increase the value of the liquid reward relative to the solid food reward. If the rats had learned about the particular outcomes that were contingent on the two responses, they should

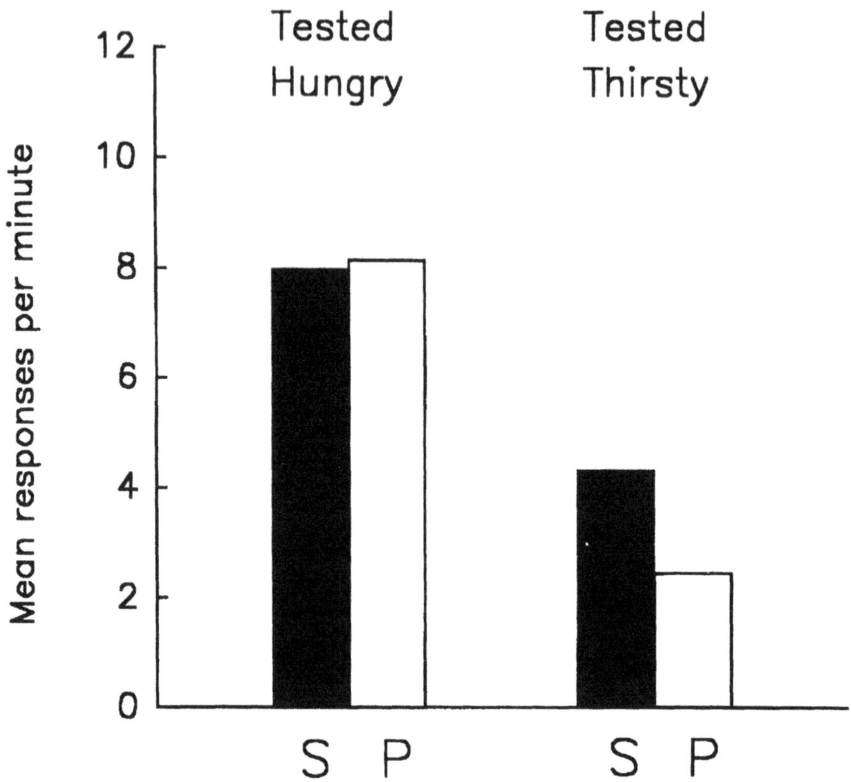

FIGURE 3.14 Mean responses per min during extinction tests conducted when the rats were either hungry (pair of bars on left) or thirsty (pair of bars on right). For each drive state, responding is shown separately when the sucrose outcome (S) was used to train the response and when the pellet outcome was used to train the response (P). Under hunger, there was no significant difference between pellet-trained and sucrose-trained responses; under thirst, sucrose-trained responses were significantly more likely than pellet-trained responses, Wilcoxon $\underline{T}$ (16) = 27, $p < .05$. (Based on data from Shipley and Colwill, in press).

demonstrate a bias for the sucrose-trained response during testing. The results of testing are shown in Figure 3.14. Under thirst, the rats displayed a significant preference for the sucrose-trained response; under hunger, there was no significant difference between performance of the two responses. These examples of sensitivity of instrumental responses to the current value of the outcomes used to train them provide unequivocal evidence that learning about those outcomes is a natural product of exposure to an instrumental contingency.

Limitations of Detection Methods

The techniques described in this paper constitute a set of powerful and effective probes for detecting the presence of associative learning. However, there are some limitations on the circumstances under which these methods can be used to identify stimulus-outcome and response-outcome associations.

The CR to a signal for a US is determined not only by properties of the US but also by features of the signal (Holland, 1977). To the degree that the influence of the signal dominates, it may be difficult to observe any unique influence of the US. In the pigeon autoshaping procedure, only localized visual signals reliably elicit conditioned pecking. Diffuse stimuli such as sounds and general illumination changes produce increases in general activity. Because it is much less likely that seed size or seed type will produce qualitative differences in patterns of locomotion, selection of diffuse stimuli would obscure outcome-dependent CRs that would have otherwise emerged with the use of localized visual cues. Thus, the selection of particular CSs may importantly determine the conditions under which the form of the CR can be used as a tool for detecting associations with a specific outcome.

The success of the transfer, contingency and outcome devaluation procedures depends upon a sufficiently detailed encoding of the outcome that renders it easily discriminable from any other outcome. However, it turns out that not all conditioning parameters lead to such encoding. In some cases, for instance, changing the value of an outcome may have no effect on responding established by that outcome. The particular circumstances that produce responding that is insensitive to subsequent changes in outcome value are only partially understood (Dickinson, 1980). Rescorla (1980, 1984a) has suggested that variations in both the temporal arrangements of the CS and US and the similarity between them alter the degree to which the subject may be able to identify the US that accompanied a particular CS. Findings of this sort have been taken as evidence that encoding of an outcome is quite flexible. Under some conditions, the sensory aspects of the outcome will participate in the association; under other conditions, the response to the outcome will serve as the associate. Since many physically different USs may elicit similar responses, the devaluation, contingency and transfer techniques will not be suitable for identifying associations in situations that favor response encoding.

Finally, application of these tools to the analysis of Pavlovian and instrumental conditioning in invertebrates provides a practical challenge to investigators. For many species, preliminary work will be required in order to identify multiple stimuli to serve as CSs and USs in Pavlovian conditioning procedures, and multiple behaviors and outcomes for use in instrumental conditioning procedures. The task of selecting potential conditioning events will be facilitated by knowledge of the properties of the subject's sensory apparatus and by familiarity with the demands placed on an individual by its natural habitat. The accomplishments to date of investigators working with simple systems, however, offer reassurances that the behavioral assays used to analyze vertebrate learning processes are in fact likely to reveal the contribution of an associative mechanism to learning in invertebrates.

Summary

The discriminative conditioning paradigm has proven invaluable as a tool for probing the coding properties of animal sensory systems. What it reveals about the cognitive structure of an animal's mind, however, is less obvious. I have discussed a serious limitation of the implications of a difference between the response to a stimulus paired with the US and one that was not paired for detecting associations. I described two alternative approaches that provide superior evidence for associative learning. One method relies on observations that the behavioral consequences of conditioning operations are contingent on the identity of the outcome; the other method evaluates that dependency with respect to the value of the outcome. Although the use of each may be restricted under some conditions, together they equip the researcher with powerful tools for detecting associations between events.

Acknowledgments

I appreciate the critical commentary of Seth Boatright-Horowitz, Robert Rescorla, Sara Shettleworth and Andrea Simmons on previous versions of this manuscript. I thank Robert Tetreault for his assistance in data collection and SBH for preparing Figures 3.1 and 3.3. Preparation of this chapter was supported by National Science Foundation grants BNS-8915342, BNS-8922551 and IBN-9407326 awarded to the author.

References

Abrams, T. W., Bernier, L., Hawkins, R. D. and Kandel, E. R. (1984) Possible roles of Ca^{++} and cAMP in activity-dependent facilitation, a mechanism for associative learning in *Aplysia*. *Soc. Neurosci. Abstr. 10:* 269.

Abrams, T. W., Eliot, L., Dudai, Y. and Kandel, E. R. (1985) Activation of adenylate cyclase in *Aplysia* neural tissue by Ca^{2+}/calmodulin, a candidate for an associative mechanism during conditioning. *Soc. Neurosci. Abstr. 11:* 797.

Adams, C. D. and Dickinson, A. (1981) Instrumental responding following reinforcer devaluation. *Q. J. Exp. Psychol. 33B:* 109-121.

Baxter, D. J. and Zamble, E. (1982) Reinforcer and response specificity in appetitive transfer of control. *Anim. Learn. Behav. 10:* 201-210.

Bitterman, M. E., Menzel, R., Fietz, A. and Schafer, S. (1983) Classical conditioning of proboscis extension in honeybees (*Apis mellifera*). *J comp Psychol. 97:* 107-119.

Block, R. A. and McConnell, J. V. (1967) Classically conditioned discrimination in the planarian, *Dugesia dorotocephala*. *Nature. 215:* 1465-1466.

Bolles, R. C. and Beecher, M. D. (1988) *Evolution and Learning*. Hillsdale, NJ: Erlbaum.

Bonardi, C. (1989) Inhibitory discriminative control is specific to both the response and the reinforcer. *Q. J. Exp. Psychol. 41B:* 225-242.

Byrne, J. H. and Berry, W. O. (1989) *Neural models of plasticity.* New York: Academic Press.

Byrne, J. H. and Crow, T. (1991) *Examples of mechanistic analyses of learning and memory in invertebrates. Learning and Memory.* Academic Press, pp. 329-358.

Byrne, J. H., Baxter, D. A., Buonomano, D. V., Cleary, L. J., Eskin, A., Goldsmith, J. R., McClendon, E., Nazif, F. A., Noel, F. and Scholz, K. P. (1991) Neural and molecular bases of nonassociative and associative learning in *Aplysia*. *Activity driven CNS changes in learning and development*. *Annals NY Acad. Sci. 627:* 124-149.

Carew, T. J., Walters, E. T. and Kandel, E. R. (1981) Classical conditioning in a simple withdrawal reflex in *Aplysia californica*. *J. Neurosci. 1:* 1426-1437.

Carew, T. J., Hawkins, R. D. and Kandel, E. R. (1983) Differential classical conditioning of a defensive withdrawal reflex in *Aplysia californica*. *Science 219:* 397-400.

Colwill, R. M. (1991) Negative discriminative stimuli provide information about the identity of omitted response-contingent outcomes. *Anim. Learn. Behav. 19:* 326-336.

Colwill, R. M. (1994) Associative representations of instrumental contingencies. In D. L. Medin (Ed.) *The psychology of learning and motivation Vol 31*. New York: Academic Press, pp. 1-72.

Colwill, R. M., Absher, R. A. and Roberts, M. L. (1988) Context-US learning in *Aplysia californica*. *J. Neurosci. 8:* 4434-4439.

Colwill, R. M. and Motzkin, D. K. (1994) Encoding of the unconditioned stimulus in Pavlovian conditioning. *Anim. Learn. Behav. 22:* 384-394.

Colwill, R. M. and Rescorla, R. A. (1985) Post-conditioning devaluation of a reinforcer affects instrumental responding. *J. Exp. Psychol. Anim. Behav. Processes 11:* 120-132

Colwill, R. M. and Rescorla, R. A. (1986) Associative structures in instrumental learning. In G. H. Bower (Ed.) *The psychology of learning and motivation Vol 20*. New York: Academic Press, pp. 55-104.

Colwill, R. M. and Rescorla, R. A. (1988) Associations between the discriminative stimulus and the reinforcer in instrumental learning. *J. Exp. Psychol. Anim. Behav. Processes 14:* 155-164

Colwill, R. M., Tetreault, R. R., Goodrum, K. and Martin, A. (1994) Appetitive conditioning in *Aplysia californica. Eastern Psychological Association 65th Annual Meeting*, Providence.

Crow, T. J. and Alkon, D. L. (1978) Retention of an associative behavioral change in *Hermissenda. Science 201:* 1239-1241.

Davey, G. C. L. and Cleland, G. G. (1982) Topography of signal-centered behavior in the rat: Effects of deprivation state and reinforcer type. *J. Exp. Anal. Behav. 38:* 291-304.

DeCola, J. P. and Fanselow, M. S. (1995) Differential inflation with short and long CS-US intervals: Evidence of a nonassociative process in long-delay taste avoidance. *Anim. Learn. Behav. 23:* 154-163.

Dewsbury, D. A. (1990) *Contemporary issues in comparative psychology.* Sunderland, MA: Sinauer Associates, Inc.

Dickinson, A. (1980) *Contemporary animal learning theory.* Cambridge: Cambridge Univ. Press.

Dickinson, A. (1989) Expectancy theory in animal conditioning. In S. B. Klein and R. R. Mowrer (Eds.) *Contemporary Learning Theories.* Hillsdale, NJ: Erlbaum, pp. 279-308.

Durlach, P. J. (1983) The effect of signalling intertrial USs in autoshaping. *J. Exp. Psychol. Anim. Behav. Processes 9:* 374-389.

Durlach, P. J. (1989) Learning and performance in Pavlovian conditioning: Are failures of contiguity failures of learning or performance? In S. B. Klein and R. R. Mowrer (Eds.) *Contemporary learning theories: Pavlovian conditioning and the status of traditional learning theory.* Hillsdale, NJ: Erlbaum, pp. 19-59.

Durlach, P. J. and Shane, D. O. (1993) The effect of intertrial food presentations on anticipatory goal-tracking in the rat. *Q. J. Exp. Psychol. 46B:* 289-318.

Farley, J. (1987) Contingency learning and causal detection in *Hermissenda*: I. Behavior. *Behav. Neurosci. 101:* 13-27.

Farley, J. and Alkon, D. L. (1985) Cellular mechanisms of learning, memory, and information storage. *Ann. Rev. Psychol. 36:* 419-494.

Fukushi, T. (1979) Properties of olfactory conditioning in the housefly, *Musca domestica. J. Insect. Physiol. 25:* 155-159.

Gallistel, C. R. (1990) *The organization of learning.* Hillsdale, NJ: Erlbaum.

Gallistel, C. R. (1992) Classical conditioning as a nonstationary, multivariate time series analysis: A spreadsheet model. *Behav. Res. Meth. Instrum. Comput. 24:* 340-351.

Gormezano, I., Kehoe, E. J. and Marshall, B. S. (1983) Twenty years of classical conditioning research with the rabbit. In J. M. Prague and A. N. Epstein (Eds.) *Progress in psychobiology and physiological psychology Vol 10.* New York: Academic Press.

Guthrie, E. R. (1952) *The psychology of learning* (2nd ed.). New York: Harper and Row.

Hawkins, R. D., Abrams, T. W., Carew, T. J. and Kandel, E. R. (1983) A cellular mechanism of classical conditioning in *Aplysia*: Activity-dependent amplification of presynaptic facilitation. *Science 219:* 400-405.

Hawkins, R. D., Carew, T. J. and Kandel, E. R. (1986a) Effects of interstimulus interval and contingency on classical conditioning of the *Aplysia* siphon withdrawal reflex. *J. Neurosci. 6:* 1695-1701.

Hawkins, R. D., Clark, G. and Kandel, E. R. (1986b) Cell biological studies of learning in simple vertebrate and invertebrate systems. In *Handbook of Physiology. The Nervous System. Higher Functions of the Brain.* Sec. 1, Vol 6 pp 25-83. Bethesda, MD: American Physiology Society.

Hawkins, R. D., Lalevic, N., Clark, G. A. and Kandel, E. R. (1989) Classical conditioning of the *Aplysia* siphon-withdrawal reflex exhibits response specificity. *Proc. Natl Acad. Sci. USA 86:* 7620-7624.

Hebb, D. O. (1949) *The organization of behavior: A neuropsychological theory.* New York: Wiley.

Henderson, T. B. and Strong, P. N. Jr. (1972) Classical conditioning in the leech *Macrobdella Ditetra* as a function of CS and UCS intensity. *Condit. Reflex. 7:* 210-215.

Holland, P. C. (1977) Conditioned stimulus as a determinant of the form of the Pavlovian conditioned response. *J. Exp. Psychol. Anim. Behav. Processes 3:* 77-104

Holland, P. C. (1990). Event representation in Pavlovian conditioning: Image and action. *Cognition 37:* 105-131.

Holland, P. C. and Straub, J. J. (1979) Differential effects of two ways of devaluing the unconditioned stimulus after Pavlovian appetitive conditioning. *J. Exp. Psychol. Anim. Behav. Processes 5:* 65-78.

Hollis, K. L. (1982) Pavlovian conditioning of signal-centered action patterns and autonomic behavior: A biological analysis of function. *Advances study behav. 12:* 1-64.

Hollis, K. L. (1984) The biological function of Pavlovian conditioning: The best defense is a good offense. *J. Exp. Psychol. Anim. Behav. Processes 10:* 413-425.

Hollis, K. L., Cadieux, E. L. and Colbert, M. M. (1989) The biological function of Pavlovian conditioning: A mechanism for mating success in the blue gourami (*Trichogaster trichopterus*). *J. Comp. Psychol. 103:* 115-121.

Hull, C. L. (1943) *Principles of behavior.* New York: Appleton-Century-Crofts

Hull, J. H. (1977) Instrumental response topographies of rats. *Anim. Learn. Behav. 5:* 207-212.

Jenkins, H. M. and Moore, B. R. (1973) The form of the autoshaped response with food or water reinforcers. *J. exp. Anal. Behav. 20:* 163-181.

Karrer, T. and Sahley, C. L. (1988) Discriminative conditioning alters food preferences in the leech, *Haemopis marmorata. Behav. Neural Biol. 50:* 311-324.

Kesner, R. P. and Olton, D. S. (1990) *Neurobiology of comparative cognition.* Hillsdale, NJ: Erlbaum.

Kruse, J. M., Overmier, J. B., Konz, W. A. and Rokke, E. (1983) Pavlovian conditioned stimulus effects upon instrumental choice behavior are reinforcer specific. *Learn. Motivat. 14:* 165-181.

LaMon, B. C. and Zeigler, H. P. (1984) Grasping in the pigeon (*Columbia livia*): Stimulus control during conditioned and consummatory responses. *Anim. Learn. Behav. 12:* 223-231.

Lederhendler, I., Gart, S. and Alkon, D. L. (1986) Classical conditioning of *Hermissenda*: Origin of a new response. *J. Neurosci. 6:* 1325-1331.

Mackintosh, N. J. (1974) *The psychology of animal learning*. London: Academic Press.
Mackintosh, N. J. (1983) *Conditioning and associative learning*. Oxford: Oxford Univ. Press.
Macphail, E. M. (1993) *The neuroscience of animal intelligence: from the seahare to the seahorse*. New York: Columbia Univ. Press.
Mitchell, D., Kirschbaum, E. H. and Perry, R. L. (1975) Effects of neophobia and habituation on the poison-induced avoidance of exteroceptive stimuli in the rat. *J. Exp. Psychol. Anim. Behav. Processes 1:* 47-55.
Mitchell, D., Parker, L. F. and Johnson, R. (1976) Absence of a generalization decrement in the poison-induced avoidance of interoceptive stimuli in the rat. *Physiol. Psychol. 4:* 121-123.
Mitchell, D., Scott, D. W. and Mitchell, L. K. (1977) Attenuated and enhanced neophobia in the taste-aversion "delay of reinforcement" effect. *Anim. Learn. Behav. 5:* 99-102.
Nelson, M. C. (1971) Classical conditioning in the blowfly (*Phormia regina*): Associative and excitatory factors. *J. comp. physiol. psychol. 77:* 353-368.
Ocurr, K. A., Walters, E. T. and Byrne, J. H. (1985) Associative conditioning analog selectively increases cAMP levels of tail sensory neurons in *Aplysia. Proc. Natl Acad. Sci. USA. 82:* 2548-2552.
Pavlov, I. P. (1927) *Conditioned reflexes*. Oxford: Oxford Univ. Press.
Pavlov, I. P. (1928) *Lectures on conditioned reflexes Vol* 2. New York: International Publishers.
Pfautz, P. L., Donegan, N. H. and Wagner, A. R. (1978) Sensory preconditioning versus protection from habituation. *J. Exp. Psychol. Anim. Behav. Processes 4:* 286-295.
Rescorla, R. A. (1967) Pavlovian conditioning and its proper control procedures. *Psychol. Rev. 74:* 71-80.
Rescorla, R. A. (1968) Probability of shock in the presence and absence of CS in fear conditioning. *J. Comp. Physiol. Psychol. 66:* 1-5.
Rescorla, R. A. (1980) *Pavlovian second-order conditioning: Studies in associative learning*. Hillsdale, NJ: Erlbaum.
Rescorla, R. A. (1982a) Effect of a stimulus intervening between CS and US in autoshaping. *J. Exp. Psychol. Anim. Behav. Processes 8:* 131-141.
Rescorla, R. A. (1982b) Comments on a technique for assessing associative learning. In M. L. Commons, R. J. Herrnstein and A. R. Wagner (Eds.) *Quantitative Analysis of Behavior Vol. 3*. Ballinger, pp. 41-63.
Rescorla, R. A. (1984a) Comments on three Pavlovian paradigms. In D. L. Alkon and J. Farley (Eds.) *Primary neural substrates of learning and behavioral change*. Cambridge: Cambridge Univ. Press, pp. 25-45.
Rescorla, R. A. (1984b) Signaling intertrial shocks attenuates their negative effect on conditioned suppression. *Bull. Psychon. Soc. 22:* 225-228.
Rescorla, R. A. (1988) Pavlovian conditioning: It's not what you think it is. *Amer. Psychol. 43:* 151-160.
Rescorla, R. A. and Cunningham, C. L. (1978) Within-compound flavor associations. *J. Exp. Psychol. Anim. Behav. Process. 4:* 267-275.
Rescorla, R. A. and Holland, P. C. (1976) Some behavioral approaches to the study of learning. In M. R. Rosenzweig and E. L. Bennett (Eds.) *Neural*

mechanisms of learning and memory. Cambridge, Mass: MIT Press, pp. 165-192.

Ricker, J. P., Brzorad, J. N. and Hirsch, J. (1986) A demonstration of discriminative conditioning in the blow fly, *Phormia regina*. *Bull. Psychon. Soc. 24:* 240-243.

Rogers, R. F., Schiller, K. M. and Matzel, L. D. (in press) Contextual conditioning in *Hermissenda crassicornis* using diffuse chemosensory cues. *Anim. Learn. Behav.*

Sahley, C. L., Gelperin, A. and Rudy, J. W. (1981) One-trial associative learning modifies food odor preferences of a terrestrial mollusc. *Proc. Natl. Acad. Sci. 78:* 640-642.

Sahley, C. L., Rudy, J. W. and Gelperin, A. (1984) Associative learning in a mollusk: a comparative analysis. In D. L. Alkon and J. Farley (Eds.) *Primary neural substrates of learning and behavioral change*. Cambridge: Cambridge Univ. Press, pp. 243-258.

Shipley, B. E. and Colwill, R. M. (in press) Direct effects on instrumental performance of outcome revaluation by drive shifts. *Anim. Learn. Behav.*

Skelton, R. W., Spetch, M. L. and Wilkie, D. M. (1980) A method for automatically recording topographical differences in pigeons' keypecking for food and water reinforcers. *Behav. Res. Meth. Instrum. 12:* 349-352.

Spetch, M. L., Wilkie, D. M. and Skelton, R. W. (1981) Control of pigeons' keypecking topography by a schedule of alternating food and water reward. *Anim. Learn. Behav. 9:* 223-229.

Stanhope, K. J. (1989) Dissociation of the effect of reinforcer type and response strength on the force of a conditioned response. *Anim. Learn. Behav. 17:* 311-321.

Stanhope, K. J. (1990) Effect of intertrial unconditioned stimulus (US) presentations upon responding to a conditioned stimulus predictive of either the same or a different appetitive US. *Learn. Motivat. 21:* 266-286.

Stanhope, K. J. (1992) The representation of the reinforcer and the force of the pigeon's keypeck in first- and second-order conditioning. *Q. J. Exp. Psych. 44B:* 137-158.

Thorndike, E. L. (1911) *Animal intelligence: Experimental studies*. New York: Macmillan.

Trapold, M. A. (1970) Are expectancies based upon different positive reinforcing events discriminably different? *Learn. Motivat. 1:* 129-140.

Trapold, M. A. and Overmier, J. B. (1972) The second learning process in instrumental learning. In A. A. Black and W. F. Prokasy (Eds.) *Classical conditioning II: Current research and theory*. New York: Appleton-Century-Crofts, pp. 427-452.

Wagner, A. R. (1978) Expectancies and the priming of STM. In S. H. Hulse, H. Fowler and W. K. Honig (Eds.) *Cognitive processes in animal behavior*. Hillsdale, NJ: Erlbaum, pp. 177-209.

Walters, E. T. (1989) Transformation of siphon responses during conditioning of *aplysia* suggests a model of primitive stimulus-response association. *Proc. Natl Acad. Sci. USA 86:* 7616-7619.

Walters, E. T. and Byrne, J. H. (1983) Associative conditioning of single sensory neurons suggests a cellular mechanism for learning. *Science 219:* 405-408.

Walters, E. T., Carew, T. J. and Kandel, E. R. (1979) Classical conditioning in *Aplysia californica*. *Proc. Natl. Acad. Sci. 76:* 6675-6679.

Walters, E. T., Carew, T. J. and Kandel, E. R. (1981) Associative learning in *Aplysia*: Evidence for conditioned fear in an invertebrate. *Science 211:* 504-506.

4

Neuroethological Studies on How Birds Discriminate Song

Jeffrey Cynx

Abstract

The ease with which zebra finches learn conspecific song discriminations is related to the bird's vocal repertoire. It is also related to two other neuroethological variables of interest: gender and season. The perceptual specialization is lateralized, under hormonal control, and relies on neural circuitry previously identified as necessary for learning song vocalizations.

Introduction

Neuroethology often entails the study of specialized capabilities. The classic studies include echolocation in bats (Moss and Simmons, Chapter 11, Haresign et al. Chapter 12, this volume), electrosensory capabilities in fish (Heiligenberg, 1991), cricket auditory communication (Huber and Thorson, 1985), and animal navigation (Walcott and Green, 1974). Instead of searching for the least common denominator of behavior or neurobiology, neuroethology recognizes the unique capabilities of each species. A species that is a champion in a behavior may display neural optimizations that are missing or obscured in other animals (Bullock, 1984; Heiligenberg, 1990). Neuroethology can be viewed more generally as an integrated approach to brain/behavior issues, dealing with the complexity of a behavior's evolution, function, development, and neural implementation. In this sense, it denies reductionism (Churchland, 1986).

The study of birdsong is a case in point. Song is a specialized communication system, perhaps the most complex learned signalling system outside of human speech. Our understanding depends on the studies of ecologists, ethologists, psychologists, neuroendocrinologists, neuroethologists, neurophysiologists, neuroanatomists, and molecular biologists. In each of these fields, successful research has taken advantage of the complexity of the behavior.

This chapter concerns one aspect of this complexity, song perception. Although one may study the acoustics of song or the molecular biology of brain nuclei involved in perception, the difficulty is understanding how song is represented in the intermediate routes between air molecules and neural molecules. The next section reviews three overlapping aspects of birdsong that concern song perception: song development, the neural bases of song, and auditory discrimination. This is followed by a discussion of five operant experiments that concern the behavioral and neural representation of song or song like structures.

A Short Review of Birdsong

Song Development

Developmental studies of birdsong have addressed the species-specificity of naturally-occurring learning (Marler, 1990; Thorpe, 1961). There are upwards of ten thousand species of birds, and at least half of these, the oscines or songbirds, learn their species-specific vocalizations (Kroodsma, 1982). This learning is guided by innate tendencies that lead to the acquisition of species-appropriate song (Marler, 1990; 1991). Zebra finches (*Taeniopygia guttata*), on which this chapter focuses, go through their song learning phase from approximately 30 to 90 days post-hatching. The learning of song is effective only during this sensitive period (Immelmann, 1969). All songbirds need to see and hear an adult tutor, and to hear themselves practice, to learn their song. When isolated from tutors or deafened during development, they produce abnormal songs as adults (Konishi, 1965; Price, 1979; Thorpe, 1961).

Adult Song Production

At about 90 days post-hatching, zebra finch song becomes stereotyped or crystallized. This stereotypy was originally thought to be impervious to change. Recent evidence suggests that this is not true. Zebra finches deafened as adults will show some changes in song structure (Nordeen and Nordeen, 1992). An adult will stop or alter song in midstream when his monitoring of song is disrupted (Cynx, 1990; Cynx and von Rad, 1995). Thus hearing and discrimination of conspecific song is critical in normal song development, and apparently also in the maintenance of adult song. Regardless, birds like zebra finches and other "closed-ended learners" in general maintain the same song throughout adulthood. In "opened-ended learners," such as the canary, songs can change dramatically from year to year (Nottebohm and Nottebohm, 1978).

Neural Basis of Song

The song system is known in some detail. Neuroanatomical tracing studies have revealed the major brain nuclei that control song production in the canary (Nottebohm, 1980; Nottebohm et al. 1982; Stokes et al. 1974). These nuclei have been found in every species of songbird exam-

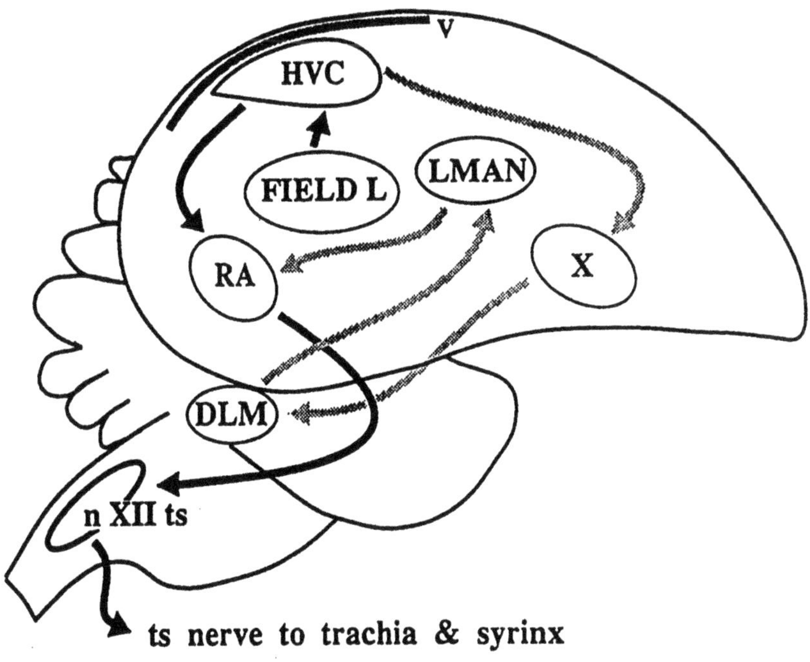

FIGURE 4.1 A sagittal section of the brain of a songbird, showing major song nuclei. Primary auditory input comes through Field L to HVC. The circuit, HVC-RA-nXIIts forms the backbone of the song production system, controlling use of the syrinx, the bird's vocal organ. Lesions anywhere along it severely disrupt song production in adults and juveniles. A second route, the recurrent pathway, also goes from HVC to RA but by way of Area X and nucleus MAN. Lesions along it fail to disrupt adult song production, but severely disrupt song learning.

ined, but only in songbirds (Figure 4.1). Lesions to nuclei, HVC, RA and nXIIts, along what is termed the main efferent pathway to the syrinx, the bird's vocal organ, cause severe deficits in adult song production (Nottebohm, et al. 1976).

There is a second pathway that goes from HVC to RA, but by way of Area X, thalamic nucleus DLM, and MAN (Bottjer et al. 1989; Nottebohm et al. 1982). Lesions along this pathway have no effect on adult song production. However, they disrupt song development (Bottjer et al. 1984; Sohrabji, Nordeen and Nordeen, 1990; Scharff and Nottebohm, 1991), indicating that this pathway plays a critical role in song learning, but not adult song production.

The neuroanatomical studies were supplemented by studies demonstrating that the volume of HVC and some other song control nuclei correlated with changes in seasonal song production (Nottebohm, 1981). These size changes reflect variation in cell activity (Gahr, 1990) and the production of new neurons in adulthood, in both open-learners (Goldman and Nottebohm, 1983) and closed-learners (Alvarez-Buylla et al. 1990; Nordeen and Nordeen, 1988).

The song system in zebra finches is sexually dimorphic. As with most species, only males sing, and males have larger song nuclei (Nottebohm and Arnold, 1976). Some neuronal connections found in males are missing in females (Konishi and Akutagawa, 1985; Williams, 1985). Both the male production of song and the development of sexual dimorphism have been shown to be under hormonal control (Arnold, 1980).

The song system, besides being modular or compartmentalized to specific brain nuclei, is also lateralized (Nottebohm, 1977). In zebra finches (Williams, 1992) and other songbirds, unilateral lesions to HVC have a profound effect on song production. This suggests that, as with humans, one side of the brain plays a dominant role in the control of this learned behavior. There are close anatomical and physiological relations between auditory and motor areas. Brain nuclei originally designated as motor control areas have been found to have units responsive to auditory stimuli, especially conspecific song (Doupe and Konishi, 1991; Katz and Gurney, 1979; Margoliash, 1983; McCasland and Konishi, 1981; Williams and Nottebohm, 1985). Exposing zebra finches to recordings of conspecific song causes a specific change in gene expression in auditory areas associated with the song system (Mello et al. 1992).

Neurobiological Orphans

The neurobiology remains puzzling in light of the known behavior. It makes one suspect that we do not know enough about the behavior to make sense of the neurobiology. Why do females zebra finches have song nuclei at all if they do not sing? Why does neurogenesis occur in adult birds that have stereotyped song? What is the role of lateralization? These findings can be thought of as neurobiological orphans, lacking functional roles or behaviors to claim them. Although much has been learned concerning the auditory pathways, the presence of auditory information, and the patterns of motorneuron firing that precede and accompany song production, the question of how signals in the auditory system are translated into commands in the motor system during learning has not been answered. Nor does anyone understand why known motor areas down to the level of motor neurons are responsive to auditory input. In general, we know more about the neurobiology of birdsong, especially song perception, than we understand. What is needed are behavioral models — especially in song perception — that can claim the neurobiological phenomena by explaining their roles. Auditory perception in birds has been an active area of experimental research for decades. There is still no complete picture. The next section reviews two of the models for song perception that have been put forward.

Discrimination of Song

Up to the mid 1970s, the precise psychophysical functions of oscine hearing were unknown. For example, what was the smallest frequency difference that could be detected? What was the smallest time difference that could be detected? Further, it was unclear whether these parametrics were similar across species and how they compared with mammalian hearing. Despite the lack of data, hypotheses had been put forward concerning the relation of hearing and production in songbirds. One hypothesis was that the avian auditory system was capable of finer temporal and/or frequency resolution than humans (Greenewalt, 1968; Pumphrey, 1948). This "Bird For All Frequencies" hypothesis was proposed to account for the rapidly modulated features in much of birdsong. It was presumed that oscines needed enhanced auditory resolution to extract the subtle features of song.

A second "Bird For Each Frequency" hypothesis was based on the knowledge that birdsong is learned through auditory feedback, and each species shows selective biases for learning its own song. This suggested that each species' peripheral auditory system is 'tuned' to species-specific settings (Schwartzkopff, 1949). The species-specific sensory adaptations would then bias song learning.

The universality of both these hypotheses was either thrown into doubt or falsified by Dooling, his colleagues, and others using animal conditioning techniques. Auditory resolution for most song birds was shown to be no greater than for humans, nor are there any differences between sexes (Dooling, 1982; Fay, 1988; also see Dooling et al., Chapter 6, Weary, Chapter 7, this volume). These psychophysical findings have proven valuable in documenting the conservative nature of the peripheral hearing system in birds and mammals. Regardless of the morphological differences in the peripheral auditory systems of songbirds and mammals, the auditory processing remains strikingly similar (differences have been found in a few bird species, for example, owls (*Tyto alba* and *Bubo virginianus*) and the budgerigar (*Melopsittacus undulatus;* Fay, 1988). These sensory studies therefore indicate that the sensory processes are necessary but not sufficient to explain species-specific song perception. This led to the asking of cognitive questions. Rather than asking whether songbirds can hear sounds, researchers ask how songbirds attend to, discriminate between, and remember sounds, including birdsong. Even if songbirds and humans use the same sensory bricks and mortar in the acoustic world, they may reach different perceptual and cognitive ends.

Song Discrimination: Five Experiments

These experiments use operant techniques to ask birds how they hear song. It seems possible that by removing an animal from its normal context and placing it on an operant procedure, motivational biases for responding to songs could be confounded with perception of the songs. A bird in the laboratory might fail to recognize song as song, or at least as a familiar sound. This is an reasonable neuroethological question that touches on the definition of "special". The term may refer to a kind of song discrimination that occurs only in the context of its natural usage of territorial defense and courtship. This discrimination of song may be radically different from discrimination in a laboratory, not only at a functional level, but also at the levels of the neural substrate. However, the validity of studying song perception in an operant setting was provided by Sinnott (1980). She trained cowbirds (*Molothrus ater*) and red-winged blackbirds (*Agelaius phoeniceus*) to discriminate between either conspecific or alien songs. Both species showed superior performance in discriminating between conspecific songs. Neurophysiological experiments also provide evidence for song recognition in the laboratory (Doupe and Konishi, 1991; Margoliash, 1983; McCasland and Konishi, 1981; Williams and Nottebohm, 1985). Thus it appears that song perception is not context-limited, at least to the extent that it has been studied in the laboratory. In fact, it may pre-empt general percep-

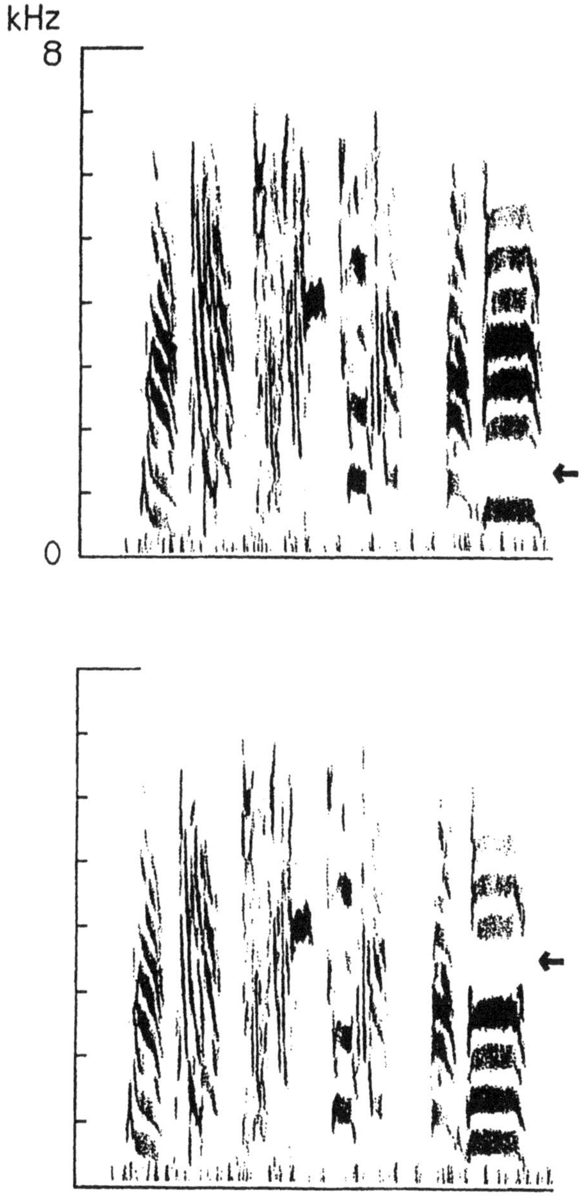

FIGURE 4.2 A typical zebra finch song. In the top panel, the second harmonic of the last syllable has been removed. In the bottom panel, the fifth harmonic of the same syllable has been removed (indicated by arrows). Each song is about 700 ms in duration. The y-axis ranges shows frequency from 0 to 8 kHz. One set of birds in Experiment 1 was trained to discriminate the harmonic syllables in isolation. A second set of birds was trained to discriminate them as shown (embedded in song).

tual processes regardless of context, as has been posited in human speech perception (Mann and Liberman, 1983). The first two experiments address this issue.

Experiment 1: Song Context

Harmonics, frequency components that stand in integer relations, are ubiquitous in animal communication. They encode imaging information in echolocating bats (Moss and Simmons, Chapter 11, Haresign et al. Chapter 12, this volume; Suga, 1990), and are integral parts of vocalizations in human speech (Ladefoged, 1982) and birdsong (Nowicki and Marler, 1988). Normal zebra finch song contains a number of harmonic structures. These harmonic components are seldom of equal amplitude. For example, the third harmonic for one zebra finch may consistently be 15 dB lower than the two surrounding harmonics. This amplitude variation is consistent across time, learned, and under neural control (Williams et al. 1989). It therefore seems possible that this variation, to birds, is a salient song property.

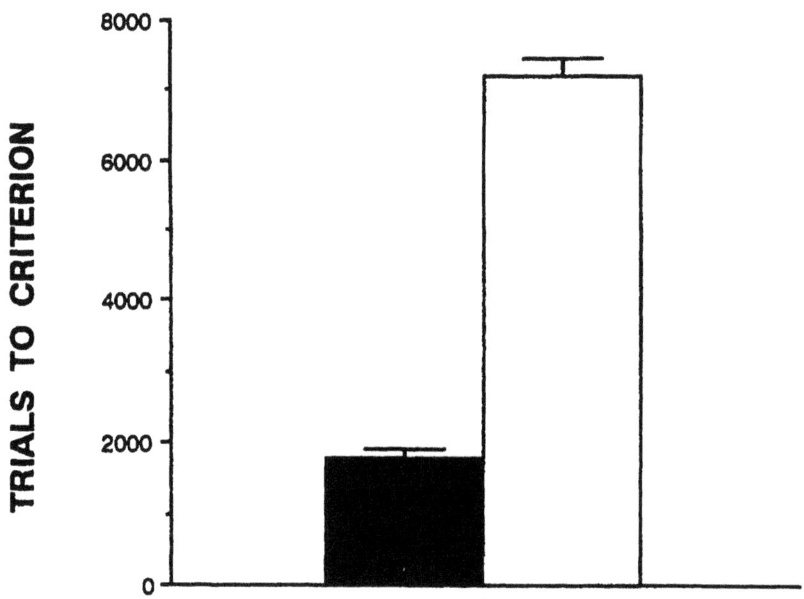

FIGURE 4.3 Results from Experiment 1. Birds trained on harmonically structured syllables embedded in song (solid black bar) required significantly fewer trials than those trained on the same syllables in isolation (open white bar). All birds were males. The error bars show SEMs.

To test this, male zebra finches were trained to discriminate between two harmonic stacks. The stimuli were made from the same syllable edited from the same song (Figure 4.2). The only difference between the two was that one had a missing second harmonic, the other a missing fifth harmonic. The operant was perch-hopping. A zebra finch started a trial by landing in one location. He then heard either one stimulus or the other. One stimulus signalled he could go to a food location. The other stimulus signalled that he should not go, but should wait through the 3-sec. interval to begin another trial. If he went incorrectly, the houselights went out for 15 seconds to tell him the move was incorrect. As the bird acquired the discrimination, his latency to responses to the two stimuli diverged, approaching a 3-second difference. In one case, he went as quickly as possible; in the other case, he waited out the 3-sec. interval. A 75% correct score — halfway between chance and errorless performance — was used as the acquisition criterion.

Other male birds were trained to perform the same discrimination, but with the syllable embedded in song (Figure 4.2). This would seemingly make the task more difficult as the salient information is now more hidden. However, there was a facilitative effect when the harmonic information was embedded (Figure 4.3).

The effect could be due either to perception or to attention. In other words, did the birds hear the sounds differently when embedded or not? To test this, the birds trained on syllables alone were transferred to the stimuli with the embedded syllables. The birds trained on the embedded syllables were transferred to the syllables alone. If they had learned to perceive different sounds, then the transfer should impair discrimination performance. If the initial difference was attentional, then there might be no deficit upon transfer since the birds perceived the same things once they had learned to attend to the salient cues. The result was that all birds were unaffected by the transfers. All birds maintained the discriminations from the onset of the transfer sessions. This indicates that all birds ended up learning the same thing. The difference in acquisition rates was linked to the bird's ability to recognize and attend to the salient cue. Some song property facilitated attending to the syllable's structure.

Experiment 2: Song Familiarity

The second experiment indicates that birds in operant experiments discriminate between songs based on their social and ecological import. Each zebra finch in an aviary has a different song, having improvised a unique song based on imitating the song of a male tutor (Immelmann, 1969; Williams, 1990). A bird's own song seems to occupy a special

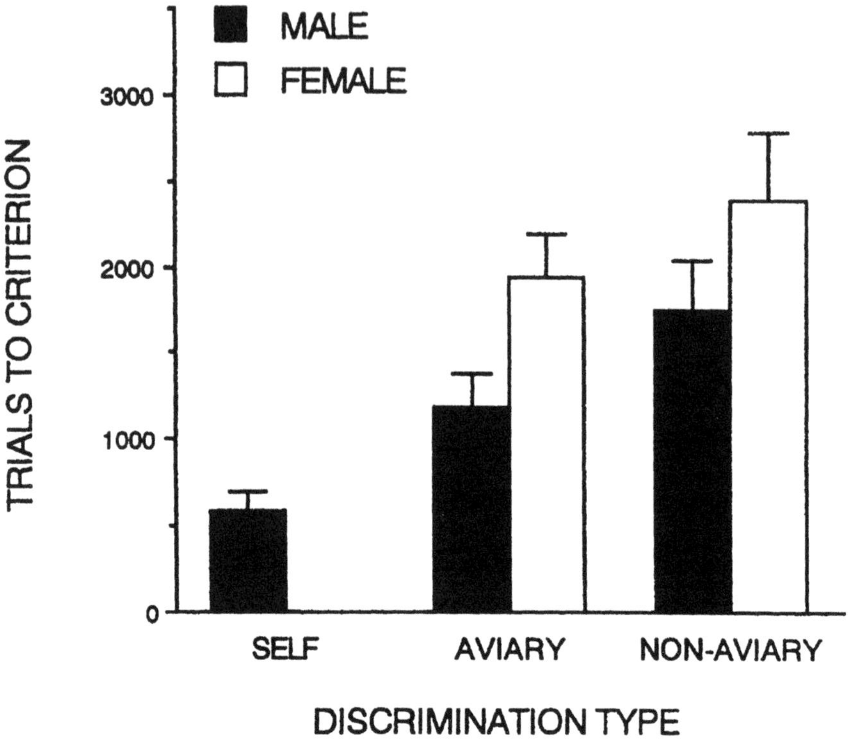

FIGURE 4.4 Mean trials to criterion for the five stimulus conditions in Experiment 2. Each bar on the abscissa represents the mean and SEM for five birds, one from each cohort. Filled bars represent males, and open bars females (females do not sing, so could not be tested with their own song). The three male groups differed significantly among themselves, but the difference between the two female groups was not significant. Males in the "aviary" condition required significantly fewer trials than the corresponding females.

status in a number of species with differing ecologies. It also has a special status in stimulating units in brain nuclei (Doupe and Konishi, 1991; Margoliash, 1983). The bird's relation to song was therefore used as a variable in training zebra finches to discriminate between songs (Cynx and Nottebohm, 1992a). Five birds — three males and two females — were trained to discriminate between the same two songs. One male bird was discriminating between its own song and a cage mate's song. Two other birds from that aviary — a male and a female — were trained on the same stimuli. Finally, a male and female from another aviary were also trained on the same stimuli. This designed ensured that each bird of each gender had a different historical relation

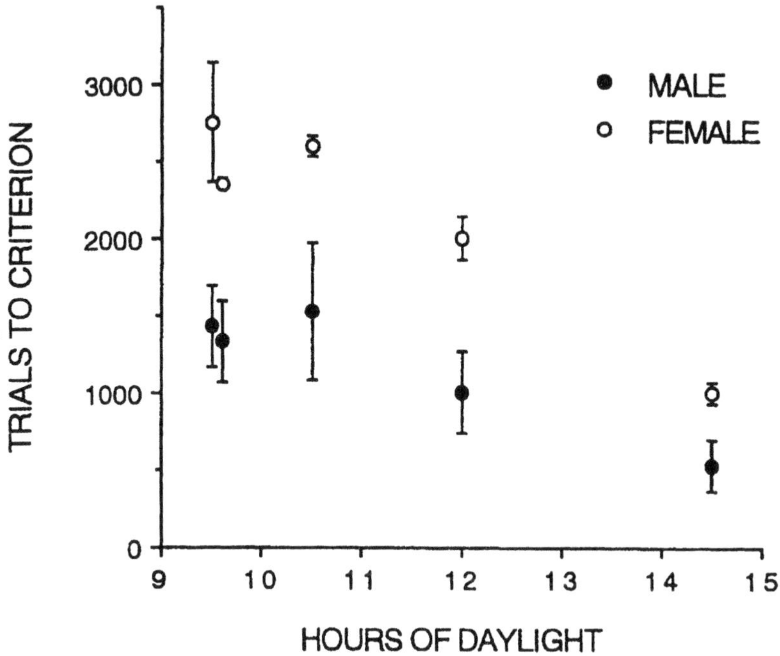

FIGURE 4.5 The mean number of trials and SEM needed for the three males (filled circles, averaging "self", "aviary", and "other aviary" conditions) and two females (open circles, averaging "aviary" and "other aviary" conditions) in each cohort to reach criterion in Experiment 2. Each cohort is plotted according to the number of daylight hours available during training. All birds were on a natural photoperiod.

to the two song stimuli. If zebra finches recognized the stimuli as songs, their respective facility in acquiring the discriminations might depend on their past experience with these songs or their gender. Five cohorts of birds were run with this design.

The results were that male birds acquired these discriminations in fewer trials if the songs were familiar, and most quickly if one of the songs was their own (Figure 4.4). Females took more trials to reach criterion, regardless of their relation to the songs. It seems doubtful that the sex difference indicates that females do not pay attention to song. Female zebra finches remember and respond to their father's or mate's song (Miller, 1979a,b). These song dimensions — which were not controlled in this experiment — may have more ecological relevance to female zebra finches than a song's familiarity.

There was an unexpected overall seasonal effect (Figure 4.5). The birds had been kept on a natural photoperiod, and those trained in the summer acquired the discriminations more quickly than those trained in the winter. I will return to this issue of seasonal variation later in the chapter.

Experiment 3: Lateralization of Song Discrimination

The earlier experiments revealed psychophysical and ecological aspects of song discrimination. Another way to study birdsong is to lesion the song system and see if performance changes. The value of this neuropsychological approach is that it ties specific behaviors with specific brain nuclei. The first of two neuropsychological experiments presented here concerns lateralization (Cynx et al. 1992).

Lateralization in the control of a learned behavior was first shown in a nonhuman species in songbirds (Nottebohm, 1977). The motor control of birdsong showed a hemispheric asymmetry. Until the early 1970's, such hemispheric asymmetries were considered the hallmark of the human brain and human language. Since that time lateralization has been found in a number of different animals, and has been tied to animal communication systems.

To test for lateralization of song discrimination in zebra finches, a technique was adapted from that used in determining laterality of vocalization discrimination in Japanese macaques (*Macaca fuscata;* Petersen et al. 1978; Heffner and Heffner, 1984). Fourteen male zebra finches were randomly paired. In each pair, either the left or right nucleus ovoidalis, a thalamic auditory nucleus that projects unilaterally to the forebrain (Kelley and Nottebohm, 1979), was partially destroyed. By lesioning one half of it, we presumed that we were largely deafening either the right or left forebrain including the telencephalic song system. Each pair then was trained to discriminate between its two songs. Birds A and B, for example, were discriminating songs A and B. The results from Experiment 2 predicted that all birds should acquire this discrimination in a few hundred trials. However, the birds with left side lesions took at least twice as many trials as birds with right side lesions to acquire the discriminations (Figure 4.6, Panel A). This experiment took place in June. We then replicated it in December. Again, birds with left side lesions took more trials. However, the overall acquisition rates also increased. We then replicated the experiment again in the next June. The seasonal effect was apparent here, as it had been in the last experiment (Figure 4.6, Panel A).

We then trained the same pairs of birds on a second discrimination. This was the task from Experiment 1 with the harmonic stack embedded in song. The results here were reversed from those with the birds' own song. In this case, birds with left side lesions required fewer trials than

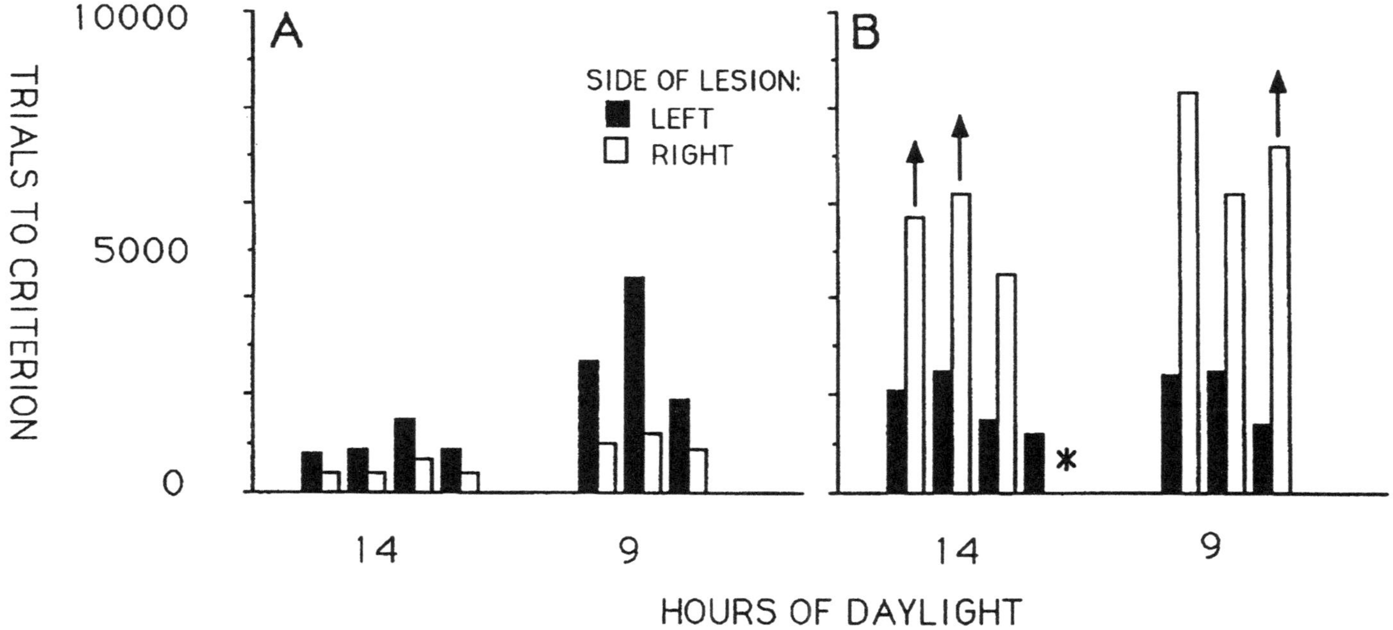

FIGURE 4.6 A) Trials to learning criterion for each zebra finch when trained to discriminate between its own song and the song of a cage mate. In each of the seven pairs, a bird received a lesion to either the left or the right nucleus ovoidalis. The numbers along the x-axis indicate the amount of daylight for the summer and winter experimental series. B) Trials to criterion for the same birds when subsequently trained to discriminate between two versions of an unfamiliar song that differed only in the harmonic profile of one syllable (Figure 2). Testing of right-side lesion birds in pairs 1, 2, and 7 in the second task (represented by arrows above bars) stopped before the birds reached learning criterion. One bird, indicated by an asterisk (Pair 4), died before the second experiment began.

birds with right side lesions (Figure 4.6, Panel B). There was also little or no evidence for the seasonal effect here. The effect may be tied to certain features of song present in the first but not the second task. The results have some speculative elegance. The left side of a zebra finch brain seems dominant for distinguishing between whole song structures, as where the right side seems dominant for a kind of feature detection of harmonic components. This distinction is a hallmark of human lateralization (Sperry, 1974).

Experiment 4: Hormonal Mediation of Song Discrimination

The seasonal effect in Experiments 2 and 3 was unexpected. Birdsong in many species is seasonal. The amount of seasonal song production is known to be determined partly by levels of testosterone or its metabolites (Pröve, 1974). One possibility is that there is steroid mediation of song discrimination as well as song production. Hormonal mediation has been reported in human absolute pitch perception (Wynn, 1971)

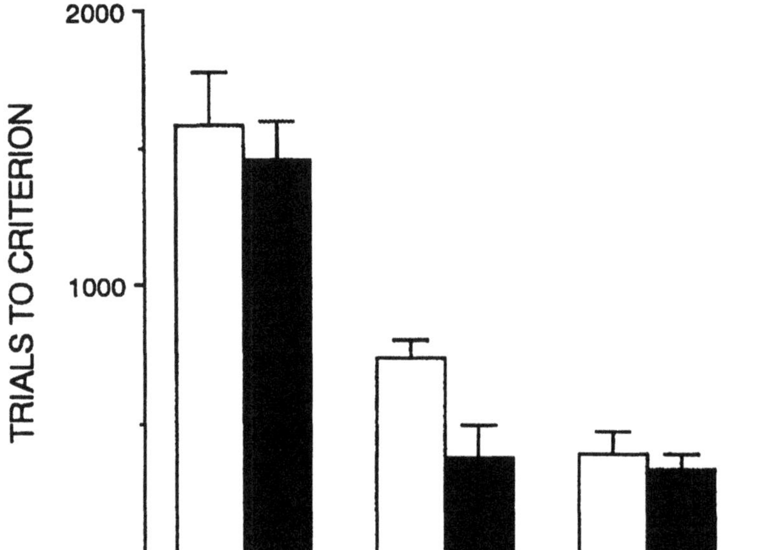

FIGURE 4.7 The results from the experiment with castrated zebra finches. Control birds are represented by the empty bars. The testosterone-implanted birds are represented by the filled black bars. The first pair of bars shows results from training each of seven pairs on a heterospecific (canary) song discrimination; the middle pair of bars shows results from training each pair on their own two (conspecific) songs; and the pair of bars on the far right shows results from training each pair on a second heterospecific (canary) song discrimination.

and in avian olfactory perception (Clark and Smeraski, 1991). However, zebra finches are not known to be a seasonal species. As opposed to temperate birds, they are able to reproduce throughout the year, largely unaffected by changes in light levels (Immelmann, 1971). This experiment tested whether there might be a previously unsuspected hormonal mediation of one aspect of song behavior.

Fourteen male zebra finches on a 12:12 light:dark cycle were castrated. This provided fourteen birds with negligible endogenous levels of testosterone during the experiment. Half of the fourteen were implanted some weeks later with silastic tubes of testosterone. The other half were implanted with empty tubes as controls. They were divided into seven pairs, each pair with one testosterone-implanted bird and a control. They then were trained on a number of discriminations. Each pair was first trained on a hetereospecific song discrimination, then on their own two songs, then on three other discriminations (not addressed in this chapter), then on a hetereospecific discrimination again. This design determined if any effect was due to general changes in learning or was tied to zebra finch song in particular.

Canary song was the hetereospecific song. Each pair was trained on the same canary discrimination, but different pairs were trained on different canary discriminations. When the zebra finches were initially trained on the canary songs, they showed no differences between testosterone-implanted and control birds (Cynx and Nottebohm, 1992b). (Figure 4.7, left). When the pairs were next trained on the zebra finch songs, there were consistent differences (Figure 4.7, middle). Testosterone-implanted birds acquired the discriminations in fewer trials than the control birds in every case. Finally, the birds were trained on another set of canary song stimuli. Again, there were no consistent differences (Figure 4.7, right). This second canary discrimination required fewer trials, presumably due to an overall learning effect.

It appears that testosterone facilitates a conspecific, but not hetereospecific song discrimination. This points to mechanisms underlying song discrimination that are selective to conspecific song while being plastic in their response to hormone levels. The results also raise a general ornithological issue of zebra finch's hormonal response to seasonal changes in light levels. Since the first report of these results, people have mentioned in lab lore that breeding is more difficult in the winter for zebra finches, and that physiological measures in song control brain nuclei may be suppressed in the winter months. What now is needed is the hormonal profile across seasons, plus some indication of the mechanism. For example, the mechanisms could be direct targeting of receptors, or it could be a feedback loop that includes both behavior and hormones (Lehrman, 1965). Birds with testosterone sing more

(Pröve, 1974), and the singing could drive the physiological changes. We also need to know what is changing in the discrimination. There should be salient features embedded in zebra finch song that can be identified. Finally, as mentioned earlier, seasonal changes have been noted in rates of neurogenesis (Alvarez-Buylla et al. 1990). These changes may be reflected in seasonal variation in the ability to update perceptual memories.

Experiment 5: The Recurrent Pathway

The last study, like the third study, concerns the functional neuroanatomy of song discrimination (Figure 4.1). The main efferent pathway goes from HVC to RA then to NXIITS. A second pathway goes from HVC to RA via Area X, thalamic nucleus DLM, then through MAN. Lesions to this second circuit have no effect on adult song performance, but disrupt song development (Bottjer et al. 1984; Nottebohm, 1980; Scharff and Nottebohm, 1991; Sohrabji et al. 1990). However, there is a confound in this interpretation. Normal song development requires auditory information from the tutor and auditory feedback from practice. What may be disrupted is not development, but perception during development. The disruption may appear developmental because the dependent variable is developmental. This circuit's innervation of RA is also of interest because of the possible involvement of NMDA receptors and long term potentiation in the song system at that point (Kubota and Saito, 1991; Mooney and Konishi, 1991). These cellular mechanisms are thought to play a crucial role in learning in other learning systems (Cotman et al. 1989; Lynch et al. 1984).

Disrupting this circuit in an adult bird impairs its ability to learn to discriminate among conspecific sounds (Cynx et al. 1991). This was determined by using a procedure similar to the last experiment. Twelve male zebra finches were paired off. In each pair, one bird was given a bilateral lesion to Area X, one of the nuclei in the "developmental" circuit. Each pair then was trained to discriminate between two heterospecific songs. As in the last experiment, there were no reliable differences between lesioned and intact animals. When subsequently trained on the zebra finch songs, all lesioned birds required more trials than intact birds to reach criterion. Finally, when trained on a second song hetereospecific discrimination, there were no reliable differences. These findings provide the first evidence that Area X, one of the two nuclei along the recurrent pathway, is involved in the adult discrimination of conspecific song.

Conclusion

Varieties of Perceptual Experience

This paper has focused on how a songbird's perceptual world is engineered. An adult bird's ability to discriminate among songs is not a general phenomenon, but is dependent on the bird's sex, its familiarity with the song, side of the brain, and the season in which the song is learned. At least the last variable seems to be under the control of testosterone. Thus song discrimination is not static, but shows a number of forms of behavioral and neural plasticity.

Acknowledgments

Fernando Nottebohm, Constance Scharff, and Heather Williams were experimental colleagues in experiments reported in this paper. Mary Lou Sotanski, Katie Gould, Liesel Gould, Sonya Hornung, Barbara Petri, Gabor Tomasz and Uta Von Rad provided laboratory assistance. Susan Brown and Daniel Weary commented on a draft of this article. The Whitehall Foundation, NIH, Simpson Trust, and Mary Flagler Cary Charitable Trust supported the research.

References

Alvarez-Buylla, A., Kirn, J. R., and Nottebohm, F. (1990). Birth of projection neurons in adult avian brain may be related to perceptual or motor learning. *Science 249*: 1444-1446.

Arnold, A. P. (1980). Sexual differences in the brain. *Am. Sci. 68*: 165-173.

Bottjer, S. W., Miesner, E. A. and Arnold, A. P. (1984). Forebrain lesions disrupt development but not maintenance of song in passerine birds. *Science* 224: 901-902.

Bottjer, S.W., Halsema, K.A., Brown, S.A., and Miesner, E.A. (1989). Axonal connections of a forebrain nucleus involved with vocal learning in zebra finches. *J. Comp. Neurol, 279*: 312-326.

Bullock, T. H. (1984). Comparative neuroscience holds promise for quiet revolution. *Science 225*: 473-478.

Churchland, P. S. (1986). *Neurophilosophy: Towards a Unified Science of the Mind/ Brain.* Cambridge, MA: MIT Press.

Clark, L. and Smeraski, C. A. (1991). Seasonal shifts in odor acuity by starlings. *J. Exp. Zool. 255*: 22-29.

Cotman, C.W., Bridges, R.J., Taube, J.S., Clark, A.S., Geddes, J.W., and Monaghan, D.T. (1989). The role of the NMDA receptor in the central nervous system plasticity and pathology. *J. NIH Res. 1*: 65-74.

Cynx, J. (1990). Experimental determination of a unit of song production in the zebra finch (*Taeniopygia guttata*). *J. Exp. Psychol. 104*: 3-10.

Cynx, J. and Nottebohm, F. (1992a). The role of gender, season, and familiarity in discrimination of conspecific song by zebra finches (*Taeniopygia guttata*). *PNAS, USA 89*: 1368-1371.

Cynx, J., and Nottebohm, F. (1992b). Testosterone facilitates conspecific song discriminations in castrated zebra finches (*Taeniopygia guttata*). *PNAS, USA 89*: 1376-1378.

Cynx, J., Scharff, C. and Nottebohm, F. (1991). Effects of Area X lesions on song discriminations in zebra finches. *Soc. Neurosci. Abstr. 17*: 1051.

Cynx, J., and von Rad, U. (1995). *Effects of delayed auditory feedback on bird song production*. Manuscript submitted for publication.

Cynx, J., Williams, H. and Nottebohm, F. (1990). Timbre discrimination in zebra finch (*Taeniopygia guttata*) song syllables. *J. Comp. Psychol. 104*: 303-308.

Cynx, J., Williams, H. and Nottebohm, F. (1992). Hemispheric differences in avian song discrimination. *PNAS, USA 89*: 1372-1375.

Dooling, R. J. (1982). Auditory perception in birds. In D. E. Kroodsma and E. H. Miller (Eds.) *Acoustic Communication in Birds, Vol. 1*. NY: Academic Press, pp. 95-130.

Doupe, A. J. and Konishi, M. (1991). Song-selective auditory circuits in the vocal control system of the zebra finch. *PNAS, USA 88*: 11339-11343.

Fay, R. R. *Hearing in Vertebrates: A Psychophysics Databook*. Chicago: Hill-Fay Associates, 1988.

Gahr, M. (1990). Delineation of a brain nucleus: Comparisons of cytochemical, hodological, and cytoarchitectural views of the song control nucleus HVC of the adult canary. *J. Comp. Neurol. 294*: 30-36.

Goldman, S. A. and Nottebohm, F. (1983). Neuronal production, migration, and differentiation in a vocal control nucleus of the adult female canary brain. *PNAS, USA 80*: 2390-2394.

Greenewalt, C. H.(1968). *Birdsong: Acoustics and Physiology*. Washington, D. C.: Smithsonian Institution Press.

Heffner, H. E. and Heffner, R. S. (1984). Temporal lobe lesions and perception of species-specific vocalizations by macaques. *Science 226*: 75.

Heiligenberg, W. (1991). The neural basis of behavior: A neuroethological view. *Ann. Rev. Neurosci. 14*: 247-267.

Huber, F., Thorson, J. (1985). Cricket auditory communication. *Sci. Am. 253*: 46-54.

Immelmann, K. (1969). Song development in the zebra finch and other Estrildid finches. In: R. A. Hinde (Ed.) *Bird Vocalisations*. London: Cambridge University Press, pp. 61-74.

Immelmann, K. (1971). Ecological aspects of periodic reproduction. In D. S. Farner and J. R. King (Eds.) *Avian Biology, Vol., 1*. New York: Academic Press, pp. 341-389.

Katz, L. C. and Gurney, M. E. (1979). Auditory responses in the zebra finch's motor system for song. *Brain Res. 221*: 192-197.

Kelley, D. B., and Nottebohm, F. (1979). Projections of a telencephalic auditory nucleus — field L — in the canary. *J. Comp. Neurol. 183*: 455-470.

Konishi, M. (1965). The role of auditory feedback in the control of vocalization in the white-crowned sparrow. *Zeit f. Tierpsychol. 22*: 770-783.
Konishi, M. and Akutagawa, E. (1985). Neuronal growth, atrophy and death in a sexually dimorphic song nucleus in the zebra finch brain. *Nature 315*: 145-147.
Kroodsma, D. E. (1982). Learning and the ontogeny of sound signals in birds. In D. E. Kroodsma and E. H. Miller (Eds.) *Acoustic Communication in Birds, Vol. 2*. New York: Academic Press, pp. 1-23.
Kubota, M. and Saito, N. (1991). NMDA receptors participate differentially in two different synaptic inputs in neurons of the zebra finch robust nucleus of the archistriatum in vitro. *Neurosci. Lett. 125*: 107-109.
Ladefoged, P (1982). *A Course in Phonetics (2d ed)*. New York: Harcourt Brace Jovanovich.
Lehrman, D.S. (1965). Interactions between the internal and external environments in the regulation of the reproductive cycle of the ring dove. In F.A. Beach (Ed) *Sex and Behavior*. New York: Wiley, 355-380.
Lynch, G., McGaugh, J. L., and Weinberger, N. M. (Eds.) (1984) *Neurobiology of Learning and Memory*. New York: Guilford Press.
Mann, V.A. and Liberman, A.M. (1983). Some differences between phonetic and auditory modes of perception. *Cognit. 14*: 211-235.
Margoliash, D. (1983). Acoustic parameters underlying the responses of song-specific neurons in the white-crowned sparrow. *J. Neurosci. 3*: 1039-1057.
Marler. P. (1990). Innate learning preferences: Signals for communication. *Develop. Psychobiol. 23*: 557-568.
Marler, P. (1993). Song-learning behavior: the interface with neuroethology. *TINS 14*: 199-206.
McCasland, J.S. and Konishi, M. (1981). Interaction between auditory and motor activities in an avian song control nucleus. *PNAS, USA 78*: 7815-7819.
Mello, C.V., Vicario, D.S., and Clayton, D.F. (1992). Song presentation induces gene expression in the songbird forebrain. *PNAS, USA 89*: 6818-6822.
Miller, D.B. (1979a). The acoustic basis of mate recognition by female zebra finches (*Taeniopygia guttata*). *Anim. Behav. 27*: 376-380.
Miller, D.B. (1979b). Long-term recognition of father's song by female zebra finches. *Nature 280*: 389-391.
Mooney, R. and Konishi, M. (1991). Two distinct inputs to an avian song nucleus activate different glutamate receptor subtypes on individual neurons. *PNAS, USA 88*: 4075-4079.
Nordeen, K. W., and Nordeen, E. J. (1988). Projection neurons within a vocal motor pathway are born during song learning in zebra finches. *Nature 334*: 149-151.
Nordeen, K.W. and Nordeen, E.J. (1992). Auditory feedback is necessary for the maintenance of stereotyped song in adult zebra finches. *Behav. Neur.Biol. 57*: 58-66.
Nottebohm, F. (1977). Asymmetries in neural control of vocalization in the canary. In: S. R. Harnad (Ed.) *Lateralization in the Nervous System*. New York: Academic Press, pp. 23-44.
Nottebohm, F. (1980). Brain pathways for vocal learning in birds: a review of the first 10 years. *Prog. Psychobiol. Physiol. Psychol. 9*: 85-124.

Nottebohm, F. (1981). A brain for all seasons: Cyclical anatomical changes in song control nuclei of the canary brain. *Science 214*: 1368-1370).

Nottebohm, F. and Arnold, A.P. (1976). Sexual dimorphism in vocal control areas of the songbird brain. *Science 194*: 211-213.

Nottebohm, F., Kelley, D.B., and Paton, J.A. (1982). Connections of vocal control nuclei in the canary telencephalon. *J. Comp. Neurol. 207*: 344-357.

Nottebohm, F. and Nottebohm, M.E. (1978). Relationship between song repertoire and age in the canary, *Serinus canarius*. *Zeit. f. Tierpsychol. 46*: 298-305.

Nottebohm, F., Stokes, T.M., and Leonard, C.M. (1976). Central control of song in the canary, *Serinus canaria*. *J. Comp. Neurol. 165*: 457-486.

Nowicki, S. and Marler, P. (1988). How do birds sing? *Music Percept. 5*: 391-426.

Petersen, M.R., Beecher, M.D., Zoloth, S.R., Moody, D.B. and Stebbins, W.C. (1978). Neural lateralization of species-specific vocalizations by Japanese macaques (*Macaca fuscata*). *Science 204*: 324-327.

Price, P.H. (1979). Developmental determinants of structure in zebra finch song. *J. Comp. Physiol. Psychol. 93*: 260-277.

Pröve, E. (1974). Der Einfluss von Kastration und Testosteronsubstitution auf das Sexualverhalten mannlicher Zebrafinken [The influence of castration and testosterone treatment on sexual behavior of male zebra finches] (*Taeniopygia guttata castonatis* Gould). *J. f. Ornithol 115*: 388-347.

Pumphrey, R.J. (1948). The sense organs of birds. *Ibis 90*: 171-199.

Scharff, C. and Nottebohm, F. (1991). A comparative study of the behavioral deficits following lesions of various parts of the zebra finch song system: Implications for vocal learning. *J. Neurosci. 11*: 2896-2913.

Schwartzkopff, J. (1949). Uber sitz und Leistung von Gehor und Vibrationssinn bei Vogeln. *Zeit. f. Vergleich. Physiol. 31*: 527-608.

Sinnott, J.M. (1980). Species-specific coding in bird song. *J. Acoust. Soc. Am. 68*: 494-497.

Sohrabji, F., Nordeen, E.J. and Nordeen, K.W. (1990). Selective impairment of song learning following lesions of a forebrain nucleus in the juvenile zebra finch. *Behav. Neural Biol. 53*: 51-63.

Sperry, R. W. (1974). Lateral specialization in the surgically separated hemispheres. In F. O. Schmitt and F. G. Worden (Eds.) *The Neurosciences: Third Study Program*. Cambridge, MA: MIT Press, pp. 5-20.

Stokes, T.M., Leonard, C.M., and Nottebohm, F. (1974). The telencephalon, diencephalon, and mesencephalon of the canary, *Serinus canaria*, in stereotaxic coordinates. *J. Comp. Neurol. 156*: 337-374.

Suga, N. (1990). Biosonar and neural computation in bats. *Sci. Am. 262*: 60-66.

Thorpe, W.H. (1961). *Bird-Song*. London: Cambridge University Press.

Weary, D.M. (1990). Categorization of song notes in great tits: which acoustic features are used and why? *Anim. Behav. 39*: 450-457.

Williams, H. (1985). Sexual dimorphism of auditory activity in the zebra finch song system. *Behav. Neural Biol. 44*: 470-484.

Williams, H., Crane, L. A., Hale, T. K., Esposito, M. A., and Nottebohm, F. (1992). Right-side dominance for song control in the zebra finch. *J. Neurobiol. 23*: 1006-1020.

Williams, H. (1990). Models for song learning in the zebra finch: Fathers or others? *Anim. Behav. 39*: 745-757.

Williams, H., Cynx, J. and Nottebohm, F. (1989). Timbre control in zebra finch (*Taeniopygia guttata*) song syllables. *J. Comp. Psychol. 103*: 366-380.
Williams, H. and Nottebohm, F. (1985). Auditory responses in avian vocal motor neurons: A motor theory for song perception. *Science 229*: 279-282.
Walcott, C, and Green, R. P. (1974). Orientation of homing pigeons altered by a change in the direction of an applied magnetic field. *Science 184*: 180-182.
Wynn, V.T. (1971). "Absolute pitch" — a bimensual rhythm. *Nature 230*: 337.

5

Ecological Specialization in the Avian Brain

Susan D. Healy

Abstract

The search for an explanation for variation in brain size has been a long one, fascinating philosophers and scientists alike (see Gould 1981 for review). Collaborations between evolutionary biologists, ethologists and neuroscientists are beginning to provide exciting new insights into how and why such variation may have come about. This chapter reviews the latest of these advances as well as providing an introduction to some of the methods, familiar to evolutionary biologists, that have made a major contribution to this progress. The chapter attempts to demonstrate the way in which behavior and its neural substrates are closely related and the fundamental role that natural selection has played in the evolution of the avian brain.

Brain Size and Body Size

Elephants have bigger brains than hummingbirds. Why? Oddly enough, we do not really know. In this particular example, it is not too difficult to see that the variation in the size of the brain is related in some way to the size of the rest of the animal's body. However, the close relationship between brain and body size has not always been so well appreciated. Around the turn of the century, for example, there was a flurry of studies comparing the brain size of different humans, yet in few of these studies was a role for body size entertained. One conclusion drawn was that women had smaller brains than did men. We know now that this difference in brain size reflects a difference in body size and not one in intellectual ability. Many of the other conclusions, such as apparent differences between different ethnic groups and between men of differing professions, where they were not completely fallacious, would not have been arrived at in these early studies if the effects of body size had been properly accounted for (Gould, 1981).

It is now taken for granted that there is an allometric relationship between brain and body size (brain weight = $a \times (\text{body weight})^b$, where a is the allometric coefficient and b is the allometric exponent). By no means are all the explanations for this relationship of an adaptive nature. Consider, for example, that increased body weight conveys a selective advantage, and such increased weight might be brought about through a growth hormone which concomitantly acts to increase brain weight. A mechanism of this nature assumes that changes in brain size, unlike those in body size, are not closely controlled by natural selection but are merely dragged along by changes in body size. Other, not dissimilar, non-adaptive explanations have been proposed but it is more commonly thought that selection does indeed play an active role in the determination of brain size. Currently, there is no highly favored explanation as to why brain size tends to be closely correlated with body size. Similarly unexplained is the observation that the relationship between brain and body size is not the same in different taxa: birds and reptiles differ from mammals, and within taxa the relationship among species is different from that among genera (Harvey and Pagel, 1991).

Relative Brain Size

Body size, closely related as it is to brain size, however, does not explain all the variation, either inter- or intraspecific. This effect can be observed statistically in regressions of brain size on body size in which the data are scattered around the regression line (for example, see Figure 5.1). The degree of scatter gives an indication of how closely brain and body size are related. When body size is very closely correlated with brain size, there will be little scatter and a significant regression. The greater the scatter the less body size accounts for variation in brain size. Over the last decade more and more examples of variation in brain size have been found that cannot be explained simply by differences in body size alone. In nearly all cases a plausible argument has been made that the observed variation in relative brain size is correlated with variation in an ecological or behavioral variable. One of the early examples is the relationship found by Clutton-Brock and Harvey (1980) between relative brain size and diet in primates. They found that frugivorous primates have relatively larger brains than do folivorous species. Indeed,

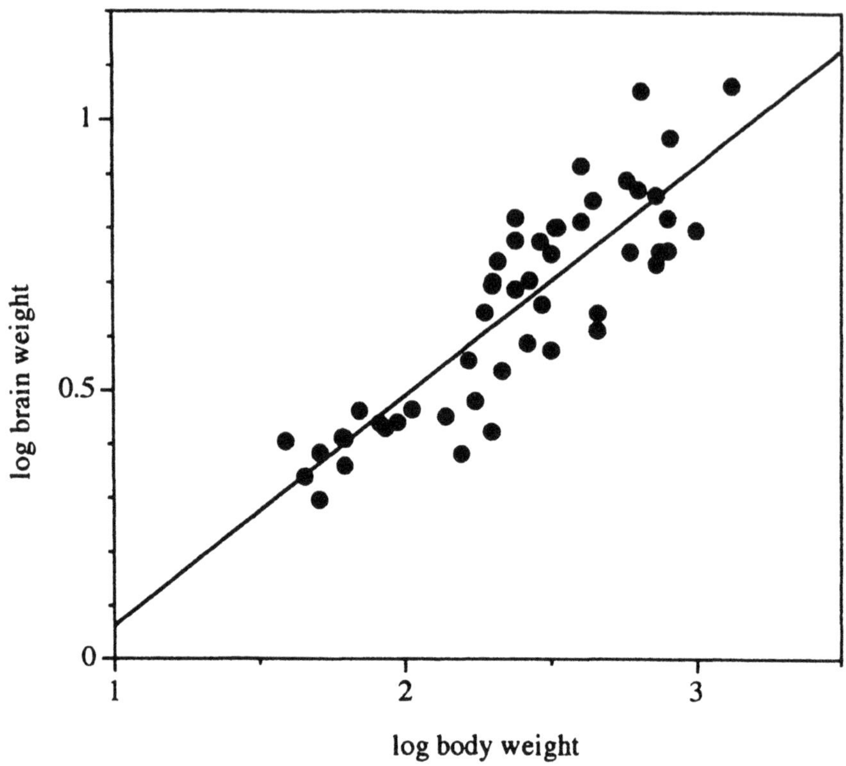

FIGURE 5.1 A regression of brain weight on body weight of 50 species from the family Sciuridae. Data from Mace et al. (1981).

in other mammalian groups relative brain size has also been shown to be correlated with diet, e.g. bats (Eisenberg and Wilson, 1978) and rodents (Mace et al. 1981). The relationship between diet and relative brain size is not one common to all animals however. In birds relative brain size is not correlated with diet, but it is correlated with mode of development, i.e. whether the species produces precocial or altricial young (Bennett and Harvey, 1985a). Birds which produce altricial young have relatively larger brains as adults than those species which produce precocial young. Conversely, this is not a relationship found in mammals (Bennett and Harvey, 1985b). In spite of the lack of a widespread explanation for the variation in relative brain size among vertebrate species, these studies, correlational though they were, began to give the impression that the agents of natural selection were closely tied to changes in brain structure and that brain structure might be more plastic than previously had been considered. It was beginning to look as if evolutionary biologists might have something useful to contribute to an understanding of the brain, previously strictly the domain of neuroscientists, psychologists and philosophers.

The Comparative Method

It may not be immediately obvious how an evolutionary viewpoint could contribute to an understanding of brain structure and function. However, by looking at brain structure and function across a range of differing species, similarities and differences, as hinted at in two-species comparisons (e.g. Jacobs et al. 1990), can be seen as evolutionary patterns. Such patterns are only correlational in nature and do not prove causation (detailed experimental work is required to be convinced of causal relationships) but where significant relationships can be discerned, insights may be gained as to how selection has acted on neural structures. The overview provided by a comprehensive cross-species analysis, for example, can act as a guide as to those species worthy of further, detailed experimental investigation. Those species for which the differences are most apparent in a comparative analysis should be those most likely to yield results at the experimental level. In any comparative study there are also species which do not appear to fit the general pattern and it is likely that these too will contribute insight as to why or how a particular species 'escaped' the mould into which all the others fit, more or less. The advantage of single species experimental investigations is that they can say a lot about how evolution has acted on neural mechanisms within a species. But they have little to say as to whether these mechanisms have been commonly employed. So, generality of neural design can be determined with comparative analyses.

In order to determine the generality of a certain relationship, phylogenetic relatedness must be dealt with. Closely related species are likely to be more similar to each other both in their ecology and their morphology than they are to more distantly related species. The aim of the comparative method is to assess the number of instances in which natural selection has independently brought about change in a feature (e.g. brain size) in response to a change in the animal's environment (e.g. diet). The greater the number of times that natural selection has converged on the same solution to a problem the less likely it is that the solution is itself a coincidence. The phylogenetic basis of comparative analyses thus allows an assessment of the origins of specific neural structures and their relative ages. Once a relationship between brain size and an ecological variable has been established to have occurred in a large number of separate instances answers may be found to questions such as "How quickly can brain size and organization change (at a species level)?" "How do developmental patterns of brain growth differ between taxonomic groups?" "What are the trade-offs, in terms of neural capabilities, between differing rates of development?".

The Question of Size

A question commonly asked of comparative brain/body analyses, such as those of Harvey et al. (1980), is 'what does size mean?' Surely if we want to know about brains, it would be better to look at the building blocks of the brain, the neurons? Will we not find out all we need to know about brain structure and function once we have determined the structure, function and dysfunction of neurons? There is no denying that such work will take us a long way to discovering how the brain works. But how is size relevant to the brain? For the rest of the body, it has long been accepted that size means something. For example, longer legs are associated with increase in speed of gait. Increases in the size of sensory structures, such as the nose or ear, are usually correlated with increased sensitivity or increased capacity. It seems more than plausible that size, in relation to the brain, also implies increased capacity of some description. An increase in brain size that is directly correlated with increased body size may be explained simply by the extra sensory processing required by a larger animal, without any change in higher cognitive processing. However, body size is also correlated with life span: larger animals tend to live longer. Animals that live longer may do better if their brains are able not only to process more sensory information but also to store information for long periods of time and process input in more complex ways. After all, an elephant will need to remember some information for the duration of its long life. Likewise it would be useful if it could update the stored information when its environment changes, which is bound to happen over the decades of the elephant's life. An elephant with a brain that had an expanded capacity for this type of processing would probably have done better than a smaller-brained conspecific.

So, large, long-lived animals have larger brains than smaller, shorter-lived species and larger brains probably enable greater processing and storage capacity of some description. What might increasing brain size mean at the cellular level? It may be that larger brains are due to increased numbers or size of neurons. On the other hand it may be that increasing the number of neurons requires an accompanying increase in the number of support cells like glia and astrocytes. Variation in whole brain size does not allow the pinpointing of the particular cell populations that have produced the changes.

Certainly comparative studies demonstrating correlations between ecological variables and whole brain size (e.g. Harvey et al. 1980; Bennett and Harvey, 1985a and b) were not able to explain very convincingly why selection might have brought about an increase in brain size, plausible though the relationships may have appeared, let alone the nature of the cellular basis for such changes. How, for example, has frugivory brought about a change in whole brain size in primates, bats and rodents? It was suggested that there is increased complexity in foraging brought about by the spatial and temporal variation in fruit supply and this added complexity has brought about selection for increased processing and storage of information by many parts of the brain (Mace et al. 1981). It is not difficult to imagine that a range of improvements in, for example, visual and olfactory processing and in motor skills might be highly advantageous to frugivorous mammals. But what is complexity? And does increasing complexity really require an enlargement of the whole brain or has a part of it simply increased so much that the change is observable in a whole brain measurement? Furthermore, how easy is it for complexity, or any other agent of natural selection, to bring about neural changes, whatever the scale?

The Cost of Neural Changes

Worthy of consideration at this point is the assumption that brains are in many ways 'expensive'. This is most easily seen in metabolic terms: with its constantly greater demands for oxygen and glucose, the brain turns over a far larger proportion of the body's blood supply than any other organ or region. This supply must be kept constantly high. Deprivation of oxygen for as little as four minutes can lead to massive, irreparable damage. Such effects are commonly seen in ageing humans who have suffered cerebrovascular accidents (strokes). Strokes are caused by a blockage or rupture in the arterial blood supply, and the larger the area of brain downstream of the blockage the more massive the damage. Brain cells are not only more expensive to run than other body cells, they are also more precious. Unlike all other cells, once lost, they cannot be replaced. Repair and replacement of function in an adult brain must be carried out by the remaining cells which conduct a patch-up job as well as they can.

Brains, then, are expensive to maintain and problematic to repair. In spite of these constraints selection has brought about increased brain size in some species. The benefit, if somewhat nebulous, of a larger brain has outweighed these more obvious costs. Even in relatively larger-brained species, however, it seems likely that, in fact, it is not the whole brain that has increased in size but rather one, or some, of its component parts. Some areas of the brain play a particular role by receiving and processing a single type of information, e.g. visual cortex. This is most obvious in those areas processing sensory information or controlling motor output. However, there is also much work attempting to describe unimodal functioning to other, 'higher' processing areas, such as the amygdala and hippocampus. Work on brain damage and repair has shown that such precise regional organization is not characteristic of the brain's operation, or at least, it is not necessary for competent function. In the visual system, for example, it has been well documented that deprivation of input to one area may result in its contralateral partner doing the work normally done by both. Even in these cases, however, repairs are not carried out by areas receiving another kind of input nor, necessarily, a neighboring region. Connectivity studies have also shown that neighboring areas are not always in direct contact, nor are areas that process the same kind of information adjacent to each other, e.g. the visual system in mammals and the song system in birds. Why are brains wired up in this less than orderly fashion? Are all vertebrate brains wired up in the same way? Is this the most efficient possible neural structure or one that has been added to and modified piecemeal?

The Brain as a Sum of its Component Parts

If brains have been tweaked and twiddled by natural selection in response to differing environmental and behavioral demands, we might expect to see variation not only in whole brain size between species, especially those whose lifestyles are very different, but also to see variation in the size and organization of the component parts. Indeed, following the results of Harvey et al. (1980) it was not long before explanations were sought at the level of particular brain regions. How different, after all, are folivorous species from frugivores and insectivores? None of the other ecological variables investigated were correlated with variation in brain size between these groups, such as activity timing (nocturnal/diurnal), stratification (arboreal/terrestrial), habitat (forest/woodland, grassland/desert) and breeding system (monogamous/polygynous). Folivorous and frugivorous primates seem to have quite similar lifestyles, except for their diet. In more concrete terms, how is folivory less complex than frugivory? Edible fruit-bearing plants may be scattered in a more heterogeneous manner than those plants from which folivores feed. Fruit eating also has a larger temporal component: both in production and in stages of ripening. Thus fruit eaters may have to search more carefully for appropriate food, judge when fruit will be ripe and, perhaps, remember from season to season the locations of suitable plants. Which neural structures would be put under more pressure in order to cope with these extra processing demands of the frugivorous habit? Or, in other words, increased speed or efficiency of processing by which neural structures would benefit fruit eaters most? Eisenberg & Wilson (1978) found that frugivorous bats, for example, have a larger relative neocortex as well as larger visual and olfactory projection areas. Enhanced visual processing might enable better detection of fruit, enhanced olfactory processing may aid discrimination of edible or ripe fruit and increased cortical areas may provide better analysis of all the incoming sensory information, more efficient manipulation of fruit (e.g. peeling) as well as providing 'space' for storage of memories for locations of fruit-bearing plants. These suggestions are, of course, hypothetical. However, it may be the case that frugivorous bats are better at certain visual, olfactory and higher processing tasks than their folivorous relatives. Determination of such behavioral differences requires experimental manipulation.

Ecology and Brain Regions: Sensory Systems

Trying to pinpoint where in the brain the processing of 'complex' information may occur is probably a waste of time. However, there are areas of the brain in which specific kinds of information are known to be processed, the most obvious of which are the sensory processing areas. It seems sensible to begin with these neural structures when looking for relationships between variation in their size or organization and variation in an animal's environment. In order to do this two different kinds of data are required, firstly, measurements of the size of brain regions and, secondly, descriptions of each species' behavior and ecology. In collecting the second kind of data it has become obvious that, even though most interest might be focused on mammals, particularly those most closely related to humans, we know much more about the habits of birds. Birds, after all, are frequently brightly colored, conspicuous and diurnal. Mammals have a tendency to be cryptically colored, secretive and nocturnal. In addition, through the mammoth efforts of Sibley and Ahlquist (1991) the phylogeny of the birds, determined by DNA hybridization, has been largely agreed upon. It is, thus, from the birds that most of the recent studies on variation in brain size have garnered results.

Activity Timing and Olfactory Bulbs

Portmann (1946; 1947) was the earliest to look at relationships between parts of bird brains (brain stem, optic lobes, cerebellum and the hemispheres) and ecological variables. Like Bennett and Harvey (1985a), his most conspicuous finding was that species producing altricial young had larger brains as adults than those producing precocial offspring. Subsequent to Portmann's findings few comparative studies were carried out and none of a widespread nature until the work of Cobb (1960), followed by Bang and Cobb in 1968 and Bang in 1971. They measured the olfactory bulb size and cerebral hemispheres of over 130 bird species and to find whether olfactory bulb size was associated with diet, after claims that some birds, such as turkey vultures, located food via smell. Cobb (1960) found a correlation, weak though it was, between olfactory bulb size and aquatic habits in birds. Bang (1971) also found a relationship between olfactory bulb size and living on or near bodies of water. These species also tended to be piscivorous/carnivorous, colonial ground-nesters. These analyses did not provide strong evidence that the size of the avian olfactory bulb was correlated with diet nor with any other ecological variable. And yet subsequent behavioral and neurophysiological data consistently showed not only that many birds can smell, they can detect odors in many different situations e.g. odor aversion learning, food finding, orientation and navigation (see review in Bang and Wenzel, 1985). Coupled with these data that show birds can detect and do use olfactory information is the observation that all birds possess olfactory bulbs and there is enormous variation between species in olfactory bulb size (Bang, 1971).

In 1990 Healy and Guilford reanalyzed the data of Cobb (1960) and Bang (1971), making three substantial modifications. Firstly, they used a recent phylogeny; secondly, they removed the effects of body weight; and finally, they tested the hypothesis that olfactory bulb size is correlated with activity timing. Neither the effects of taxonomic association or body weight had been considered in the previous analyses nor had activity timing been examined. Across the 124 species they surveyed, Healy and Guilford (1990) found 13 cases in which the activity timing of two sister taxa differed, and in 12 of these the nocturnal birds had larger relative olfactory bulbs than their diurnal relatives (Figure 5.2). Healy and Guilford did not find olfactory bulb size correlated with other ecological variables such as diet, nest type, development, nest dispersion, or migration. It is not yet clear what advantage enlarged olfactory bulbs confer on nocturnal birds although it seems likely that for animals living in an environment in which visual information must be more difficult to obtain, increasing the ability to detect food, conspecifics and predators

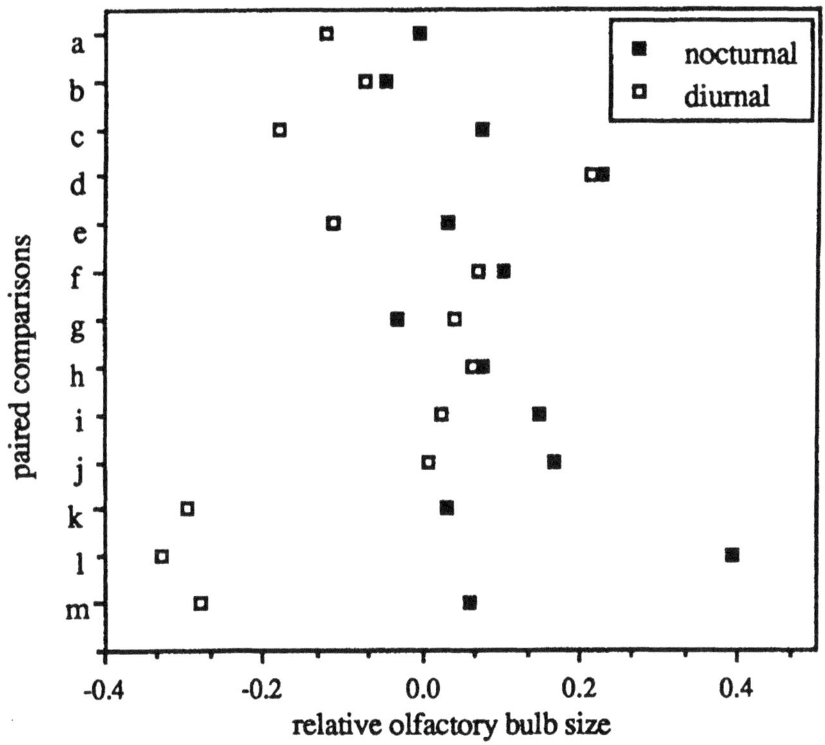

FIGURE 5.2 The relative size of the olfactory bulb in nocturnal/crepuscular and diurnal birds. The horizontal axis represents the relative size of the olfactory bulb after removing the effects of body and brain size by multiple regression. The groups represented and sample sizes are given in Healy & Guilford (1990). Solid symbols, nocturnal/crepuscular groups; open symbols, diurnal groups.

by olfaction must be advantageous. Interestingly, although there was a very marked difference in the size of the olfactory bulb between nocturnal and diurnal species there was no difference in whole brain size. This lead Healy and Guilford to speculate that some trade-off had occurred among the neural structures. Is it possible that the olfactory areas could only have been enlarged at the expense of another area, for example, the visual processing areas? This idea was not tested due to a lack of relevant data. More recently, Gittleman (1991) showed that the association between nocturnality and olfactory bulb size is not confined to birds. He found that nocturnal carnivores have larger olfactory regions than do their diurnal relatives.

Birdsong and its Neural Substrates

It is from birds again, however, that the two most established bodies of work on relationships between ecological or behavioral variables and variation in brain regions have been produced. The foremost of these, on the neural substrates of bird song, is the product of several laboratories in the United States and began in the mid 1970's (Nottebohm et al. 1976). Prior to this, however, investigations into the behavioral nature of bird song, its development, innate predispositions and interspecific variation had laid the essential foundations for the neural work that followed (e.g. Konishi, 1963; Marler, 1957; see also chapter by Cynx).

Birdsong: The Behavior

Song is produced largely by males and most frequently during the breeding season. Not only may the syllables used in the construction of song differ between individuals within a species, the syllable type, length and complexity of songs varies enormously between species. The males of some species sing the same song over and over again whereas males of other species may have extremely large repertoires and only rarely repeat themselves. Some species sing very specific songs irrespective of their geographical location, whilst others have songs that are location-specific and culturally transmitted, known as 'dialects' (e.g. Marler and Tamura, 1962). Why is there such phenomenal variation in bird song? After three decades of intensive work we are still unable to answer the question fully. Nor is the variation in birdsong confined to the sound itself. The development of song also varies greatly between species. For most birds there is a 'sensitive period' within the first few months of life during which they are exposed to song and this song(s) will form the basis for their own singing in the following breeding season. Birds exposed to song outside this sensitive period will find it difficult or impossible to learn. There is interspecific variation in the time at which the sensitive period begins as well as in its duration. Song sparrows and zebra finches, for example, have a sensitive period that begins within two-three weeks of hatching and lasts four to eight weeks (Immelman, 1969; Marler, 1987). Other species are able to copy songs throughout their first year and still others continue to acquire new songs their entire life (e.g. canaries). Not only do some species have temporal constraints imposed on their song learning, there is evidence for restrictions on song content. Most young birds learn a species-specific song even though they are usually surrounded by a chorus of different songs, indeed, they often learn from only one male (Marler, 1991). Experimental treatments have shown that some birds can sing species-specific song even when isolated during early development (Price, 1979) although the species-specific song of deafened birds will gradually degrade (Nottebohm et al. 1976). This latter result would suggest that auditory feedback plays a major role in song acquisition and production. Birdsong, then, is usually produced by males, often acquired only during a sensitive period, and varies enormously in structure among species (see reviews in Konishi, 1985; Slater, 1989; Marler, 1991; chapter by Cynx).

Birdsong: The Neurobiology

The syrinx is the organ which produces song. Its muscles are innervated by projections from the hypoglossal nucleus in the brainstem (nXIIts) which in turn receives input from nucleus robustus archistriatalis (RA) and the dorso-medial nucleus of the intercollicular complex (DM) in the telencephalon. RA receives input from the high vocal centre (HVC), field L and LMAN (lateral portion of the magnocellular nucleus of the anterior neostriatum). HVC also projects to Area X. A number of neat lesion experiments have demonstrated the roles of these areas in the acquisition, perception and production (for review see DeVoogd, 1994). LMAN is crucial to song acquisition in juveniles and grows rapidly within the first few weeks following hatching of both males and females. Connections to RA grow rapidly and then regress within about two months, to a greater degree in females (Nordeen et al. 1992). The size of LMAN itself also decreases (Bottjer et al. 1985), due partly to changes in the size of neurons and of their dendritic trees (Nordeen et al. 1987). These size changes have been shown to be coincident with the sensitive period and thus with early acquisition of song.

RA and HVC are both involved in the acquisition and production of song, perhaps even in song discrimination (Brenowitz, 1991). They also increase in size during this time, reaching greater volumes in males, but do not show the regression pattern of LMAN. In some species, their size actually fluctuates seasonally, coincident with song production by the adult male. The discovery of this seasonal brain change in the 1980's (Nottebohm, 1981) was very exciting as it opened up the possibility that in some way the behavioral changes were causing those in the brain, a feat not previously thought possible. Subsequently, the story has become much more complicated, not least because more species have been investigated.

The original brain/behavior changes were shown in the canary. The volumetric changes in canary RA come about through variation in the length of dendrites and dendritic spine density. Both of these neuronal features are greatly increased in the spring (Hill and DeVoogd, 1991).

The seasonal variation in HVC appears to be caused by changes in the rate of cell death and recruitment of new cells (Kirn et al. 1994). Male canaries incorporate new songs into their repertoire each year whereas in other species, like the zebra finch, song learnt in the first year is that sung in succeeding years. In the zebra finch, although there is sexual dimorphism in the size of RA and HVC, there is no seasonal fluctuation in volume of these two nuclei. The detailed comparisons of these two species suggested that here was evidence for a causal relationship between song acquisition and production and major neurobiological changes. In addition, the discovery of a substantial degree of neurogenesis occurring in the adults provided the mechanism for such neural changes and the substrate for acquisition of new songs. In the summer song production ceases and many neurons from both RA and HVC are lost. The loss of so many cells suggests that it is of no benefit, or is costly, to maintain them when song is constructed anew in the succeeding winter and spring. The song system of zebra finches, on the other hand, does not show this degree of neurogenesis or loss of neurons. Neurogenesis and volumetric changes in the song system seemed to be causally related to seasonal song acquisition, and this conclusion seemed to be confirmed by data from red-winged blackbirds (*Agelaius phoeniceus*) and white-crowned sparrows (*Zonotrichia luecophrys*) (Kirn et al. 1989; Baker et al. 1984). Not all of the data, however, fit this conclusion quite so neatly. For example, rufous-sided towhee (*Pipilo erythrophthalmus*) males do not learn new songs after the first year but do show seasonal fluctuation in HVC size (Brenowitz et al. 1991), as do Gambel's white-crowned sparrows (*Z. leucophrys gambelii*) which sing throughout the year (Smith et al. 1992). These data would suggest that investigation of the song system in a greater variety of species, especially those which differ in the various features of song acquisition and production, would be useful in the near future.

With the accumulation of more species, relationships between the variation in the size and structure of the song system will become apparent. One such pattern has been found in a comparative analysis across 45 species of songbird (DeVoogd et al. 1993). The experimental studies have shown that HVC is crucially involved in the learning and production of song. DeVoogd et al. (1993) showed that variation in the relative size of HVC was positively correlated with variation in repertoire size: species with larger repertoires have a relatively larger HVC volume (Figure 5.3). HVC volume is not, however, correlated with a measure of song complexity, the number of syllable types per song. The differences found between the species were large enough that it is likely that they were caused by differences in the number of neurons and the extent of axonal and dendritic branching although, currently, there are insufficient data to confirm this supposition. The volume of Area X, which plays a role in acquisition of song (e.g. Scharff and Nottebohm, 1991), was also measured but did not correlate with repertoire size.

It would appear, then, that selection has acted specifically on the various subregions of the song system, rather than having had a more general effect. Such specificity, even with respect to these quite tiny regions, provides a good idea as to how finely tuned selective effects can be, such that selection can pinpoint quite precisely the regions in which changes are most beneficial. In addition, perhaps the occurrence of neurogenesis is a result of selection for an appropriate mechanism for the acquisition of new information which, after use, would be best dumped. Endlessly increasing the brain size (and thus weight) of a small, flying bird would surely be a costly matter. For those species for which changing songs each year is beneficial, then getting rid of the neural substrates of the used songs would free space for the forthcoming breeding season and avoid an accumulation of unwanted synaptic connections.

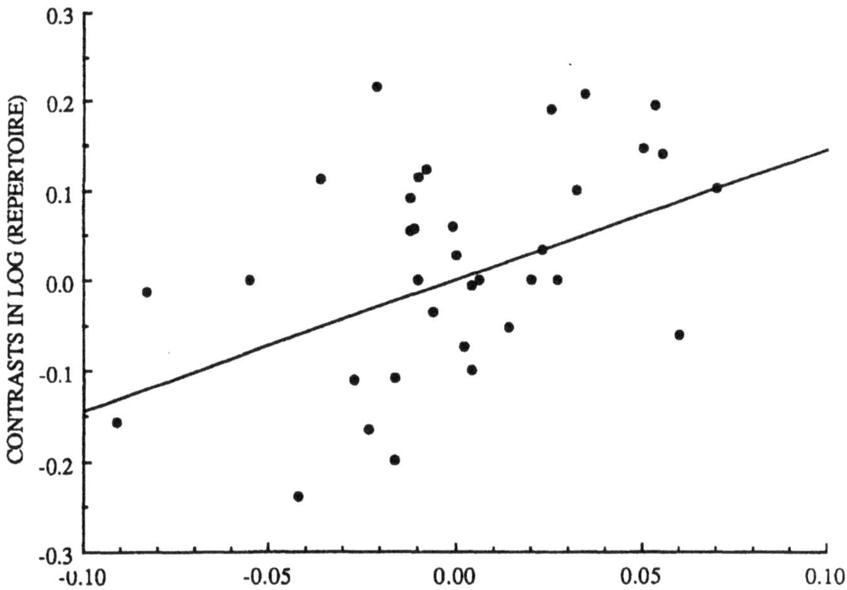

FIGURE 5.3 The correlation between HVC volume and song repertoire size. The horizontal axis represents the relative volume of the hippocampus after removing the effects of brain size by regression. The groups represented, sample sizes and analytical method are given in DeVoogd et al. (1989). Solid symbols, species; open symbols, genera; crosses, families.

Ecology and Brain Regions: Higher Processing Areas — Food Storing

As with the work on birdsong, the investigation into the relationship between food storing and the hippocampus began with observations of a rather remarkable behavior. Food storing is not a very common behavior but it is taxonomically widespread, the best known food-storing species probably being honey bees and squirrels. All manner of food stuffs are stored, in different kinds of locations and are retrieved over varying time periods (see Vander Wall, 1990 for review). The stored food may be used to feed young (e.g. honey bees), other relatives (e.g. acorn woodpeckers), or the hoarders themselves. The amount of food stored varies too: cats will store a single prey carcass whilst squirrels and jays store many thousands of seeds. Why animals store food continues to be a source of debate (e.g. McNamara et al. 1990; Brodin, 1994) but a common theme seems to be that food storing may enable animals to cope with stochasticity in food acquisition. For some animals this may be a daily problem and for others, one that occurs seasonally.

Storers fall into two general categories: larder hoarders and scatter hoarders. Larder hoarders store all of their food in one or a few locations, usually near the nest or burrow. Scatter hoarders, on the other hand, store each item singly and storage sites are widely spread. The larder hoarder has no difficulty in relocating the food but does have the problem of defending it. Scatter hoarders have no problem defending their hoards, but how do they retrieve their dispersed stores? For those scatter-hoarding species that store hundreds or thousands of items retrieval poses a major problem. In the 1950's and 1960's it was thought, with rare exceptions like Haftorn (1956), that storers must retrieve their food either at random or, perhaps, by smell. It was not until the 1980's that the possibility that these animals might use memory for relocation was seriously investigated. Balda and Kamil working on corvids, and Krebs, Sherry and Shettleworth working on parids began accumulating evidence that, in these birds at least, memory was indeed the mechanism for accurate food retrieval. Various features of the memory used in food retrieval have now been described. Not only can the birds remember a prodigious number of different locations, they can remember the quality of the food item (returning first to preferred food items) and even that they have emptied a site of food. These birds do not return to sites from which they have removed the stored item. Added to the already impressive memory task these animals set themselves is the fact that they may not retrieve this food for several days, weeks or even months (for reviews see Sherry, 1989; Balda & Kamil, 1989; Shettleworth, 1990). Food storers seem to have set themselves an awesome memory problem.

The Hippocampus

The hippocampus is an area of the mammalian brain which over the past decade, particularly, has excited much interest, primarily because of the massive impairment to memory caused by hippocampal damage. The possibility that the avian structure of the same name was also involved in memory processing came about following the results of experiments involving hippocampal lesions in food storers. Black-capped chickadees (*Parus atricapillus*) with lesions to the hippocampus are unable to relocate their stored food accurately, although they are still able to store it, and do attempt to retrieve (Sherry and Vaccarino, 1989; Rice, 1992). It seems as if lesioned birds are impaired in their ability to process spatial information, yet are as able as normal birds to perform other memory tasks that do not involve memory for locations. So, a putative role for the avian hippocampus is that it processes spatial information. Support for this hypothesis was provided by two comparative studies which between them looked at hippocampal volume in over 50 passerine species (Krebs et al. 1989; Sherry et al. 1989). Both studies found that the size of the hippocampus is relatively larger in food-storing birds than it is in closely related non-storers (Figure 5.4). Both of these studies examined the effects of other ecological variables (e.g. mode of development, nesting dispersion, diet), none of which significantly explained any of the variation in hippocampus size. Interestingly, migration, the ecological variable thought most likely to be require spatial information processing, was also not correlated with variation in hippocampal volume (Healy et al. 1991). Presence of food storing, then, is correlated with an enlarged hippocampus.

The family Corvidae provided an ideal opportunity to examine further the closeness of this brain/behavior relationship because species of corvids vary in the degree of food storing. Jays (*Carrulus glandarius*), for example, store many thousands of seeds and may not retrieve them for several months; magpies (*Pica pica*) and crows (*Corrus corone corone*) store tens or hundreds of food items, retrieving them within a few hours or days; and jackdaws (*Corrus monedula*) store little or not at all. Amongst six species of European corvid, Healy and Krebs (1992) found that relative hippocampal volume is larger in those species which store more food or store it for longer (Figure 5.5). A similar relationship has been found amongst North American corvids (Basil et al. in press) and also amongst species within the Paridae (Healy and Krebs, in press; Hampton et al. 1995).

One of the questions that arose from this work concerned the timing of this hippocampal enlargement. Was the food storer's hippocampus relatively larger throughout development even though food storing does not begin until a few weeks after fledging (Haftorn, 1992; Clayton, 1992)?

Or was hippocampal enlargement tied more intimately to the onset of the behavior? Two within-species comparisons have shown that the enlarged hippocampus seen in adult food storers only appears after the onset of food storing (Healy & Krebs 1993; Healy et al. 1994a). In food-storing magpies and marsh tits (*P. palustns*) prefledging juveniles have similarly sized brains as the adults and yet their hippocampal volumes are significantly smaller. Hippocampal volume in prefledging juveniles and adults of close non-storing relatives, jackdaws and blue tits (*P. caeruleus*), respectively, does not show this age-related change (Figure 5.6). The observed changes in hippocampal volume are associated with a higher neuron density in the juveniles than in the adults, and the non-storing adults had significantly fewer hippocampal neurons than the storing adults.

Most recently, an even more surprising result has been found: in a study with hand-raised food storers it was found that the hippocampus

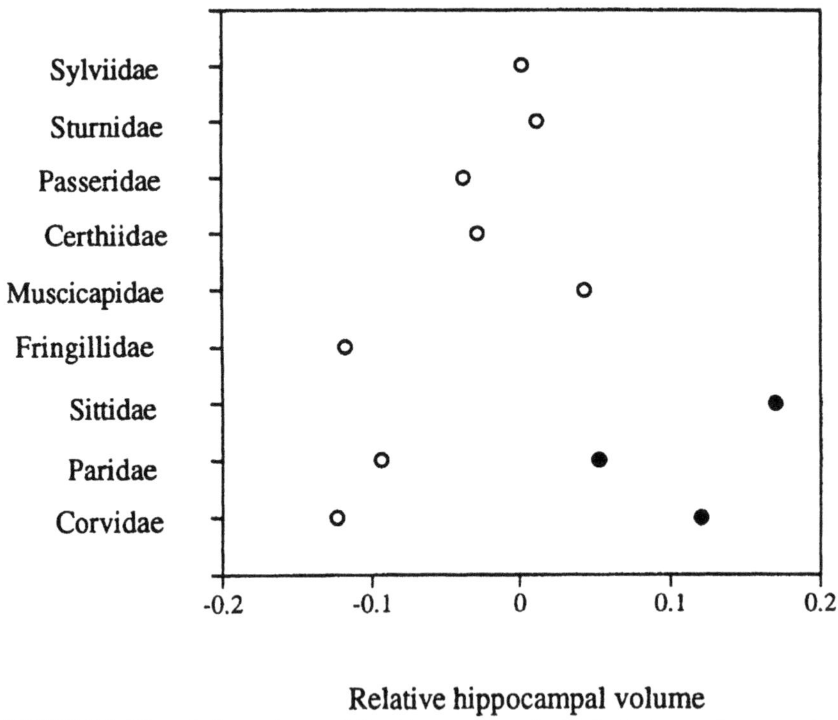

FIGURE 5.4 The relative volume of the hippocampus in food-storing and non-storing passerine birds. The horizontal axis represents the relative volume of the hippocampus after removing the effects of body and brain size by multiple regression. The groups represented and sample sizes are given in Krebs et al. (1989). Solid symbols, food-storing species; open symbols, non-storing species.

of birds deprived of food-storing experience was *smaller* than that of similarly-aged conspecifics with food-storing experience (Clayton & Krebs, 1994). This result suggests that, through behavioral experience, large scale changes are possible in the brains of adult animals. What is bringing about these volumetric changes is not clear, but one suggestion is that the experience of food storing is, in some way, inhibiting cell death through apoptosis. Apoptosis has itself recently become a focus for attention and is commonly referred to as 'programmed cell death' as it frequently occurs during events such as embryogenesis and metamorphosis (Martin, 1993). It does, however, also occur in adult brains although the role it plays is, as yet, uncertain.

Changes in the hippocampus of marsh tits appear to be caused by the experience of food storing. The hippocampus, however, does not control food storing itself as hippocampal lesions do not prevent birds from

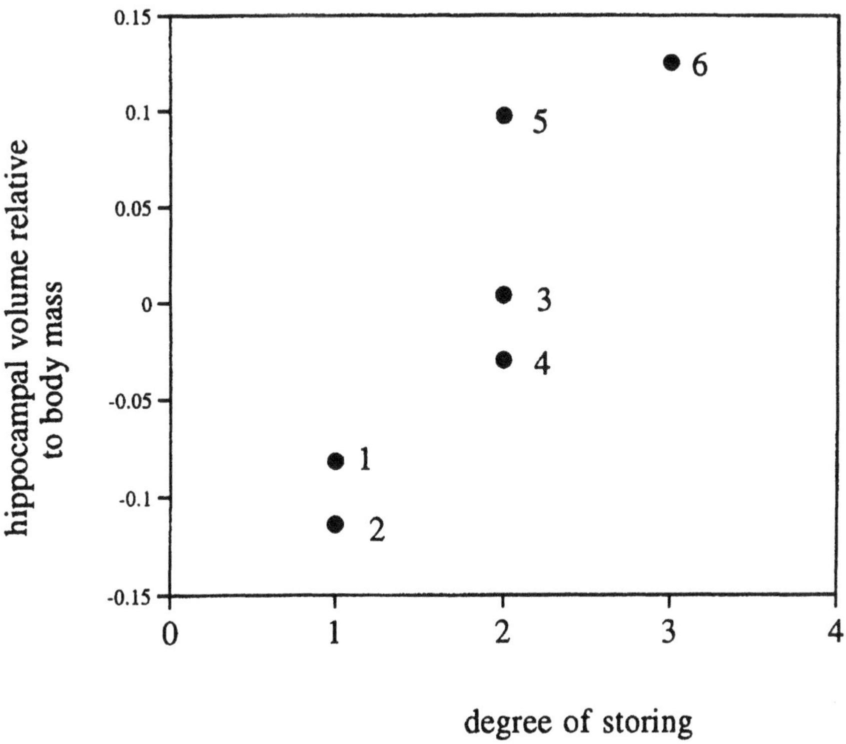

FIGURE 5.5 The volume of the hippocampus relative to the body mass plotted as a function of category of storing behavior (scored as 1, 2, and 3 for low, medium and high storing). The species studied and sample sizes are given in Healy & Krebs (1992).

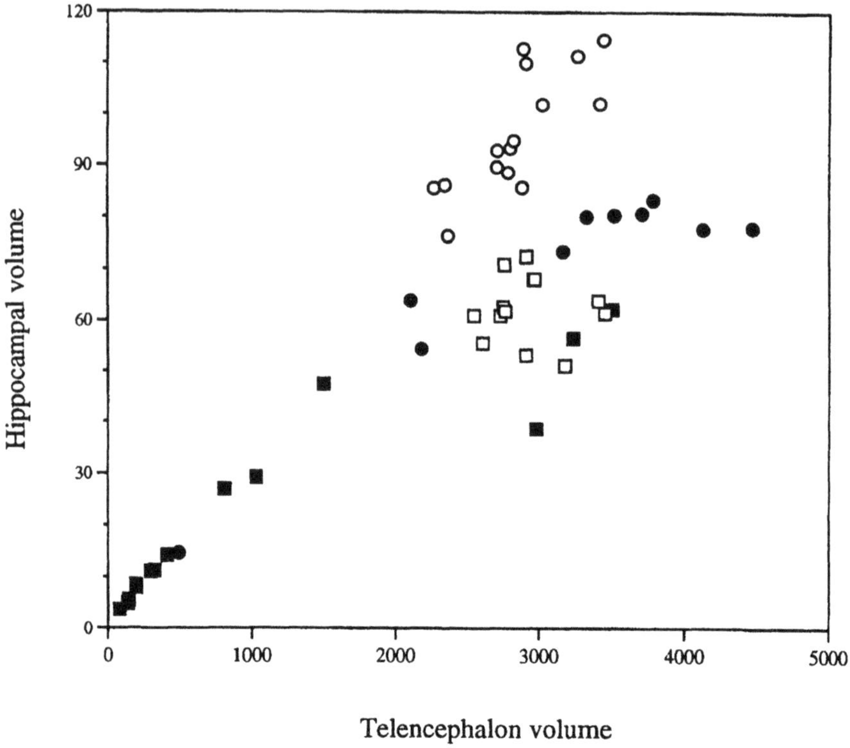

FIGURE 5.6 Hippocampal volume as a function of volume of the rest of the telencephalon (correlated with age). Solid squares: 5-25 day old jackdaws; open squares: >320 day old jackdaws, solid circles: 5 -25 day old magpies;open circles: > 320 day old magpies. From Healy & Krebs (1993).

storing (Sherry and Vaccarino, 1989; Rice, 1992). Because the lesioned birds are incapable of accurate retrieval, it has been assumed that the hippocampus is processing information about the storage locations and thus spatial information. Given the consistent and clear difference between storers and non-storers in hippocampal volume it has been surprisingly difficult to demonstrate a difference in memory capacities between the two groups. Some rather ingenious experiments have been devised (e.g. Hilton and Krebs, 1990) in order to make comparisons, mostly to little avail. Recently, Brodbeck and Shettleworth (1995) and Clayton and Krebs (1994b) showed that food storers had a preference for spatial information when the alternative was the site's local color pattern whereas the non-storers had no strong preference. When tested for recall ability of the two types of cue in a separate experiment, Healy (in press) did not find differences in memory performance between storing and non-storing tit species.

The comparative experiments to date do not provide consistent support for the idea that the enlarged hippocampus enables the food storer to remember a few locations for a longer time than non-storers. Then what feature of the memory required for food retrieval is it that requires the larger hippocampus? Is it attention to detail so the animal remembers the close environs of each storage location? Is it that the food storer encodes the more distal cues into a 'map' and this map can simply contain more detail than that of the non-storer remembering just its territorial boundaries? Does a larger hippocampus enable associative memory with a finer discriminative capacity? Something of this kind is suggested by the results of Clayton and Krebs (1994b) and Krebs et al. (1990): storers are able to discriminate between sites visited previously that either did or did not contain food whilst non-storers appear not to make such distinctions. Finally, one of the most remarkable features of food-storing memory, the sheer number of locations that are recalled, has not seriously been assessed. Results from an experiment by Olson (1991) provide tantalizing evidence: nutcrackers (*Nucifraga columbiana*), a species heavily dependent on food storing, outperformed scrub jays (*Aphelocoma coerullescens*), a species that is far less reliant on stored food, in their ability to remember multiple sites. Although only three sites were sequentially presented, vastly fewer than the number of locations these birds can remember, it may be that the selective advantage of the larger hippocampus is additional processing for a large number of locations.

Sex Differences

The ecological significance of the hippocampus can be investigated further by looking for other situations in which there appears to be an increased demand for spatial memory. Sherry et al. (1993) have found one such situation. Brown-headed cowbirds (*Molotunis ater*) are North American brood parasites. The female visits as many as 40 potential host nests before she returns to lay her own eggs. The male takes no part in this activity. This nest choice procedure by the female would seem to require an impressive memory capacity and, indeed, hippocampal volume is greater in the female than the male. This sex difference was not observed in two other closely related icterid species which are not brood parasites, the red-winged blackbird and the common grackle (*Quiscalus quiscula*).

Migration

A second case in which there may be an increased spatial memory load concerns migration, the ecological variable previously found not to be correlated with variation in hippocampal volume (Healy et al. 1991). In a study that is still in progress, preliminary data (Healy et al. 1994b) do point towards an association between migration experience and hippocampal volume. In late summer, garden warblers (*Sylvia borin*) migrate from their breeding grounds in central Europe to overwinter in tropical Africa. There is evidence that they tend to use the same breeding and overwintering sites each year as well as the same stopover sites during the migration (e.g. Loevei, 1988). Healy et al. (1994b) have suggested that they are able to return to these same sites by remembering the local landmarks associated with each of them and have proposed that these memories impose increased spatial information processing by experienced birds. To test this hypothesis, hippocampal volumes of birds that had migrated at least once were compared with those of young birds about to set out on their first migration. It was found that older birds have relatively larger hippocampal volumes than do younger birds. As this result may have been brought about through some effect of increasing age, birds from both groups (experienced migrants and inexperienced birds) were kept in captivity for 12 months before a second batch of measurements were made. Relative hippocampal size of inexperienced birds was still found to be smaller than that of the experienced birds. Hippocampal volumes of a closely related non-migratory warbler, the Sardinian warbler (*S. melanocephala*), were also measured and no difference between juvenile and adult birds was found.

These results would appear to be at odds with that found earlier by Healy et al. (1991) and it is not yet clear why this is so. It may be that the migration variable tested, migration distance, was an inappropriate one in terms of an increased demand for spatial information processing, especially as the resident species in the analysis included food-storing species. As we know these species have an enlarged hippocampus it is perhaps not surprising that the migrant species hippocampus size did not differ from that of the resident species. In addition, there may be qualitative or quantitative differences in the spatial memory requirement between food storing and migration which would result in differences in hippocampal volume.

Trade-offs Within the Brain

However the apparent discrepancy in the migration/hippocampal volume story is resolved, evidence is gradually accumulating that the size of the avian hippocampus is associated with variation in demand for the processing of spatial information. In none of the cited examples was variation in brain size correlated with variation in the relevant ecological variable, demonstrating again how finely selection has manipulated the neural structures underlying behavior. If food storers have an enlarged hippocampus but not an enlarged brain then it seems as if the benefits of increased spatial memory that accrue to the bird from the enlarged hippocampus must have been costly in some other way, specifically in the decrease of size and function in another part of the brain. It is not yet known, either in behavioral or neural terms, where the trade-off has been made. The first good evidence in support of neural trade-offs was provided by Barton et al. (1995). They found that, among the primates, nocturnal species tend to have larger olfactory structures than diurnal species. Conversely, the diurnal species had larger striate visual cortices than their nocturnal relatives.

It has proven to be more straightforward to demonstrate relationships between sensory processing areas and behavior and ecology and so it is likely to be easier to acquire evidence for neural trade-offs in the higher processing areas. Localization of function to a given brain region is required and this may be difficult, if not, in some cases, impossible. Notwithstanding this problem, the finding of neural trade-offs, as in the case of Barton et al. (1995), cries out for the demonstration of the correlated trade-off in behavioral terms. This will not necessarily be easy as again, whilst it is possible to quantify sensory and motor capacities, it has long proven difficult to demonstrate convincing interspecific differences in higher processing capabilities. One method that may prove successful is that of taking comparisons between species that differ as little as possible except in that variable of interest and this is a strategy being adopted in comparative psychology (Kamil, 1988). This type of experimentation will be an important part in the description of brain/behavior relationships.

Conclusions

Nearly two decades ago comparative analyses presented ecological and behavioral variables as plausible explanations for variation in brain size. Now evidence is accumulating to show that whole brain size is a rather crude measure and that selection has manipulated brains in a rather more delicate fashion. Brains are a composite of many parts, some of which have clearly defined functions. It is now obvious that this composition is not simply a by-product of an animal's size nor is it a hangover from the evolutionary past. Brains have been stretched and squashed in a very precise manner by natural selection in response to a wide variety of ecological and behavioral demands. So whilst comparing elephants and hummingbirds is probably not the best place to begin looking at variation in brain size it seems, after removing the obvious effects of taxonomy and body size, explanations will be found in their lifestyle.

Acknowledgments

Thanks to Tim Guilford who gave valuable, long-distance advice and criticism, to Sara Shettleworth for useful comments and to Helen and Denis Healy for their hospitality. Work by the author cited in this chapter was supported financially by SERC, St John's College, Oxford and the Royal Society.

References

Baker, M.C., Bottjer, S.J. and Arnold, A.P. (1984) Sexual dimorphism and lack of seasonal changes in vocal control regions of the white-crowned sparrow brain. *Brain Res., 295*: 85-89.

Balda, R.P. and Kamil, A.C. (1989) A comparative study of cache recovery by three corvid species. *Anim. Behav., 38*: 486-495.

Bang, B.G. (1971) Functional anatomy of the olfactory system in 23 orders of birds. *Act. Anat. (Suppl.), 58*: 1-76.

Bang, B.G. and Cobb, S. (1968) The size of the olfactory bulb in 108 species of bird. *Auk, 85*: 55-61.

Bang, B.G. and Wenzel, B.M. (1985). Nasal cavity and olfactory system. In A. S. King & J. McLelland (eds.) *Form and function in birds 3*. New York: Academic Press, pp. 195-225.

Barton, R.A., Purvis, A. and Harvey, P.H. (1995) Evolutionary radiation of visual and olfactory systems in primates, bats and insectivores. *Proc. R. Soc. B.,* 348: 381-392.

Basil, J.A., Kamil, A.C., Balda, R.P. and Fite, K.V. (1995) Differences in hippocampal volume among food storing corvids. *Brain, Behav. Evol.*, in press.

Bennett, P.M. and Harvey, P.H. (1985a) Relative brain size and ecology in birds. *J. Zool. (Lond.), 207*: 151-169.

Bennett, P.M. and Harvey, P.H. (1985b) Brain size, development and metabolism in birds and mammals. *J. Zool. (Lond.), 207*: 491-509.

Bottjer, S.J., Glaessner, S.L. and Arnold, A.P. (1985) Ontogeny of brain nuclei controlling song learning and behavior in zebra finches. *J. Neurosci., 5*: 1556-1562.

Brenowitz, E.A. (1991) Altered perception of species-specific song by female birds after lesions of a forebrain nucleus. *Science, 251*: 303-305.

Brenowitz, E., Nalls, N., Wingfield, J. and Kroodsma, D., (1991) Seasonal changes in avian song nuclei without seasonal changes in song repertoire. *J. Neurosci.* *11*: 1367-1374.

Brodbeck, D.R., and Shettleworth, S.J. (1995) Matching location and color of a compound stimulus: Comparison of a food-storing and a non-storing bird species. *J. Exp. Psych.: Anim. Beh. Proc.* 22: 64-77.

Brodin, A. and Ekman, J. (1994) Benefits of food hoarding. *Nature*, *372*: 510.

Clayton, N.S. (1992) Ontogeny of food-storing and associated retrieval in marsh tits, *Parus palustris*. *Behavior*, *122*: 11-25.

Clayton, N.S. and Krebs,J. R. (1994). Hippocampal growth and attrition in birds affected by experience. *Proc. Natl. Acad. Sci. USA*, *91*: 7410-7414.

Clayton, N.S., and Krebs, J.R. (1994b) Memory for spatial and object-specific cues in food-storing and non-storing birds. *J. Comp. Physiol.*, *174*: 371-379.

Clutton-Brock, T.H. and Harvey, P.H. (1980) Primates, brains and ecology. *J. Zool. (Lond.)*, *190*: 309-323.

Cobb, S. (1960) Observations on the comparative anatomy of the avian brain. *Perspect. Biol. Med.*, *3*: 383-408.

DeVoogd, T.J. (1994) The neural basis for the acquisition and production of song. In J.A. Hogan and J.J. Bolhuis (Eds.) *Causal mechanisms of behavioral development*. Cambridge: University of Cambridge Press, pp. 49-81.

DeVoogd, T.J., Krebs, J.R., Healy, S.D. and Purvis, A. (1993) Relations between song repertoire size and the volume of brain nuclei realted to song: comparative evolutionary analyses amongst oscine birds. *Proc. Roy.Soc. (B)*, *254*: 75-82.

Eisenberg, J.F. and Wilson, D.E. (1978) Relative brain size and demographic strategies in didelphi marsupials. *Am. Nat.*, *118*: 1-15.

Gittleman, J.L. (1991) Carnivore olfactory-bulb size - allometry, phylogeny and ecology. *J. Zool.*, *225*: 253-272.

Gould, S.J. (1981) *The mismeasure of man*. New York: W. W. Norton & Co., Inc.

Haftorn, S. (1956) Contribution to the food biology of tits especially about storing of surplus food. Part III. The Willow-Tit (*Parus atricapillus L.*). *Det Kgl. Norske Vidensk. Selsk. Skr.*, *3*.

Haftorn, S. (1992) Ontogeny of food storing in titmice. *Ibis*, *134*: 69-71.

Hampton, R.R., Sherry, D.F., Shettleworth, S.J., Khurgle, M. and Ivy, G. (1995) Hippocampal volume and food storing in three species of parids. *Brain, Behav. Evol.*, *45*: 54-61.

Harvey, P.H. and Pagel, M. 1991 *The comparative method in evolutionary biology.* Oxford: Oxford University Press.

Harvey, P.H., Clutton-Brock, T.H. and Mace, G.M. (1980). Brain size and ecology in small mammals and primates. *Proc. Natl. Acad. Sci. USA*, *77*: 4387-4389.

Healy, S.D. (1995) Memory for objects and positions: delayed-non-matching-to-sample in tits. *Q. J.Experimental Psychology. 48B: 179-191.*

Healy, S.D. and Guilford, T. (1990) Olfactory bulb size and nocturnality in birds. *Evolution*, *44*: 339-346.

Healy, S.D., Krebs, J.R. and Gwinner, E. (1991) Hippocampal volume and migration in birds. *Naturwissen.*, *78*: 424-426.

Healy, S.D. and Krebs, J.R. (1992) Food storing and the hippocampus in corvids: amount and size are correlated. *Proc. Roy.Soc., 248*: 241-245.

Healy, S.D. and Krebs, J.R. (1993) Development of hippocampal specialisation in a food-storing bird. *Behav. Brain Res., 53*: 127-131.

Healy, S.D., Clayton, N.S. and Krebs, J.R. (1994a) Development of hippocampal specialisation in two species of tit (*Parus spp.*) *Behav. Brain Res., 61*: 23-28.

Healy, S.D., Gwinner, E. and Krebs, J.R. (1994b) Hippocampus size in migrating garden warblers: effects of age and experience. *J. für Ornith., 135*: 74.

Hill, K.M. and DeVoogd, T.J. (1991) Altered daylength affects dendritic structure in a song-related brain region in red-winged blackbirds. *Behav. Neural Biol., 56*: 240-250.

Hilton, S.C. and Krebs, J.R. (1990) Spatial memory of four species of Parus: performance in an open-field analogue of a radial maze. *Q. J. Exp. Psych., 42B*: 345-368.

Immelman, K. (1969) Song development in the zebra finch and other Estrildine finches. In R. A Hinde (Ed.) *Bird vocalizations.* Cambridge: Cambridge University Press, pp. 61-74.

Jacobs, L.F., Gaulin, S.J.C., Sherry, D.F. and Hofman, G.E. (1990) Evolution of spatial cognition - sex specific patterns of spatial behavior predict hippocampal size. *Proc. Natl. Acad. Sci., 87*: 6349-6352.

Kamil, A.C. (1988) A synthetic approach to the study of animal intelligence. In D.W.Leger (Ed.) *Comparative perspectives in modern psychology.* Lincoln, Nebraska: University of Nebraska Press, pp. 257-308.

Kirn, J.R., Clower, R.P., Kroodsma, D.E. and DeVoogd, T.J. (1989) Song-related brain regions in the red-winged blackbird are affected by sex and season but not repertoire size. *J. Neurobiol., 20*: 139-163.

Kirn, J., O'Loughlin, B., Kasparian, S. and Nottebohm, F. (1994) Cell death and neuronal recruitment in the high vocal center of adult male canaries are temporally related to changes in song. *Proc. Natl. Acad. Sci. USA, 91*: 7844-7848.

Konishi, M. (1963) The role of auditory feedback in the vocal behavior of the domestic fowl. *Z. Tierpsychol., 20*: 349-367.

Konishi, M. (1985) Birdsong: from behavior to neuron. *Ann. Rev. Neurosce., 8*: 125-170.

Krebs, J.R., Sherry, D.F., Healy, S.D., Perry, V.H. and Vaccarino, A.L. (1989) Hippocampal specialization of food-storing birds. *Proc. Natl .Acad. Sci. (U.S.A), 86*: 1388-1392.

Krebs, J.R., Healy, S.D. and Shettleworth, S.J. (1990) Spatial memory of Paridae: comparison of a storing and a non-storing species, *Parus ater* and *Parus major*. *Anim. Behav., 39:* 1127-1137.

Loevei, G.L. (1989) Passerine migration between the Palaearctic and Africa. In D. M. Power (Ed.) *Current ornithology 6* . London, New York: Academic, pp 143-174..

McNamara, J.M., Houston, A.I. and Krebs, J.R. (1990) Why hoard? The economics of food storing in tits, Parus spp. *Behav. Ecol., 1*: 12-23.

Mace, G.M., Harvey, P.H. and Clutton-Brock, T.H. (1981) Is brain size an ecological variable? *Trends Neurosci., 3*: 193-196.

Marler, P. (1957) Specific distinctiveness in the communication signals of birds. *Behavior, 11*: 13-39.
Marler, P. (1987) Sensitive periods and the roles of specific and general sensory stimulation in birdsong learning. In J. P. Rauschecker and P. Marler (Eds.) *Imprinting and cortical plasticity*. New York: John Wiley & Sons, pp. 99-136.
Marler, P. (1991) Differences in behavioral development in closely related species: birdspng. In P. Bateson (Ed.) *The development and integration of behavior.* Cambridge: Cambridge University Press, pp. 41-70..
Marler, P. and Tamura, M. (1962) Song variation in three populations of white-crowned sparrow. *Condor, 64*: 368-377.
Martin, S.J. (1993) Apoptosis: suicide, execution or murder? *Trends Cell Biol., 3*: 141-144.
Nordeen, E.J., Nordeen, K.W., and Arnold, A.P. (1987) Estrogen accumulation in zebra finch song control nuclei: implications for sexual differentiation and adult activation of behavior. *J. Neurobiol., 18*: 569-582..
Nordeen, E.J., Grace, A., Burek, M. and Nordeen, K.W. (1992) Sex-dependent loss of projection neurons involved in avian song learning. *J. Neurobiol., 23*: 671-679.
Nottebohm, F. (1981) A brain for all seasons: cyclical anatomical changes in song-control nuclei of the canary brain. *Science, 214*: 1368-1370.
Nottebohm, F., Stokes, T.M., and Leonard, C.M. (1976) Central control of song in the canary, *Serinus canarius*. *J. Comp. Neurol., 165*: 457-486.
Olson, D. (1991) Species differences in spatial memory among Clark's nutcrackers, pigeons and scrub jays. *J. Exp. Psych.: Anim. Beh. Proc., 17*: 363-376.
Portmann, A. (1946) Etudes sur la cerebralisation chez sur oisaux. I. *Alauda, 14*: 2-20.
Portmann, A. (1947) Etudes sur la cerebralisation chez sur oisaux. II. Les indices intracerebraux. *Alauda, 15*: 1-15.
Price, P. (1979) Developmental determinants of structure in zebra finch song. *J. Comp. Physiol. Psychol., 93*: 260-277.
Rice, H N. (1992) *Food-caching, simple association and spatial memory in the Black-capped chickadee.* Unpublished MA thesis, University of Western Ontario, Canada.
Scharff, C. and Nottebohm, F. (1991) A comparative study of the behavioral deficits following lesions of various parts of the zebra finch song system: implications for vocal learning. *J. Neurosci., 11*: 2896-2913.
Sherry, D.F. (1989) Food storing in the Paridae. *Wilson Bulletin, 101*: 289-304.
Sherry, D.F. and Vaccarino, A.L.. (1989) Hippocampus and memory for food caches in black-capped chickadees. *Behav. Neurosci., 103*: 308-318.
Sherry, D.F., Vaccarino, A.L., Buckenham, K. and Herz, R.S. (1989) The hippocampal complex of food-storing birds. *Brain, Behav. Evol., 34*: 308-317.
Sherry, D.F., Forbes, M.R L., Khurgel, M. and Ivy, G.O. (1993) Females have a larger hippocampus than males in the brood-parasitic brown-headed cowbird. *Proc. Natl. Acad. Sci., 90*: 7839-7843.
Sibley, C.G. and Ahlquist, J.E. (1991) *Phylogeny and classificaton of birds: a study in molecular evolution.* Yale University Press, New Haven, Conneticut.

Slater, P.J.B. (1989) Bird song learning: causes and consequences. *Ethol., Ecol. Evol.,* 1: 19-46.

Smith, G.T., Brenowitz, E., Nalls, B. and Wingfield, J. (1992) Testosterone changes song control region volume in photostimulated male Gambel's sparrows. *Proc. 3rd int'l Cong. Neuroethol.* 292.

Vander Wall, S. (1990) *Food Hoarding in Animals,* Chicago University Press.

PART TWO

Specializations for Perception of Biologically Relevant Stimuli

6

The Perceptual Foundations of Vocal Learning in Budgerigars

Robert J. Dooling, Susan D. Brown, Kazuchika Manabe, and Elizabeth F. Powell

Abstract

Most of what we know about vocal learning in birds comes from studies of oscine songbirds with much less known about vocal learning in non songbirds such as parrots. Here, we describe a series of studies on vocal learning, vocal development and auditory and visual perception in the domesticated budgerigar or parakeet (Melopsittacus undulatus). Call learning in these birds occurs in both males and females, continues throughout adulthood, and involves specialized auditory and visual processes. Compared with many songbirds, the constraints on vocal learning in this species appear to be quite broad with social factors playing a crucial role. Budgerigars are an especially tractable laboratory subjects in such that they can be readily trained with operant conditioning techniques to modify the characteristics of their vocal signals. This opens up a new approach to the study of sensory and motor interactions in vocal learning and development.

Introduction

Some of the most fascinating avian behaviors are those involving the perception, production, and learning of complex vocal signals. Vocal learning is rare among vertebrates and, apart from birds, has only been shown unequivocally in humans. Almost half of the roughly 9000 extant species of birds are believed to rely on learning for the development of at least part of their vocal repertoire (Kroodsma and Miller, 1982). Vocal learning has been shown in oscine songbirds (*Order Passeriformes*), parrots (*Order Psittaciformes*), hummingbirds (*Order Apodiformes*, Wiley, 1971), and toucans (*Order Piciformes*, Wagner, 1944) and probably arose independently in each of these groups (Kroodsma and Miller, 1982; Nottebohm, 1972).

Most of what we know about vocal learning in birds comes from studies of oscine songbirds. Within this group of birds, the focus has been almost exclusively on the long, complex, songs produced predominately by males of the species. We now know that the timing of learning and the variety of acoustic signals that can serve as models varies considerably among species. For some species, song learning occurs primarily during a very restricted period early in development. Ethologists and neurobiologists have sometimes characterized this form of vocal learning as age-dependent or sensitive-period learning. On the other hand, open-ended or age-independent learning can occur throughout life. Examples include the acquisition of new songs or song elements just before each breeding season as in canaries or starlings (Nottebohm, 1984).

In contrast to developmental or seasonal learning of relatively long, species-specific song, the learning or sharing of shorter vocalizations or calls can occur throughout adulthood. Though much less attention has been paid to this kind of adult vocal learning, a number of investigators report evidence of sharing and convergence of calls in some oscine songbirds (Mundinger, 1979; Mammen and Nowicki, 1981; Nowicki, 1989). Adult vocal learning also occurs in parrots which have been shown to learn a variety of sounds including human speech and to use them in rather complex social situations (Pepperberg, 1985). This chapter reviews experiments on adult vocal learning in a small Australian parrot called the budgerigar (*Melopsittacus undulatus*) - more commonly known as the domesticated parakeet.

This review also focuses on a particular vocalization, the contact or distance call, and the perceptual capabilities that subserve the development and learning of this call. The perceptual issues are common to all vocal learning, not only in birds but also, of course, in humans. Are birds specialized for the perception of species-specific vocalizations? How does a bird learn and use only the correct, species-specific vocal signals while eschewing the myriad of incorrect sounds to which it is exposed during development and through adulthood? Do other perceptual or social factors guide, constrain, and canalize the vocal learning process?

The simple fact that budgerigars can mimic human speech argues against rigid constraints imposed by the peripheral sensory apparatus or the motor system. The auditory system of budgerigars is exquisitely competent at perceiving a variety of complex sounds including human speech. Likewise, the avian syrinx in general, and the parrot syrinx in particular, appears to be extraordinarily versatile. Thus, the fact that budgerigars develop a species-typical vocal repertoire and learn vocalizations more readily from other budgerigars than from humans or other bird species argues for an important role for social factors. A vocal learning system that operates over the organism's entire lifespan must be associated with a host of complex, integrative perceptual processes. Among these one would expect to find evidence of: (1) a special relationship between auditory perception and species-specific vocalizations, (2) specific perceptual predispositions for conspecific vocalizations evident early in life, and (3) perceptual preferences that retain a degree of plasticity through adulthood such that perceptual categories can be modified by appropriate social conditions. Here we review evidence for such processes in budgerigars and suggest some directions for future research.

Auditory Perceptual Specializations for Vocal Signals

A recent perceptual experiment sought to answer two questions concerning the budgerigar's ability to discriminate among species-specific vocal signals. The first question asks whether budgerigars process species-specific calls more efficiently (i.e., faster) than they do complex vocalizations of other avian species. The second question asks whether budgerigars discriminate among budgerigar vocalizations more efficiently than other avian species process budgerigar vocalizations. An affirmative answer to both questions provides strong evidence that budgerigars are specialized for the perception of species-specific vocalizations.

For this test, we used contact calls taken from three different species of birds: budgerigars, canaries (*Serinus canaria*), and zebra finches (*Taeniopygia guttata*) (Figure 6.1). This stimulus set consisted of contact calls recorded from four individual birds of each species (not the test birds). The contact calls of all three species are short (about 150-300 msec long) with most of the energy falling in the frequency region of 2-5 kHz. Otherwise, the calls of the three species are quite different acoustically. Budgerigar contact calls contain rapid frequency modulation but virtually no harmonic structure. Canary calls have much less frequency modulation, a simple harmonic structure if any, and a relatively high fundamental frequency. Contact calls of zebra finches, on the other hand, are rich in harmonic structure with a relatively low fundamental frequency and little, if any, frequency modulation.

Five budgerigars, four canaries, and seven zebra finches were used as subjects. We trained and tested these birds using a standard operant conditioning procedure with response latency used as the dependent

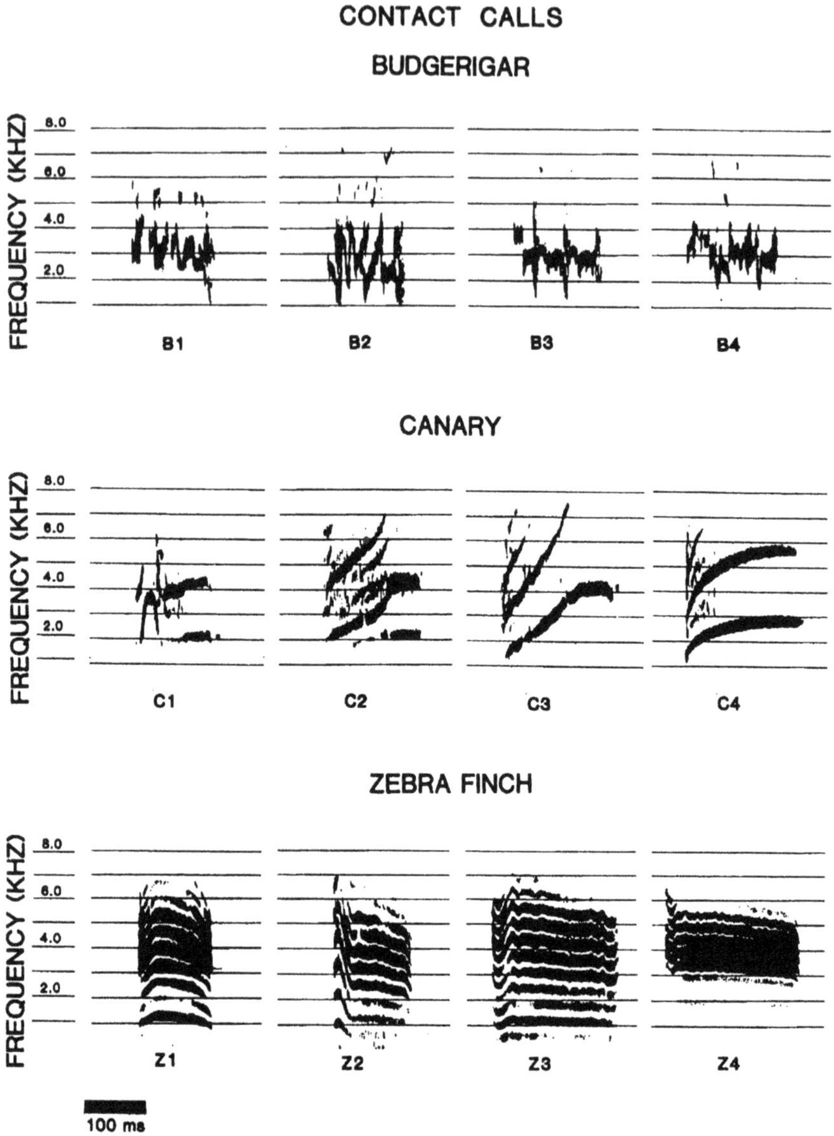

FIGURE 6.1 Sonagrams of contact calls from four budgerigars, four canaries, and four zebra finches (From Dooling et al. 1992).

variable. Detection of a new or different call in a repeating series of a single call was rewarded with food (Dooling et al. 1992). The assumption underlying this procedure is that the birds responded more quickly when the discrimination was easy and more slowly when the discrimination was difficult - an assumption that can be empirically validated using simple pure tones that vary along a single dimension (Dooling et al. 1992). So, perceptually similar calls result in longer response latencies while dissimilar calls are associated with shorter response latencies. At the conclusion of testing, average response latencies were computed for all possible pairwise combinations of calls. These triangular matrices of response latencies (i.e., similarity matrices) were analyzed using a procedure known as multidimensional scaling (MDS, the specific program and model used was SINDSCAL; see Shepard, 1980 for a general discussion). MDS produces a spatial map in which stimulus similarity is represented by spatial proximity. Calls that are perceived as similar are grouped close together in multidimensional space; those that are perceived as different are spaced apart. This analysis was computed separately for data from each of the three species to assess the perceptual categories of these stimuli for each species.

Two-dimensional spatial arrangements or maps of these calls produced by MDS were generated for each species, and these two-dimensional spatial maps for each species are shown in Figure 6.2a.

For all three species of test birds, the calls of budgerigars, canaries, and zebra finches are separated in the spatial maps generated by MDS. These results show that all three species find the acoustic differences between the categories of calls much more discriminable than the acoustic differences among calls of the same category. This provides an operational definition of a perceptual category. Often one strives to interpret these perceptual maps. For canaries, the MDS plot shows that the canary call containing considerable frequency modulation (C1) falls nearest to the group of budgerigar calls, which are also characterized by rapid frequency modulation. This suggests that frequency modulation is an important cue that canaries might use to distinguish between canary and budgerigar calls.

That each species groups these calls by species is perhaps not too surprising since even casual human listeners can readily recognize and classify these sounds as species typical calls produced by canaries, budgerigars, and zebra finches. A more important issue is how well these species discriminate among conspecific calls compared with calls from the other species. We address this issue by comparing the response latencies for discriminating among calls of the same category (i.e., among budgerigar calls, among canary calls, and among zebra finch calls). The relative response latencies for these intra-category discriminations are shown for each species in Figure 6.2b. Of the five budgerigars, four canaries, and seven zebra finches, all but on female canary showed shorter response latencies to discriminate among conspecific calls than among calls drawn from either of the other two species.

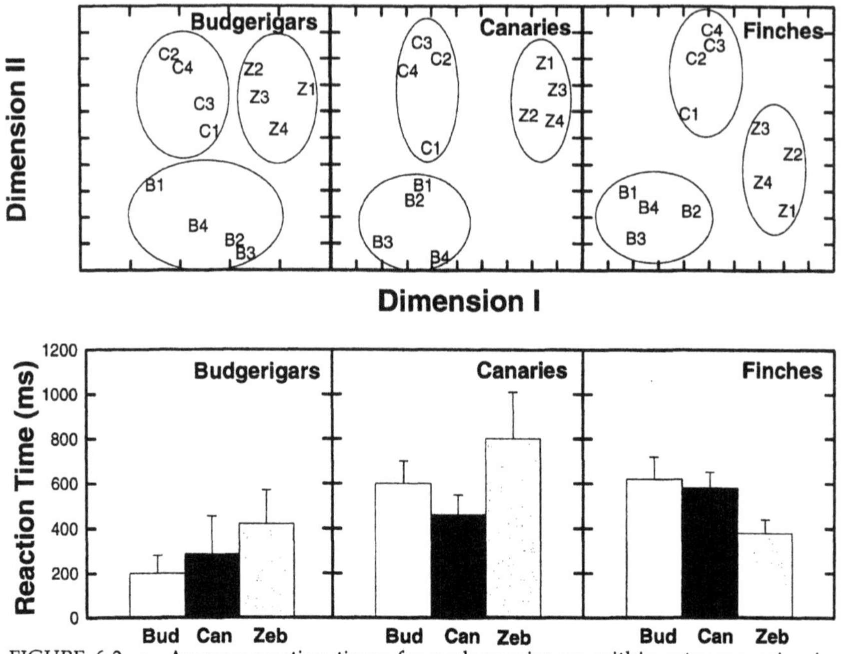

FIGURE 6.2 a. Average reaction times for each species on within-category pair-wise discriminations. Each species shows faster reactions times to conspecific calls (From Dooling et al. 1992). b. Three separate 2-dimensional spatial maps produced by SINDSCAL. Each spatial map represents an analysis of reaction times for one species discriminating among the set of 12 contact calls (shown in Figure 1).

To assess how each species discriminated among calls of the same species, we obtained each bird's normalized or relative response latency for each comparison by subtracting that birds' minimum response latency obtained over the entire experiment. For each bird, an average relative response latency was then calculated for each set of within-category discriminations (e.g. the average latency from all trials involving comparisons among canary calls). Results show that budgerigars discriminated among conspecific calls faster than among zebra finch calls or canary calls. Budgerigars also clearly discriminated among budgerigar calls faster than did canaries or zebra finches. The more efficient discrimination of conspecific calls by budgerigars cannot be due simply to peculiarities of the stimuli. Only budgerigars showed a perceptual advantage for budgerigar calls suggesting the advantage for perceiving budgerigar calls is unique to the species. The alternative explanation, that budgerigar calls are inherently more discriminable than other bird calls, is not supported by these data. In other words, budgerigars are specialized for the perception of species-specific calls.

The general case for species-specific perceptual specializations is also made by results from the other two species. Both canaries and zebra finches also discriminated among conspecific calls faster than among heterospecific calls. These findings, in three different species tested on the same stimuli with the same procedures, argue that specialization for the perception of species-specific vocal signals is a pervasive phenomenon.

How do such specializations arise? Is their development age-dependent? One possibility is that the basic hearing capabilities of each species are different and that species-specific contact calls have evolved to match these innate auditory capabilities. At least at a superficial level, the evidence for this possibility is not compelling given what we know about the basic hearing capabilities of these three species including audiograms, frequency and intensity difference limens, and masking thresholds (Fay, 1988; Okanoya and Dooling, 1987). Though there are species differences, on none of these measures can we explain the pattern of species differences observed in the perception of species-specific contact calls. To be sure, psychoacoustic tests with more complex acoustic stimuli, especially those that can be modeled after naturally occurring sounds, may yet reveal species differences at the level of basic hearing mechanisms that lead to enhanced processing of species-specific vocalizations.

At the other extreme, perhaps the development of species-specific perceptual sensitivities in these birds initially involves a set of broad, basic auditory capabilities that are similar across species. These general perceptual proclivities might then be quickly and profoundly influenced by early and selective exposure to conspecific sounds (i.e., during the nestling or fledgling phase of development), leading to perceptual preferences for the learning, production and use of conspecific sounds.

Studies of selective vocal learning in songbirds offer a variation on this second theme which is that the young bird brings to the song learning situation early perceptual sensitivities for certain, perhaps even rather simple, aspects of species-specific song. These early sensitivities serve to focus the young songbird's attention on the full complexity of species-specific song patterns which are subsequently committed to memory (Marler, 1990; 1991; Dooling and Searcy, 1980). There are strong parallels between this latter view of the ontogeny of vocal learning and other forms of perceptual plasticity (Kovach, 1983a,b) and programmed learning such as imprinting (Gould, 1982; ten Cate, 1994; Bolhius, this volume). Since budgerigars show vocal plasticity throughout life, such emerging perceptual preferences must not be immutable. Instead, certain conditions such as appropriate social situations, must be capable of modifying higher order perceptual processes throughout adulthood. There is evidence of just such perceptual plasticity in adult budgerigars.

The Perception of Learned Contact Calls

The foregoing shows that budgerigars (and probably other avian species as well) discriminate among species-specific contact calls more efficiently than among functionally similar calls of other bird species. Contact calls in budgerigars are particularly interesting because they are learned very quickly in certain social situations (Brown et al. 1988; Dooling et al. 1987; Farabaugh et al. 1994).

The fact that these calls can be learned so quickly allows a test of the plasticity of perceptual processes in adult birds. Do birds that learn to produce new calls show evidence of perceiving these calls in a special way? Brown and her colleagues (Brown et al. 1988) tested whether budgerigars that live together and develop similar contact calls show a perceptual advantage for discriminating newly learned calls.

In this experiment, contact calls were collected from three budgerigars who were housed together for three months in a small breeding cage in a large aviary. The three birds developed very similar contact calls that were indistinguishable upon sonographic analysis. Four exemplars of each bird's contact call were then used as test stimuli using the procedure described above. The three birds who were housed together, and produced these contact calls (i.e., the cagemates), were used as one group of subjects. A second group of three budgerigars (i.e., noncagemates) was also used as subjects. This group was housed in a separate, adjacent cage within clear sight and earshot of the stimulus birds but did not learn to produce the calls of the three cagemates.

When the response latencies of these two groups of birds were analyzed, several findings emerged as shown in the dendograms in Figure 6.3. First, cagemates discriminated among the calls of the three different individuals. Second, two out of three cagemates showed a significant tendency to discriminate more efficiently among their own calls than those of the other two cagemates. Third, cagemates as a group were faster at discriminating between the set of cagemate calls than were other birds (i.e., noncagemates) (Brown et al. 1988). Dendograms from a cluster analysis of the response latency data from the two groups of birds are shown in Figure 6.3 (bottom).

We conclude that cagemates have learned to extract the essential differences among cagemate calls required for individual recognition while ignoring variation in irrelevant acoustic characteristics. In other words, perceptual categories for vocal signals can be learned in this species

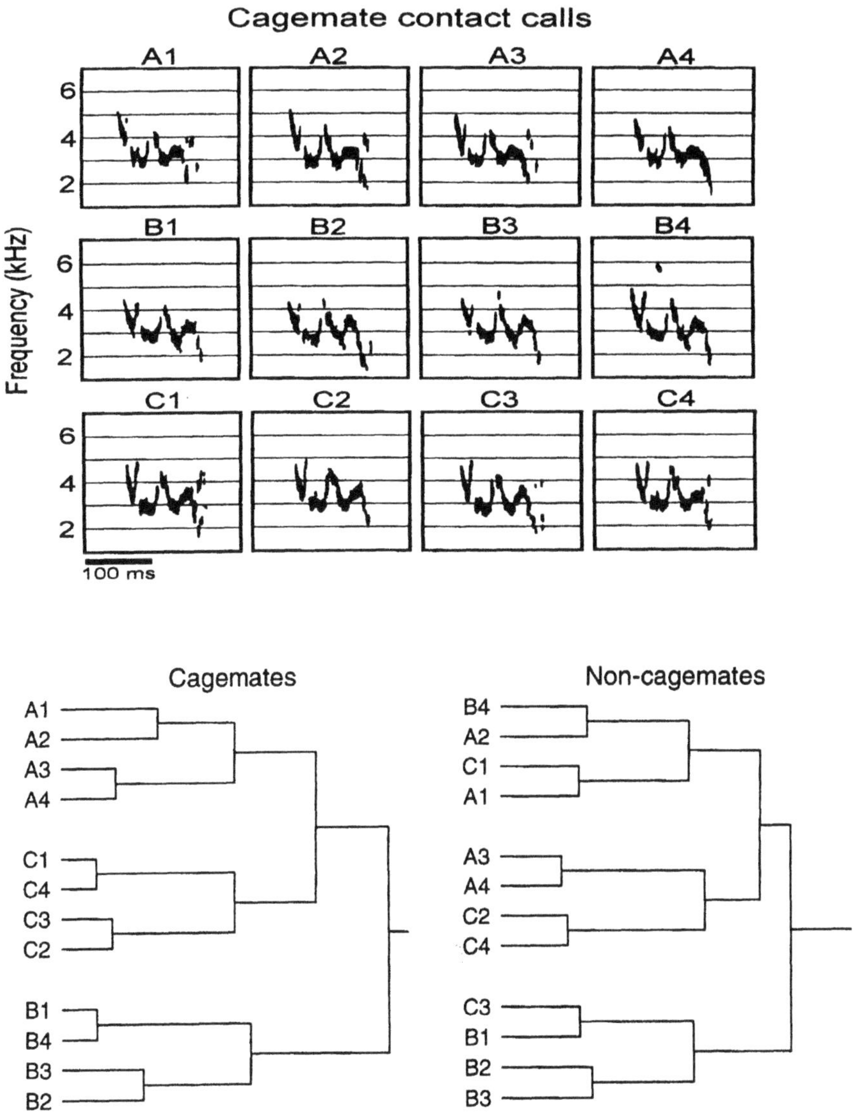

FIGURE 6.3 Top: Sonagrams of four contact calls from each of three cagemates. Each bird's contact call bears a strong resemblance to that of its cagemate. Bottom: Cluster dendograms from an analysis of the half-matrices of response latencies for cagemates and noncagemates tested on cagemate calls (From Brown et al. 1988).

throughout adulthood. The high degree of similarity among these calls also shows that these categories can depend on some rather subtle stimulus characteristics that probably include minute frequency differences in the center of the call (Brown et al. 1988). We conclude that the auditory perceptual system of the adult budgerigar is sophisticated, versatile, and plastic as opposed to being immutably dedicated to the perception of a particular set of species-specific acoustic cues established in early development. These results also suggest that the production and/or the use of contact calls in the appropriate social context is what alters a bird's perception of these calls. Thus, social factors appear to guide both the learning to produce contact calls and the learning to perceive them. A more recent study on contact call learning reinforces the importance of social interactions.

Social Factors in Adult Vocal Learning

We know that providing live tutors facilitates the learning of song by songbirds (see, for example, Marler and Peters, 1988; Petrinovich, 1988; Williams, 1990), and we saw above that social interactions are an important part of adult call learning in budgerigars. Teasing out the key elements of social interaction is difficult. In a recent study, contact calls from six adult male budgerigars, each from a different breeder, were recorded for several days to be sure that no two birds shared any of the same contact call types (Farabaugh et al. 1994). Individual budgerigars sometimes have more than one contact call type but the typical case is that either one or two call types are dominant - accounting for 95-100% of all contact call types produced. Once recorded, the birds were placed in a large cage separated by a thin cloth partition. Three birds were placed on one side of the partition and three on the other side so that birds on each side could hear but not see birds on the other side. During the next eight weeks, contact calls were recorded every few days from each bird. Within one week, we had evidence that one bird shared a contact call with another bird on the same side of the partition. At the end of eight weeks, three birds on one side of the partition were sharing the same dominant contact call types, and three birds on the other side of the partition were sharing different dominant contact call types. At this point, the experiment was stopped. Sharing of contact calls across the partition was practically nonexistent (the dominant call of one group of birds was recorded only rarely from two of the birds in the other group) (see Figure 6.4). Birds learned almost exclusively from birds that they interacted with rather than from birds they could only hear. This study refines the role of social interactions in the learning of contact calls by budgerigars and supports the argument that adult vocal learning depends, not surprisingly, on dynamic social cues. The challenge now is to isolate and manipulate these cues so that their role in vocal learning can be assessed as the study described below illustrates.

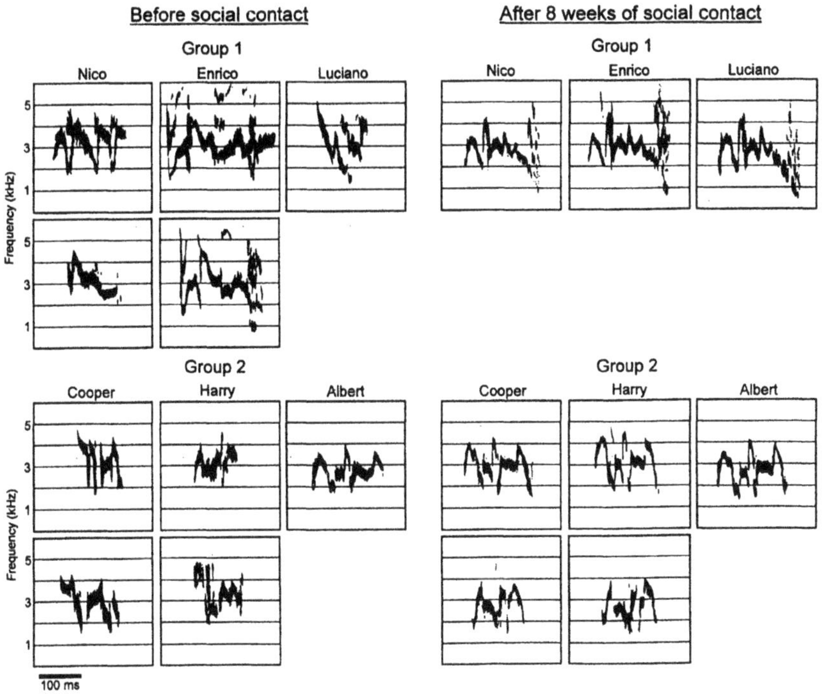

FIGURE 6.4 Top: Sonagrams of six male budgerigars dominant contact calls before social contact with each other. Bottom: Sonagrams of dominant contact calls from the same budgerigars after being housed in two groups of three birds for eight weeks (From Farabaugh et al. 1994).

Perception of Species-Specific Faces by Budgerigars

So far we know that: (1) budgerigars are specialized for the perception of species-specific vocalizations, (2) they can learn perceptual categories for vocal signals, and (3) this learning proceeds most quickly with social experience compared with simple exposure to the sounds. This experience probably includes a combination of social, visual, and motor factors. Can these processes be studied in the laboratory? As one approach, we tested whether budgerigars can discriminate among facial stimuli and whether they show special sensitivities for features of budgerigar faces as they obviously do for budgerigar voices.

As a social signal, the budgerigar face contains a wealth of information. The sex, age, breeding status of females, and individual identity of a bird can all be determined by characteristics of the face. For example, adult male budgerigars have deep blue ceres while the ceres of adult females vary from pale to dark brownish pink. Variations in female cere color are correlated with breeding condition. The stimuli used in this experiment were photographic slides of male and female budgerigars and male and female zebra finches. Male and female zebra finches are also easily distinguished; males have black stripes on the throat and chest, and white-spotted, rust-colored patches below their wings, all of which females lack. The most distinguishing characteristic between male and female zebra finches is in the face. Males have a large rust-orange cheek patch that females do not. The zebra finch faces chosen for this study can be easily distinguished by these cheek patches. Besides the zebra finches, photographs of four budgerigars with wild-type coloration, two males and two females, were selected. The budgerigar faces were chosen for their relative similarity to each other. A black and white print made from a composite of these images is shown as Figure 6.5.

Four budgerigars that were previously trained to discriminate among visual stimuli (Brown and Dooling, 1992) were tested using these faces. The response latencies generated during this task were analyzed using MDS as described above (see Brown and Dooling, 1992 for experimental details). Figure 6.6 shows the two-dimensional spatial arrangement resulting from MDS analysis of reaction times of four budgerigars tested using the set of budgerigar and zebra finch faces.

The first dimension separates budgerigar faces from zebra finch faces. The budgerigar faces are scattered along the second dimension with the faces of female budgerigars near the top of the plot and the faces of male budgerigars toward the bottom. Though the differences between male and female zebra finch faces are obvious, males have a prominent orange cheek patch that the females lack, zebra finch faces are grouped closely together indicating that these differences were not salient to the budgerigars. If differences among budgerigar faces are more salient to budgerigars than the differences among zebra finch faces, the response latencies to discriminate between pairs of budgerigar faces should be shorter than those to discriminate between pairs of zebra finch faces. This was indeed exactly the case (Brown and Dooling, 1992).

The results of this experiment show that budgerigars clearly distinguish conspecific faces from those of another, quite dissimilar, avian species. Furthermore, the relatively subtle differences among conspecific faces are more salient than the more obvious (at least to humans) differences among the faces of zebra finches. These data strongly suggest that budgerigars are specialized or biased for the perception of facial stimuli in the same sense that they are specialized for the perception

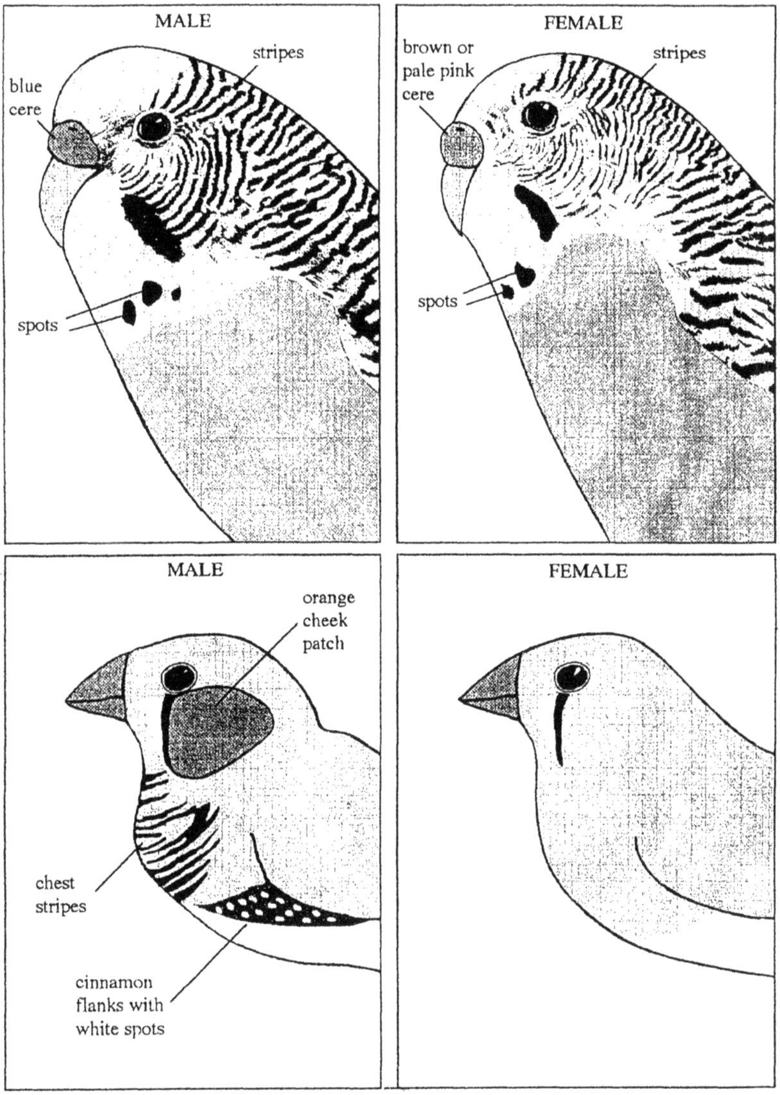

FIGURE 6.5 Black and white schematic representations budgerigar and zebra finches faces used as discriminative stimuli. Actual stimuli were color slides projected on glass response keys (From Brown and Dooling, 1993).

of vocal stimuli. In both cases this specialization takes the form of more efficient (i.e., faster) processing of certain biologically relevant stimuli. The fact that enhanced processing is evident even with static visual images of faces suggests that we have identified an important aspect of

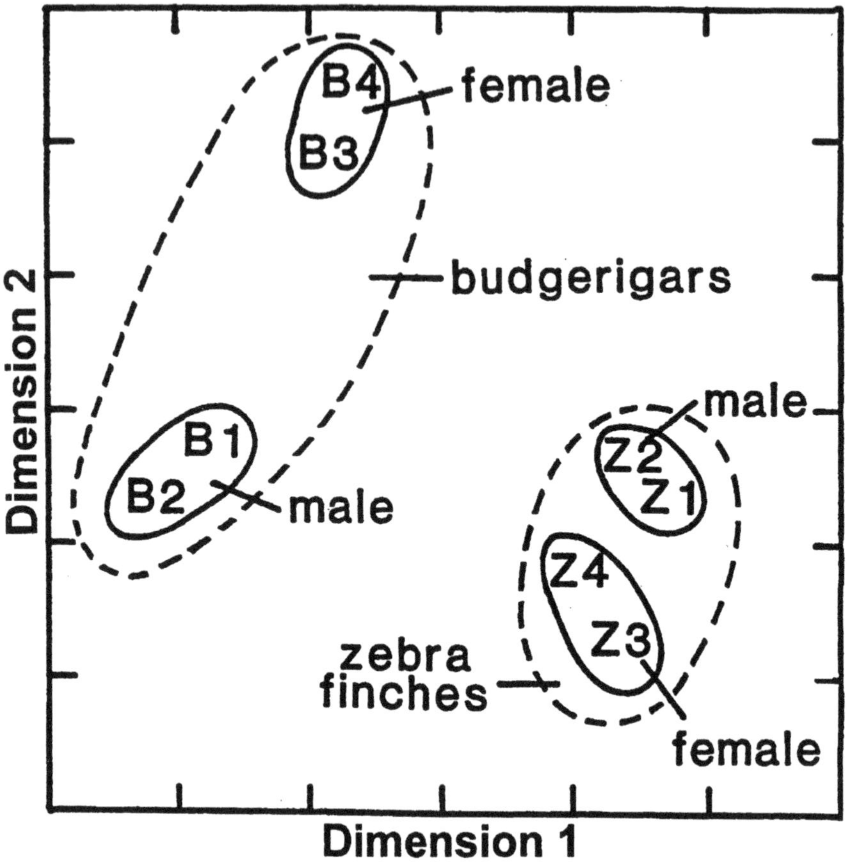

FIGURE 6.6 A 2-dimensional spatial map produced by a SINDSCAL analysis of the reactions times of budgerigars tested on budgerigar and zebra finch faces (From Brown and Dooling, 1993).

the constellation of social features that promote and canalize vocal learning in budgerigars.

Contact Call Development in Budgerigars

How does contact call learning in budgerigars develop? A recent study by Powell followed the development of contact calls in budgerigars from hatching through to young adulthood (Powell, 1993). A newly-hatched budgerigar, not yet able to see or hear well, begins to make soft, high-pitched food-begging calls that elicit feeding from its parents.

Over the next 30 days, until the time of fledging, this food-begging call undergoes a sequence of dramatic acoustic transformations. At the time of fledging, the food-begging call undergoes one final change. A brief section (150-200 msec) of the final food-begging call becomes the bird's first contact call (see Figure 6.7) and is used in the appropriate context following fledging.

Ongoing experiments have begun to reveal the maturational and social events in the nestbox that accompany these acoustic changes (Powell, 1993). The young nestling's eyes open at about 6-8 days followed a day or so later by its ears opening. Around this time, the character of food begging begins to change from somewhat passive to active and much more intense as the young bird develops motor control. Food-begging calls become loud, broadband, frequency modulated, and highly variable around 2 to 3 weeks of age.

The behavior of parents and nestlings in the box is remarkable for both its complexity and its intimacy. Both parents feed and preen both each other and the nestlings, and the nestlings often preen each other and their parents. It is quite clear from the preceding experiments on

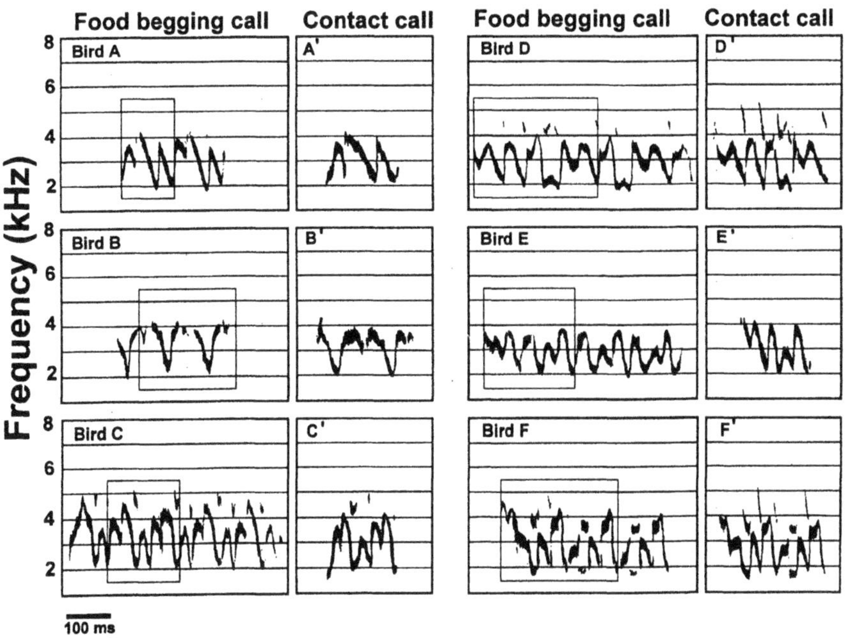

FIGURE 6.7 Sonagrams of food-begging calls produced by nestling budgerigars shortly before fledging and each bird's first contact call which is a shortened version of the food-begging call recorded in the nest box (From Powell, 1993).

vocal learning in budgerigars as well as from studies showing the effectiveness of live tutors in song learning in songbirds (see, for example, Petrinovich, 1988) that social interactions hold the key to understanding vocal learning. When in budgerigar development do these social interactions become important in vocal learning? Casual observation of behavior in the nestbox reveals quite a potpourri of social interaction, and it is hard to imagine that such activity is unrelated to the development of a normal vocal repertoire in this species. We know that developing calls of hand-reared and isolate-reared birds can be distinguished from those of normally-reared birds. Birds reared deafened or in isolation through adulthood develop both abnormal behavior and abnormal vocalizations (Dooling et al. 1987; Powell, et al. 1992). While such data implicate social factors in call development in young budgerigars, the specific aspects or timing of social experiences that might be critical remain unclear.

It is likely that parent-offspring interactions are reciprocal. Stamps and her colleagues (Stamps et al. 1985; 1987) have shown that the age, health, and sex ratio of nestlings and the rate of calling influence. There is a wealth of evidence from other avian species that begging affects parental feeding (see, for example, Mondloch, 1995; Harper, 1986). It is intriguing, but entirely speculative, as to whether parental feeding affects the characteristics of vocalizations in young budgerigars. The vocalizations and behaviors of young nestlings clearly have important consequences. Among other things, they lead to feeding and social contact. One can imagine that nestlings form some very important associations while in the nestbox. At a minimum, they probably learn to associate their own vocalizations with food, with social contact, and with the physical characteristics (especially facial features) of adult budgerigars. Though much work remains to be done, these early social interactions and their related associations may play a pivotal role in the development of perceptual preferences that guide and maintain contact call learning throughout adulthood. A demonstration that simple positive reinforcement (i.e., food) can be used to control vocal learning in adults, as described below, would open the possibility for operation of a similar mechanism in early development.

Non-social Learning of Contact Calls

Historically, vocal learning in birds has been considered a special form of learning, much like imprinting, in that it appears programmed to occur at a specific time, under a special and rather specific set of circumstances (Gould, 1982; ten Cate, 1994). In both forms of learning, the exact nature of the reinforcement that facilitates the learning remains obscure. Though recent efforts have been modestly more successful (Adret, 1993), numerous attempts over the years to use operant conditioning to control vocal learning has been an exercise in futility - probably, in part, because the true richness of the visual and social reinforcers as they operate in a bird's natural environment is difficult, if not impossible, to reproduce in the laboratory. Undaunted, Manabe and his colleagues (Manabe et al. 1995; Manabe and Dooling, 1995) have now shown that the plasticity in budgerigar contact calls can be brought under control by food reinforcement. We trained budgerigars in small operant conditioning chambers, much as in the perceptual experiments described above, to produce vocalizations for food. In these experiments, each session lasts for about 10 minutes during which the bird produces 50-100 calls, and the birds are tested in two sessions per day. The bird's vocalizations are monitored with a sensitive microphone, analyzed online using a digital signal processing methods, and compared with a previously produced vocalization or to a stored template call.

In the first experiment, budgerigars were trained to produce vocal variety. First, the bird was trained simply to vocalize for food. Then, reinforcement was contingent on whether the call produced was sufficiently different from the previous call (1-back condition). After the bird learned this task, it was then required to produce a call on each trial that was sufficiently different from the previous two different calls (2-back condition). This paradigm leads to the production of a minimum of two (1-back) or three (2-back) calls.

Figure 6.8 shows a multidimensional scaling analysis of contact calls produced by a budgerigar trained in a 1-back and then in a 2-back condition. The top left panel shows contact calls produced by the bird in the last training session in a 1-back condition (filled triangles). Notice there are two groups of solid triangles showing that the bird produced two different calls. Also plotted in this panel are the contact calls of the first session in the 2-back condition (open circles). Already there is variation especially in the group of calls labeled 1B2. The top right panel shows the same open circles for reference superimposed on the contact calls produced during the last session of a 2-back condition (filled circle). Sonagrams of the calls produced during this experiment are shown on the bottom Figure 6.8. In the 1-back condition, the bird produces two different calls - a lower pitched call with broad band component in the initial part of the call (1B1) and a higher pitched call (1B2). The contact call cluster 1B2 became more variable in the very first session in which the bird was required to produce calls that were different from the previous two different calls (first 2-back condition). This call cluster continued to diverge until, in the final training session in the 2-back condition,

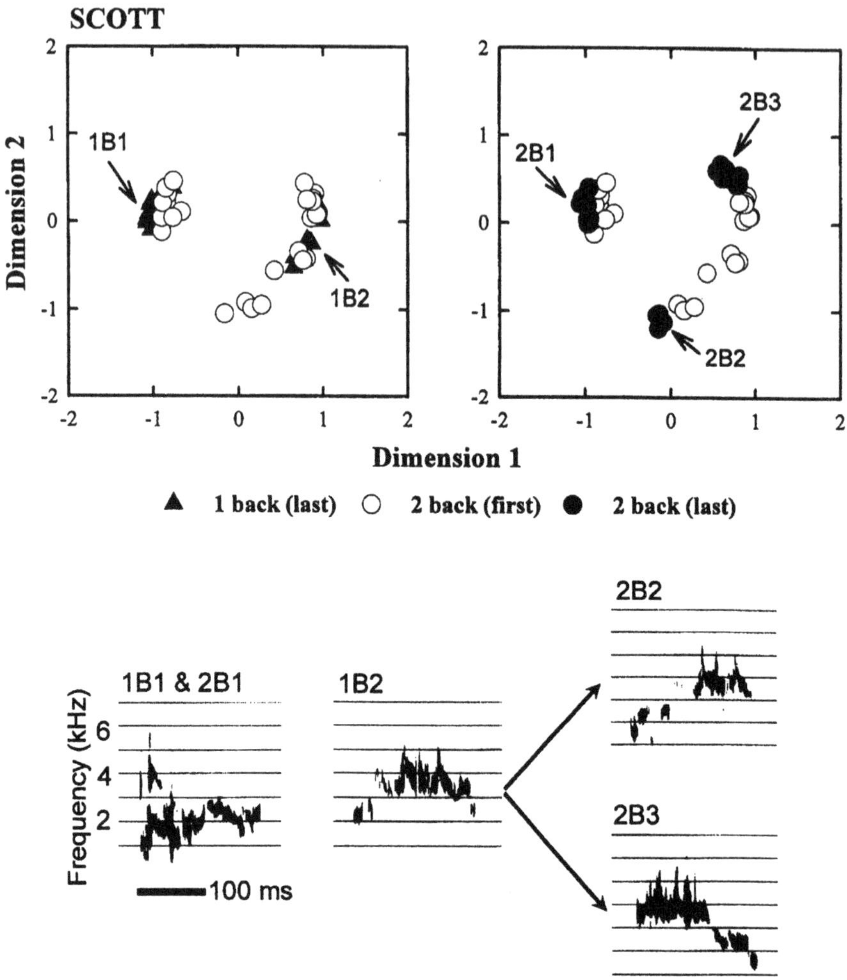

FIGURE 6.8 A 2-dimensional spatial map produced by a SINDSCAL analysis of similarity between calls in 1-back and 2-back conditions. Sonagrams of each cluster are presented on the bottom (Manabe, Staddon and Cleaveland, In preparation).

two different calls (2B2 and 2B3) emerged from the 1B2 cluster. One of the new calls is similar to the original 1B2 sound with a lower pitched segment in the initial part of the call. The second call is similar to the original 1B2 sound in that it has a lower pitched segment in the last part of the call. These results show that budgerigars can learn to produce new calls in a matter of a few minutes without social interaction and without reference sound.

In natural situations, budgerigars routinely learn calls from one another. Therefore, in a second experiment, the budgerigars' ability to mimic an acoustic template was assessed. Here a contact call was played to the bird and the bird was required to match the acoustic characteristics of this template with its own vocal production in order to obtain food reinforcement. The bird was first trained to "match" its own call within one second of hearing it. Once this task was learned with reasonable precision, the acoustic template was abruptly switched to the contact call of another individual and the bird was again reinforced for producing a "match" to this new call within one second of hearing it. The top panel in Figure 6.9 shows the day by day changes in the bird's vocal production as a function of which acoustic template was in effect. Each data point is the average of all 50-150 calls produced that day which is the average from the two ten-minute sessions each day. The solid line shows changes which occur as the bird adjusts its output to match the acoustic template. The dotted line shows the daily changes that occur when the original template is put back into effect. The bottom panel a shows sonagrams of the original call (which also served as the original acoustic template), the newly modified call after training and the new acoustic template call. These results show that birds can learn to mimic a new contact call in an amazingly short time of a few days when trained only 20 minutes per day. One could imagine that in a more natural and more social situation, call learning among budgerigars might proceed very quickly indeed.

Conclusion

Vocal learning in budgerigars is an extraordinarily sophisticated and complicated process. The behavioral experiments described above barely scratch the surface at understanding the perceptual processes that initiate, guide and maintain vocal learning in this species. As is the case with human language learning, it is of great interest to know whether vocal learning is the result of general cognitive and perceptual processes, or whether it is subserved by a host of special processes or modules dedicated specifically to the task of developing and maintaining a learned vocal repertoire. The case for special processes comes from several observations. First, budgerigars preferentially learn budgerigar vocalizations rather than other sounds though compared to other birds the constraints on mimicry are exceedingly broad. Second, budgerigars discriminate among species-specific calls more efficiently than they do among calls from other species. Third, there is a wealth of psychoacoustic evidence that budgerigars have an unusual degree of frequency resolv-

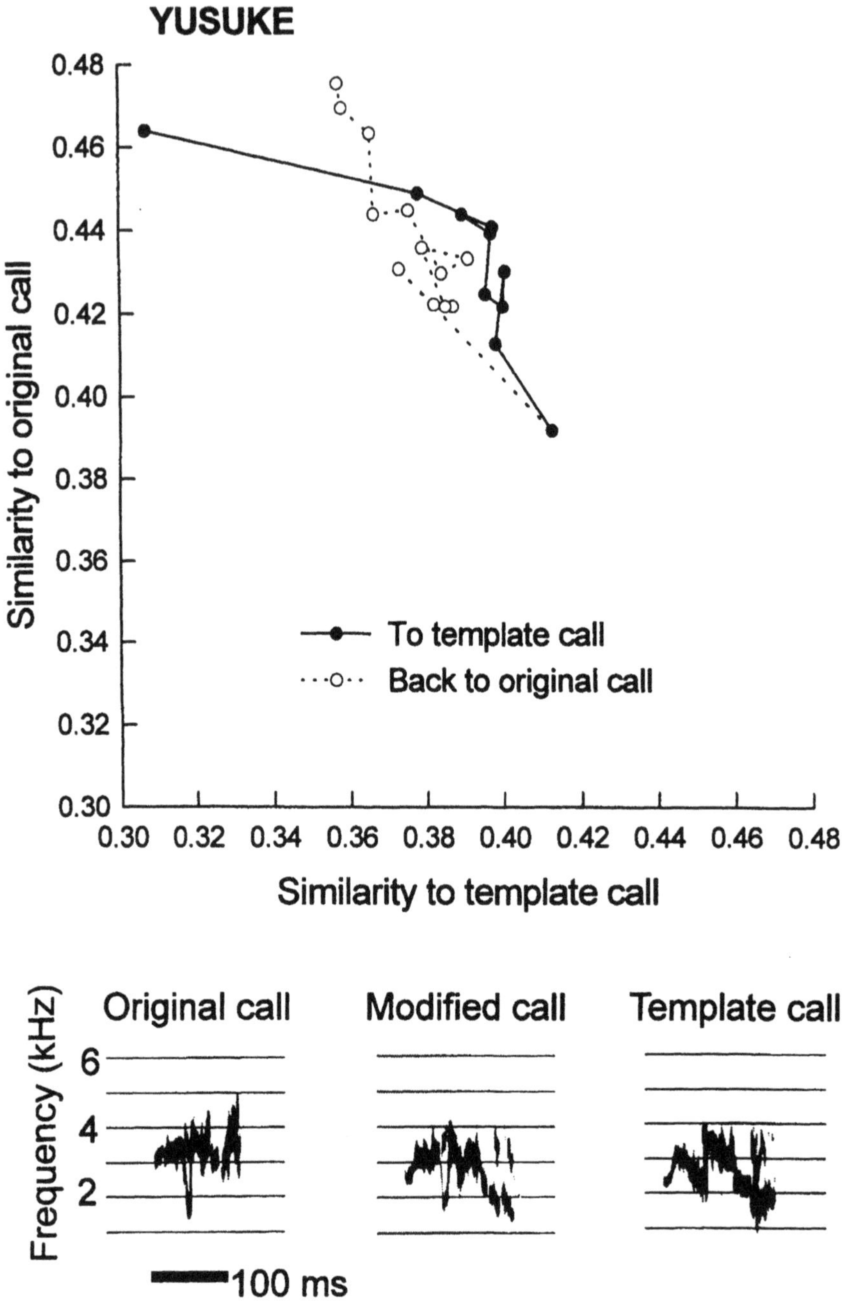

FIGURE 6.9 A similarity between a bird's call and the template call and sonagrams of original call, modified call and the template call (Manabe and Dooling, 1995).

ing power in the spectral region of 2-4 kHz - the spectral region important for acoustic communication signals. Other avian species that have been tested in a similar manner appear to lack this specialization. Such specialization could form the basis for special or enhanced processing of conspecific calls over heterospecific calls in budgerigars.

On the other hand, it is interesting that canaries and zebra finches also show enhanced processing of their species-specific calls without showing any unusual hearing or discrimination abilities in the spectral region of their vocal signals. The fact that all three species showed enhanced processing of conspecific vocal signals, argues both that higher order perceptual processes modified by experience and learning may be involved and that these processes may be quite general.

Without a doubt, social experience normally plays a pivotal role and certain kinds of experiences are more potent than others. We have identified some of the social factors important for call learning in budgerigars. Of course, the term "social factors" covers a multitude of variables and it will be a challenge to identify, quantify, and manipulate these variables. Our first foray into specifying such variables with precision was motivated by the observation that budgerigars, from the first few days after hatching through adulthood, are highly social and readily engage in a wide range of intimate behaviors. We examined the perception of conspecific and heterospecific faces by budgerigars and the results suggest that budgerigars are probably specialized for the perception of conspecific face-like stimuli and this specialization may depend in part on social and perceptual learning.

Most of the experiments conducted so far have been in adult birds. In attempting to understand how vocal learning arises, we have investigated changes in the characteristics of vocalizations during development in budgerigars. The results of these experiments show that adult-like contact calls emerge from food-begging calls used by young birds in the nestbox. Once contact calls have emerged, both the perception and production of these contact calls remain highly plastic throughout adulthood. We have shown that the formation of perceptual categories for contact calls, differing in remarkably subtle ways, probably depends on the production and/or social use of the vocalizations. Simply being exposed to vocalizations alone, something which is not likely to occur in a natural situation, is not very effective in promoting learning.

It is somewhat ironic that the very things that make the study of vocal learning in birds so extraordinarily interesting - its richness, complexity and social requirements - are the very things that make it so challenging to study. In a sense, the study of vocal learning in birds - especially seasonal learning in songbirds - has been hampered by the need to adequately satisfy these "conditions". In this sense, budgerigars may have something special to offer in that we have now shown that vocal learning can be efficiently and precisely controlled simply with food reinforcement and that this complex learning can proceed very quickly, not unlike other more common forms of vertebrate learning.

In aggregate, we have described several different behavioral approaches to understanding perceptual specializations and vocal learning in budgerigars which, especially when combined, may offer new insights into several persistent "problems" of vocal learning in birds. For instance, the question of a motor basis for perception of species-specific vocalizations can be directly addressed by training birds on tasks requiring discrimination of vocalizations they have never heard, training them to produce these vocalizations, and then retesting them for changes in perceptual efficiency. Training birds to produce with precision a specific vocalization - and only that vocalization - to a visual stimulus such as an LED, affords the opportunity to rigorously explore the role of immediate auditory feedback on adult vocal production by temporary deafening through ear-plugging or noise exposure. What is the effect, on vocal learning and on face perception in adult birds, of altering the facial features of breeding budgerigars so that nestlings are provided with incorrect facial stimuli but in a correct, dynamic social setting during development? Can the course of vocal development in young budgerigars be altered through selective reinforcement with food?

Comparative experiments should determine the extent to which adult plasticity and the control of vocalizations by food reinforcement is unique to budgerigars, or perhaps psitticines, or whether it is true as well of songbirds known to be great mimics such as the starling and mockingbird. What are the roles of sensory and motor nuclei in the kind of rapid call learning we have demonstrated in budgerigars? While the neuroethological basis of vocal learning in songbirds has a fairly long history, comparative neurobiological investigations of vocal learning in budgerigars are more recent (Ball, 1994; Striedter, 1994). As the evidence we present suggests, budgerigars may provide a particularly good model for investigating the brain mechanisms mediating social interactions, arousal, and motivational aspects of vocal learning in birds. Comparative experiments, in which different species of birds are tested under identical conditions, will reveal specialized processes that subserve vocal learning and will contribute to our understanding of the evolution of language in humans.

Acknowledgments

We thank Satoshi Amagai for comments on an earlier version of this manuscript and Tracy Kidd and Micheal Dent for help in animal care and Figure preparation. This work was supported by NIH grant DC-00198 and MH-00982 to RJD.

References

Adret, P. (1993). Operant conditioning, song learning, and imprinting to taped song in the zebra finch. *Anim. Behav., 46:* 149-159.

Ball, G.F. (1994). Neurochemical specializations associated with vocal learning and production in songbirds and budgerigars. *Brain, Behav. and Evol. 44:* 234-246.

Blough, D.S. (1985). Discrimination of letters and random dot patterns by pigeons and humans. *J. Exp. Psychol: Anim. Behav. Proc., 11:* 261-280.

Brown, S.D. and Dooling, R.J. (1992). Perception of conspecific faces by budgerigars (Melopsittacus undulatus) I. Natural faces. *J. comp. Psychol., 106:* 203-216.

Brown, S.D. and Dooling, R.J. (1993). Perception of conspecific faces by budgerigars (Melopsittacus undulatus) I. Synthetic faces. *J. comp. Psychol., 107:* 48-60.

Brown, S.D., Dooling, R.J., and O'Grady, K. (1988). Perceptual organization of acoustic stimuli by budgerigars (*Melopsittacus undulatus*). III. Contact calls. *J. comp. Psychol., 102:* 236-247.

Dooling, R.J., Brown, S.D., Klump, G., and Okanoya, K. (1992). Auditory processing of conspecific and heterospecific vocalizations in birds: Evidence for special processes. *J. comp. Psychol., 106:* 20-28.

Dooling, R.J., Brown, S.D., Park, T.J., Soli, S.D., and Okanoya, K. (1987a). Perceptual organization of acoustic stimuli by budgerigars (*Melopsittacus undulatus*): I. Pure tones. *J. comp. Psychol., 101:* 139-149.

Dooling, R.J., Gephart, B.F., Price, P.H., McHale, C., and Brauth, S.E. (1987b). Effects of deafening on the contact call of the budgerigar (*Melopsittacus undulatus*). *Anim. Behav., 35:* 1264-1266.

Dooling, R.J., Park, T.J., Brown, S.D., Okanoya, K., and Soli, S.D. (1987c). Perceptual organization of acoustic stimuli by budgerigars (*Melopsittacus undulatus*): II. Vocal signals. *J. Comp. Psychol., 101:* 367-381.

Dooling, R.J. and Searcy, M.H. (1980). Early perceptual selectivity in the swamp sparrow. *Dev. Psychobiol., 143:* 499-506.

Farabaugh, S.M., Linzenbold, A., and Dooling, R.J. (1994). Vocal plasticity in budgerigars (*Melopsittacus undulatus*): evidence for social factors in the learning of contact calls. *J. Comp. Psychol., 108:* 81-92.

Fay, R.R. (1988). *Hearing in vertebrates: A Psychophysics databook.* Illinois: Hill-Fay Assoc.

Gould, J. L. (1982). *Ethology: The mechanisms of animal behavior.* New York: W.W. Norton.

Harper, A.B. (1986). The evolution of begging: Sibling competition and parent-offspring conflict. *Amer. Nat. 128:* 99-114.

Hulse, S. (1989). Comparative psychology and pitch pattern perception in birds. In Dooling, R.J. and Hulse, S.H. (Eds.) *The Comparative Psychology of Audition: Perceiving complex sounds.* Hillsdale, NJ: Lawrence Erlbaum, pp. 331-349.

Kovach, J.K. (1983a) Constitutional biases in early perceptual learning: I. Preferences between colors, patterns, and composite stimuli of colors and patterns in genetically manipulated and imprinted quail chicks. *J. Comp. Psychol., 97:* 226-239.

Kovach, J.K. (1983b) Constitutional biases in early perceptual learning: II. Visual preferences in artificially selected, visually naive and imprinted quail chicks. *J. Comp. Psychol., 97:* 240-248.

Kroodsma, D.E. and Miller, E.H. (Eds.) (1982) *Acoustic Communication in Birds.* New York: Academic Press.

Liberman, A.M. (1982). On finding that speech is special. *Amer. Psychol., 37:* 148-167.

Mammen, D.L. and Nowicki, S. (1981). Individual differences and within-flock convergence in chickadee calls. *Behav. Ecol. and Sociobiol., 9:* 179-186.

Manabe, K., Kawashima, T., and Staddon, J. E. R. (1995). Differential vocalization in budgerigars: Towards an experimental analysis of naming. *J. Exp. Anal. Behav., 63:* 111-126.

Manabe, K., Staddon, J. E. R., and Cleaveland J. M. (In Preparation). Control of vocal repertoire by reward in budgerigars.

Manabe, K. and Dooling, R. J. (1995). *Vocal learning in budgerigars (Melopsittacus undalatus) using food reward.* Paper presented at Acoustical Society of America, Washington DC.

Marler, P. and Peters, S.S. (1989). Species differences in auditory responsiveness in early vocal learning. In R. J. Dooling and S. Hulse (Eds.), *The Comparative Psychology of Audition: Perceiving Complex Sounds.* Hillsdale, N.J. Lawrence Earlbaum, pp. 243-273.

Marler, P. (1991). *Differences in behavioral development in closely related species: birdsong.* In The Development and Integration of Behavior, P. Bateson (Ed.) Cambridge: Cambridge University Press, pp. 41-70.

Marler, P. and Peters, S. (1988). Sensitive periods for song acquisition from tape recordings and live tutors in the swamp sparrow, *Melospiza georgiana. Ethol., 77:* 76-84.

Mondloch, C.J. (1995). Chick hunger and begging affect parental allocation of feeding in pigeons. *Anim. Behav., 49:* 601-613.

Mundinger, P.C. (1970). Vocal imitation and individual recognition of finch calls. *Science, 168:* 480-482.

Mundinger, P.C. (1979). Call learning in the Carduelinae: ethological and systematic considerations. *Sys. Zool., 28:* 270-283.

Nottebohm, F. (1972). The origins of vocal learning. *Amer. Nat., 106:* 116-140.

Nottebohm, F. (1984). Bird song as a model in which to study brain processes related to learning. *Condor, 86:* 227-236.

Nottebohm, F. (1991). Reassessing the mechanisms and origins of vocal learning in birds. *Trends Neurosci., 14:* 206-211.

Nowicki, S. (1989). Vocal plasticity in captive black-capped chickadees: the acoustic basis and rate of call convergence. *Anim. Behav., 37:* 64-73.

Okanoya, K. and Dooling, R.J. (1987). Hearing in passerine and psittacine birds: A comparative study of absolute and masked auditory thresholds. *J. Comp. Psychol., 101:* 7-15.

Okanoya, K. and Dooling, R.J. (1988). Obtaining acoustic similarity measures from animals: A method for species comparisons. *J. Acoust. Soc. Amer., 83:* 1690-1693.

Pepperberg, I.M. (1985). Social modeling theory: a possible framework for understanding avian vocal learning. *Auk, 102:* 854-864.

Petrinovich, L. (1988). The role of social factors in white-crowned sparrow song development. In T.R. Zentall and B.G. Galef (Eds.) *Social Learning: Psychological and Biological Perspectives* . Hillsdale, NJ: Lawrence Erlbaum Associates, pp.255-278.

Powell, E. F. (1993). *Perception of developing vocalizations in the budgerigar (Melopsittacus undulatus).* M.S. Thesis, University of Maryland, College Park.

Powell, E. F., Dooling, R.J., and Farabaugh, S. (1992) *Development of vocal learning in the budgerigar.* Paper presented at the Society for Neuroscience, Anaheim, CA.

Shepard, R.N. (1980). Multidimensional scaling, tree-fitting, and clustering. *Science, 210:* 390-398.

Sinnott, J.M. (1980). Species-specific coding in birdsong. *J. Acoust. Soc. Amer., 68:* 494-497.

Stamps, J., Clark, A., Kus, B. and Arrowood, P. (1985). Parent-offspring conflict in budgerigars. *Behav., 94:* 1-40.

Stamps, J., Clark, A., Kus, B. and Arrowood, P. (1987). The effects of parent and offspring gender on food allocation in budgerigars. *Behav., 101:* 177-199.

Streidter, G.F. (1994). The vocal control pathways in budgerigars differ from those in songbirds. *J. Comp. Neural. 343:* 35-56.

ten Cate, C. (1994). Perceptual mechanisms in imprinting and song learning, In J.A. Hogan and J.J. Bolhuis (Eds.) *Causal Mechanisms in Behavioral Development.* Graet Britain: Cambridge University Press, pp. 116-146.

Williams, H. (1990). Song learning in the zebra finch: do fathers supply the models for their sons' songs? *Anim. Behav. 39:* 745-757.

Wagner, H.O. (1974). Notes on the history of the Emerald Toucanet. *Wil. Bull., 56:* 65-76.

Wiley, R.H. (1971). Song groups in a singing assembly of Little Hermits. *Condor, 73:* 28-35.

7

How Birds Use Frequency to Recognize Their Songs

Daniel M. Weary

Abstract

The problem of how birds recognize their songs has received a great deal of attention, perhaps more than any other system of signalling in non-human animals. This work has identified a number of features important in recognition, but relatively few seem general to more than a few species. One exception may be the frequency characteristics of the songs. I review a series of experiments showing how several species of birds use the frequency of notes, and the relationships among these frequencies, to recognize their songs. The results show that song birds, like most humans, can recognize and remember songs using the frequency relationships between adjacent notes. The results also suggest that birds can recognize individual notes by their frequency, an ability that may be rare among human listeners.

An Introduction to Bird Song

For scientists who study behavior, there are at least two good reasons to work on song. One is that it is a loud, obvious display. When birds sing they are trying to be noticed, if not by peering scientists, then by potential competitors and mates. The other reason is that song is quantifiable. Songs can be recorded, and recordings analyzed with equipment that is widely available. Many sounds can be described using just three measures: frequency, duration and amplitude.

The songs that birds sing are also highly variable both within and among species. The species used in the work that I will be describing serve as excellent examples. The songs of three of these species are illustrated in Figure 7.1. Black-capped chickadee (*Parus atricapillus*) song is acoustically very simple. It consists of two notes, which are basically tones, with a drop in frequency between them. The great tit (*Parus major*) sings a repertoire of several songs, but each is relatively simple in structure. It usually consists of pairs of notes which are sung repetitively. On the other end of the spectrum is the veery (*Catharus fuscescens*). Veeries sing repertoires of complex songs with many changes in both frequency and amplitude.

Given this enormous variability, how do birds recognize their songs? Two chapters in this volume illustrate certain specializations in how birds perceive their calls (Cynx, Dooling et al.). In this chapter, I review four studies of song recognition in birds which I think illustrate certain generalities. First I describe a field study on veeries and a laboratory study on great tits which show that frequency is an important feature in recognition. I then review another field study on veeries and a laboratory study on black-capped chickadees illustrating specific mechanisms by which birds use frequency to accomplish this task. Before I start, however, I will try to explain why birds sing and why it is important for them to be able to recognize their songs.

The Functions of Song

Birds obviously invest a great deal of time and energy both in learning and producing song. Given that they are making this investment, what are they getting out of it, or in other words, what are the biological functions of song?

Bird song is thought to have two main functions. One is to attract mates. This usually means that the male sings so as to attract females. Evidence for this function comes from at least three different types of studies. In the field, workers have noted that males of certain species spend more time singing before they have found a mate. Once they are paired the song rate shows a decline (e.g. Catchpole, 1973). The other experimental approach is to actually play songs to females. When a female song bird is sexually stimulated it tends to exhibit what is known as a 'copulation solicitation display'. This display, which normally includes a vibration of the wings and a raising of the tail, has been observed in several different species in response to song, and workers have used this response to determine female preferences for certain songs or combination of songs (see Searcy, 1992). In another type of experiment, females have been shown to be more likely to approach and enter nest-boxes which have song played from them, than other boxes from which no song is played (Eriksson and Wallin, 1986; Mountjoy and Lemon, 1991).

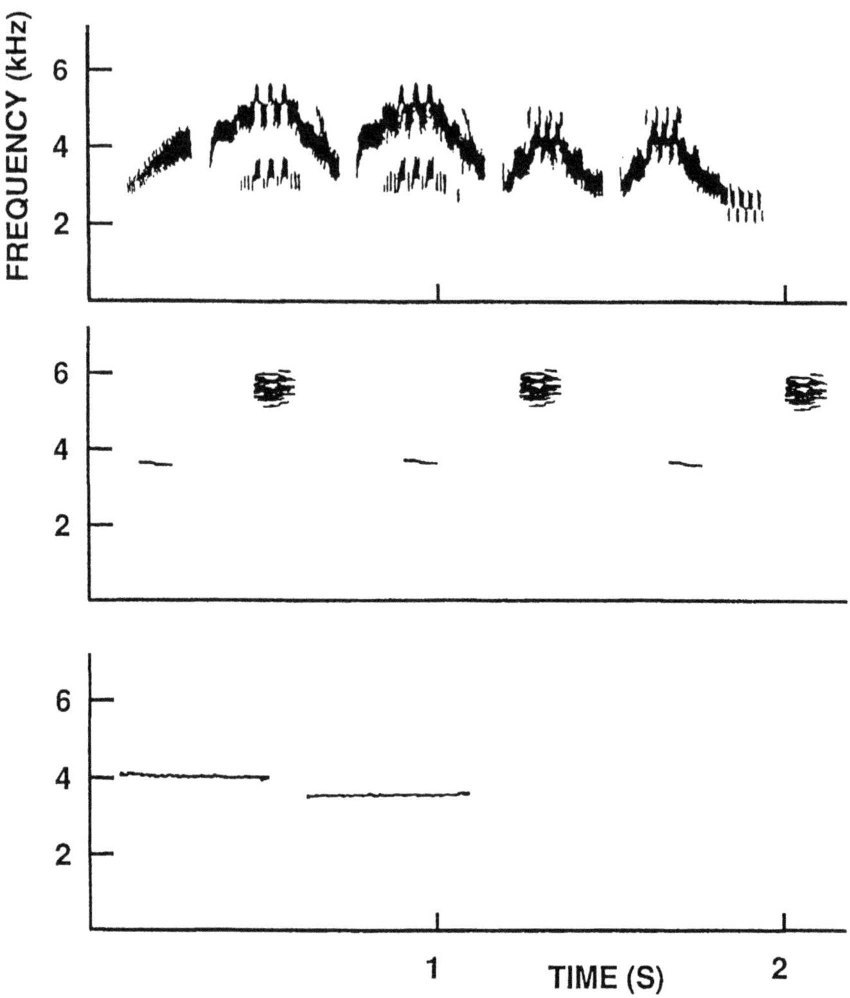

FIGURE 7.1 Sonagrams of songs from three species of song birds (from top to bottom: veery, great tit, black-capped chickadee).

The other main function of song is defence of the territory. This territorial function of song has been tested using removal and playback experiments. In the removal experiments the male is taken away from the territory and is replaced with a loudspeaker. In certain territories songs are played while in others the speakers remain silent. The territories with the silent speakers tend to be the first to be reoccupied by new males, while the territories with song being played are occupied later (e.g. Krebs et al. 1978).

Song playback to males also illustrates the territorial function of song. When researchers play tape recorded songs within the territory of a male, the male usually responds in a very aggressive fashion. He will fly near the speaker, and increase his rate of singing considerably. There are many examples of this type of experiment (see Falls, 1992).

Species and Individual Recognition

To achieve any of its functions the song must be recognizable to the listener. This means that the listener has to recognize that the song is being sung by a member of its own species. In terms of mate attraction, neither sex has much to gain by mating with a different species. Even closely related species are usually unable to produce viable young. Due to high mortality in many species of song birds, choosing a mate of the wrong species in one season can mean a lifetime reproductive success of zero. Thus, there is good reason to suspect that females should be highly discriminating when choosing a male on the basis of song. Females can also use song to reduce the time they have to spend searching for a mate. By being discriminating, females can avoid wasting time spent in approaching males of other species.

There are also good reasons for males to be discriminating. In defending his territory against members of other species the male will have to expend time and energy, and may risk injury. The males may also have little to gain from such interspecific territoriality. Males of other species make unlikely cuckolds. There is considerable experimental evidence to show that birds do discriminate their own species from others (Becker, 1982).

In addition to being able to recognize that the song is being sung by a conspecific, it may often be biologically important for the listener to determine which individual is singing. For almost any social interactions there may be good reasons to be able to recognize individuals. In terms of territorial defence, it has been suggested that birds can save time and effort by learning to recognize their territorial neighbors. Established neighbors presumably pose little threat, so in learning to recognize them the male will not have to respond each time one of them sings. Instead, he can save his efforts for intruding males which pose a more immediate threat. As expected from this reasoning, the territorial males of many species do in fact show a much stronger response to playback of songs from unfamiliar males, than they do to playback of songs from their neighbors (see Falls, 1982).

Song Recognition

The material which I reviewed above demonstrates why song is important to birds, and why it might be important for them to use song to recognize members of their own species in general, as well as recognize specific individuals. Other song recognition tasks may also be biologically important. For example, in species where individuals each sing more than one song it might be important for listeners to distinguish among these song types (Horn, 1992). Also, listeners may use song to distinguish between classes of singers like males of high and low genetic quality (Lambrechts, 1992).

For song to serve any of these functions the listeners must be able to recognize songs accurately. What are the acoustic features that listeners use to accomplish this task? The exact combination of important features seems to vary from species to species. For example, a review of how birds use song to recognize members of their own species identifies many different features as being important (Becker, 1982). Despite these between-species differences, I argue in this chapter that one feature, frequency, might be particularly important in song recognition for a number of species. The experiments I review also suggest some specific mechanisms by which birds might use frequency to recognize their songs. Humans can recognize a tune whether the notes are sung high or low (i.e. they use the constant frequency relationships or "relative pitch"), but they have trouble recognizing particular notes in isolation (i.e. "absolute pitch") (Deutsch, 1982). The experiments on birds show that they use the frequency of notes and the relationships among these frequencies to recognize their songs, and therefore indicate that both relative and absolute frequency might be important.

The Use of Frequency in Song Recognition

Veery Species Identification

As I described earlier, bird songs can be acoustically complex. This means there exists a range of different features that birds could use to recognize songs. Within the veery song illustrated in Figure 7.1, what kind of information could listeners use to recognize that the song has been sung by a member of their own species? There is a temporal organization of the song into its different notes of different durations and inter-note intervals. Then there are the large scale changes in frequency and amplitude that occur with time. Finally we have the fine scale amplitude and frequency modulations. In order to determine the importance of these different features, we have to be able to manipulate them, and the most straightforward way to do this is to work with computer-synthesized songs. Birds can respond to these synthetic songs as if they were natural: when we playback synthetic and natural songs to territorial male veeries they show strong and statistically indistinguishable responses to both (Weary et al. 1986).

Once we have synthesized a credible song, we can then manipulate it in a variety of ways, and play these manipulated songs to territorial males. In one test, we performed two types of manipulation (Weary et al. 1986). In one song we removed all the changes in frequency with time, so all notes were presented at the mean frequency of the song, but with the same amplitude changes as naturally occur. In another we removed all the changes in amplitude with time, and kept the natural changes in frequency. When we played these manipulated songs to territorial males we found a clear gradation in the territorial response (Figure 7.2). Subjects responded similarly to the normal synthetic song and the song without variation in amplitude. However, the subjects showed a much weaker response to the test song in which frequency was held constant. In fact, the response to the amplitude constant song was sta-

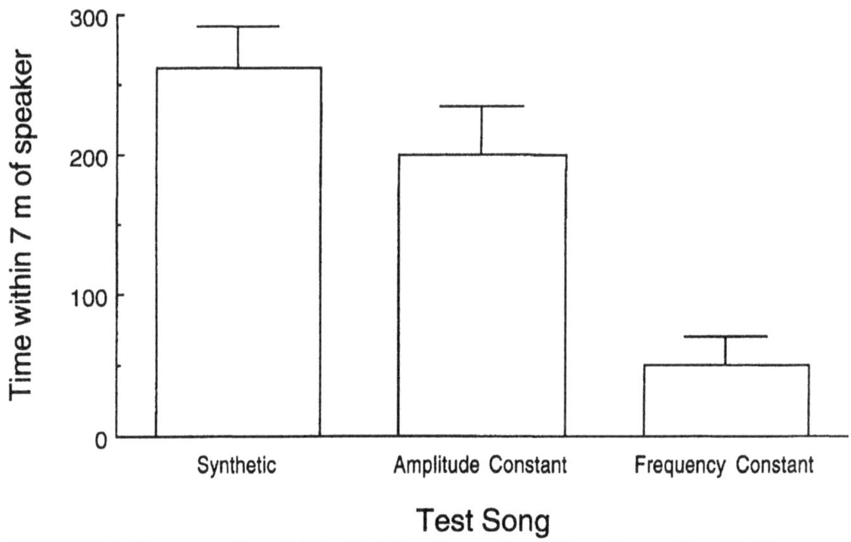

FIGURE 7.2 Responses by wild territorial male veeries to three types of synthetic songs. One song was left unmanipulated (synthetic), in another all the changes in amplitude with time were removed (amplitude constant), and in a third all changes in frequency were removed (frequency constant). The response measure illustrated is time (s) spent within 7 m of the speaker during a 5 minute trial: the more time spent close to the speaker, the stronger the response. Each song was played to five different subjects, and responses shown are averages (mean ± S.E.).

tistically indistinguishable from that to the control, while the response to the frequency constant song was significantly lower.

These results from veeries indicate that frequency changes in general are important for song recognition, at least relative to changes in amplitude. However, they do not tell us how general the importance of frequency might be. With this in mind, the following experiment examines the song features used in recognition by another, quite different species of bird.

Great Tit Operant Song Recognition

Great tit songs are acoustically more simple than those of veeries (Figure 7.1). They generally consist of repetitions of identical notes or pairs of notes, and these notes tend to be tonal. In the song illustrated, the difference between the notes can be described using five features. There is the rather obvious difference in frequency, a difference in duration, a difference in amplitude, and the buzzy structure of the upper note which is a result of rapid amplitude and frequency modulation, or AM and FM as I shall refer to them.

When a test song is played to a territorial male and the male responds strongly, we can be quite certain that the male has recognized the song as belonging to his own species. However, what can we interpret when a test evokes only a weak response? One possibility is that the male has not recognized the song as species specific. The other is that the male has recognized the song, but does not consider it to be a territorial threat. In other words, we cannot fully distinguish recognition from response (Ratcliffe and Weisman, 1992; Weary, 1992; Johnsrude et al. 1994).

One way around this problem is to take the song out of its biological context. If the experimenter, and not the bird, decides what the appropriate response is, then it should be possible to eliminate this type of bias. With this in mind, I chose to use an operant procedure to examine song perception in great tits.

Captive male great tits were trained to discriminate between two synthetic notes modelled on natural song notes recorded from wild birds. When subjects visited an electronically operated feeder following the presentation of one note, they received a food reward. Feeder visits following presentation of the other note were not rewarded. Once the subjects were reliably discriminating between the two training notes, they were then presented with unrewarded test notes.

The rules used to formulate the test notes were simple. The two training notes differed in the five parameters discussed above. In order to test the effect of each of these parameters independently, as well as any interaction effects, I synthesized all combinations of the parameters. As each parameter could be set at one of two levels, and there were five parameters, this resulted in a total of 32 notes, or in other words the 2 training notes and 30 test notes.

How did the captive great tits respond to these test notes? Figure 7.3 shows the proportion of visits to the feeder that subjects made to test stimuli, relative to the level of each of the five note parameters. Frequency was clearly a very important feature in the birds' recognition of the test notes. Notes in which the frequency was the same as the positive training note were responded to at a high rate, while those in which the frequency was the same as the negative training note were responded to infrequently. Relatively, the other note parameters had a much weaker influence on the birds' behavior. Statistically, all of the parameters significantly affected response, but frequency accounted for 29% of the variance in my measure of response, while the other parameters combined could only account for less than 10% (Weary, 1990).

Using this operant procedure and a species with a very different song from veeries, frequency once again emerged as a critical feature in song recognition. Thus, there seems to be at least some generality in the importance of frequency. Also, other tests of song perception in the laboratory and in the field have illustrated the importance of frequency for different species (Dooling et al. 1987; Nelson, 1989a).

The work that I have reviewed so far does not show us how birds are using frequency. Indeed, it is possible that the veeries and great tits do

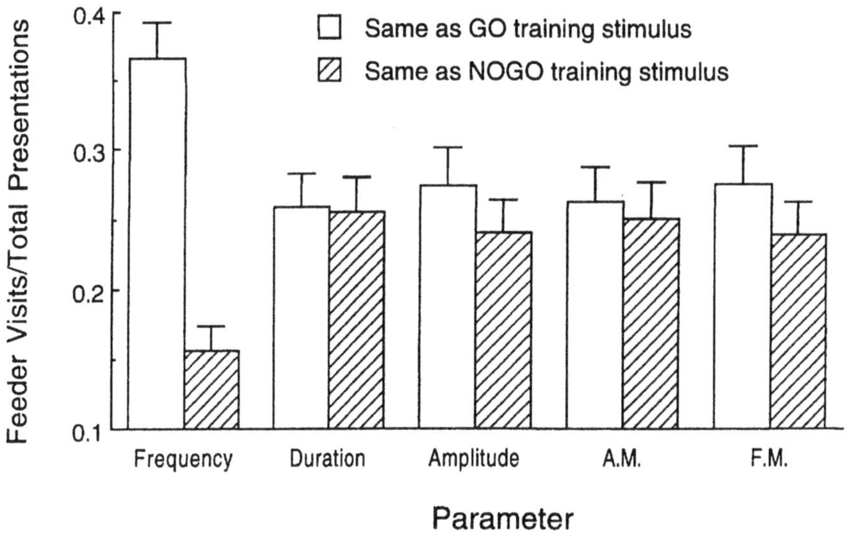

FIGURE 7.3 Response (mean ± S.E.) to the test notes averaged across four great tits which served as subjects. In a test note, the value of a parameter could either be the same (open bars) or different (hatched bars) to that in the positive (GO) training note.

not use frequency in the same way. The experiments on veeries indicated that the frequency changes that occur between notes are used by the birds to recognize the song. The entire frequency of the stimulus was not manipulated, only the relationships among the frequencies within the stimulus. For the great tits, however, the individual stimuli were at different frequencies and these were presented in isolation with substantial pauses between presentations. Thus, the great tits did not seem to be using the relationships among frequencies to recognize the test stimuli, but rather some aspect of the absolute frequency of the stimuli. This means that the two species may be using different mechanisms to identify the test songs in the two different cases.

Relative and Absolute Frequency Perception

Studies of human music perception might be relevant in formulating hypotheses about how birds recognize their songs. In music, the perception of frequency has been divided into two processes: relative and absolute pitch perception. When we use the frequency relationship between adjacent notes to recognize a sound, we call this relative pitch perception (Deutsch, 1982). Relative pitch can be subdivided in two ways. When we recognize a tune using qualitative changes in frequency, or in other words whether the second note is higher or lower than the first, we call it pitch contour perception. In contrast, pitch interval perception refers to a quantitative evaluation of the frequency change. Specifically, we identify pairs of notes using the ratio of their two frequencies. Both pitch contour and pitch interval perception are commonly observed in studies of human music perception (Deutsch, 1982).

While relative pitch is defined on the basis of the relationship between notes, absolute pitch is defined by the absence of such relationships. When animals identify individual notes without comparing them to an external standard they are using absolute pitch perception (Ward and Burns, 1982). For example, if an animal can identify a note when it is presented in isolation it is probably using absolute pitch perception. Studies of humans have shown that very few people are good at using absolute pitch to recognize notes (Miyazaki, 1989). Absolute pitch implies more than just fine frequency discrimination: it implies fine frequency discrimination for a number of different frequencies. For example, a person with excellent absolute pitch may be able to name the notes played to her on a piano.

Clearly, one of the differences between relative and absolute pitch perception has to do with how long the animal has to remember the frequency. In relative pitch perception the length of pitch memory only has to be as long as the intervals between notes, usually less than a second. With absolute pitch perception notes are identified after long intervals, even when other notes unrelated to the task have been presented to the subject in the intervening period (Ward and Burns, 1982).

There is now evidence that birds may be using these pitch recognition mechanisms to recognize their songs. The evidence for relative pitch perception is particularly convincing. First I will discuss some field evidence, using as an example another study on veeries. Then I will summarize a recent laboratory experiment on relative pitch perception in black-capped chickadees.

Evidence for Relative Pitch Perception: Veery Field Data

There are two methods workers have used to examine relative pitch perception in the field. One is to examine what birds sing, the other is to use song playback to see how they respond to song. In other words, we examine how birds both produce and recognize their songs. There are recognition and production data for four species whose songs vary from simple to complex: black-capped chickadees, Carolina chickadees (*Parus carolinensis*), white-throated sparrows (*Zonotrichia albicollis*), and veeries. I will describe the work on veeries.

The species recognition experiments on veeries that I described earlier suggested that some aspect of frequency was important to the birds. The first step in determining if they were using relative pitch was to record a large sample of males and to measure the frequency characteristics of the songs. In a typical veery song there are five notes (see Figure 7.1): an introductory whistle followed by two higher notes of similar frequency and then two lower notes of similar frequency (Weary et al. 1991). I measured the mean frequency of the two notes where the major change in frequency occurs (notes III and IV). Figure 7.4 illustrates this data. The y-axis shows the frequency at which the bird sang the higher note (note III), while the x-axis indicates the frequency of the lower note (note IV). Each bird is represented by a single point.

This plot allows us to draw several conclusions about how the birds are using relative pitch to sing their song. The first is that the distributions overlap: some veeries sing their note III at 4100 Hz, while others sing their note IV at the same frequency. However, each individual sings his note IV at a lower frequency than note III. This last result indicates that individuals are using at least frequency contour perception to produce their songs.

Are the veeries also using a quantitative aspect of the relationship? The results suggest that they are. The correlation coefficient of this relationship is 0.95, indicating that the birds are not just following the quali-

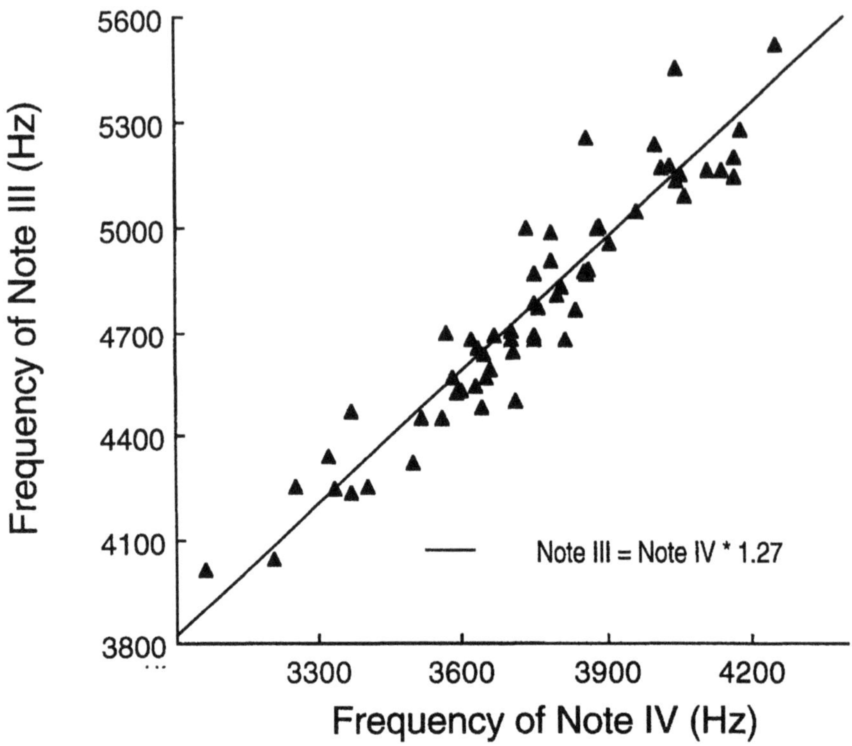

FIGURE 7.4 The relationship between the frequency of note III (phrase B) and note IV (phrase C) for 61 wild territorial male veeries.

tative rule of 'second note lower', but rather the quantitative rule of 'second note precisely so much lower'. Moreover, the results indicate that each of the birds is using the same rule. To determine the precise nature of this relationship we can simply divide the mean frequency of the higher note by that of the lower. The resulting frequency ratio equals 1.27. Thus, the solid line on the plot is based on the formula Note III = Note IV * 1.27. The fact that the points fall very close to this line suggests that the birds may be using pitch interval perception to produce their songs.

In summary, these results indicate that relative pitch is important. In addition, they suggest that the birds are sensitive to more than just the frequency contour. The birds sing the lower note at what appears to be a consistent ratio of the higher note.

Finding that the birds produce a constant frequency ratio does not demonstrate that they perceive and use this feature in song recognition.

It is at least possible that the frequency relationship observed in the production data is simply the result of a motor constraint. To demonstrate that veeries are actually using the relative frequency to recognize their songs, wild territorial males were played manipulated songs (Weary et al. 1991). Three types of test song were played: ones with normal frequency ratios, ones with a much smaller ratio of 1 (i.e. no change in frequency), and ones with frequency ratio double that which naturally occurs. Differences between versions of test songs allowed us to control for the absolute pitch of individual notes.

Subjects showed the strongest response to songs with normal frequency ratios. As illustrated in Figure 7.5, the subjects spent more time close to the playback speaker during playback of these songs, than during playback of songs with either no change in frequency between the notes, or songs where the change in frequency was twice that which occurs naturally. Given that the test songs were normal in all other respects, these results indicate that the frequency relationship between notes is important in song recognition by veeries. But the results can actually tell us more than just that relative pitch is important, they also indicate the specific mechanism in operation. If veeries had been using just pitch contour perception to recognize songs, then response should have been similar to any test song where the fourth note was discernibly lower than the third; in this experiment, these test songs would have been those with the normal and larger frequency ratios. What in fact occurred was that the songs with exaggerated ratios received a much diminished response. We thus concluded that recognition was based on more than just a difference, it was based on a specific degree of difference. Based on the song production results, and what is known about human music perception, we suggested that the frequency ratio was the salient feature.

Similar studies on two other species of birds, black-capped chickadees and white-throated sparrows, have also shown evidence for relative pitch perception (Hurly et al. 1990; 1991; 1993; Weisman and Ratcliffe, 1989; Weisman et al. 1990). Recent evidence from black-capped chickadees indicates that the frequency relationships within a note are also important (Shackleton et al. 1992). We therefore have evidence for two types of generality in song perception. First, that frequency seems to be an important feature for several species, and second, that a specific mechanism for frequency perception, relative pitch, appears to be used by a number of species.

The results for veeries, and those which I mentioned for white-throated sparrows and black-capped chickadees, support the idea that relative pitch is of general importance in song recognition by birds. However,

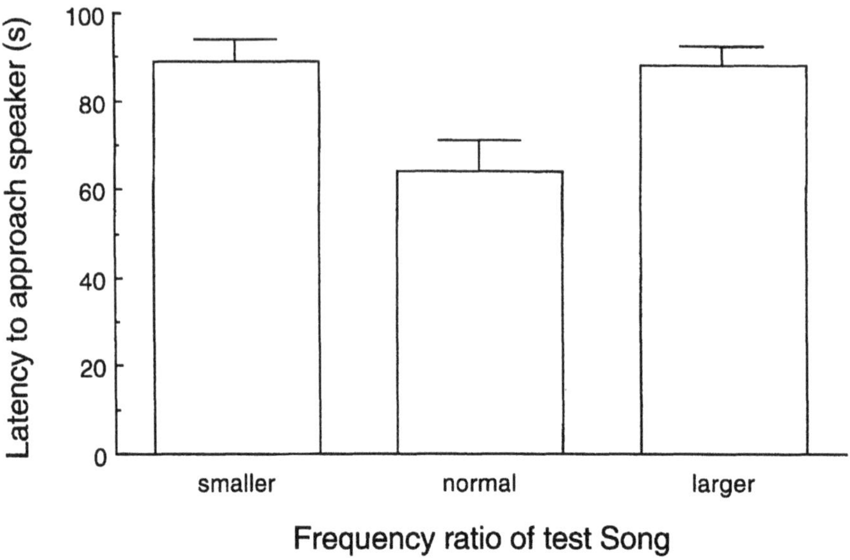

FIGURE 7.5 Responses by wild territorial male veeries to three types of test songs. Songs were composed so as to have a normal ratio between the frequencies of note III and IV (normal), or to have a ratio of 1 (smaller), or a ratio double to that which normally occurs (larger). The response measure illustrated is latency (s) to approach within 7 m of the speaker during a 5 minute trial: the shorter the latency, the stronger the response. Each song was played to each of 28 different subjects, and responses shown are averages (mean ± S.E.).

there is other evidence which is less supportive: song production data from Carolina chickadees indicates that relative pitch may be of only limited importance (Lohr et al. 1991).

All of the studies described above use the territorial response to playback of song to evaluate the use of relative pitch in song recognition. All except one of these experiments were performed in the wild. Weisman and Ratcliffe (1989) tested black-capped chickadees in the laboratory, but the birds were kept in large flight cages which they seemed to defend much like a real territory (see Ratcliffe and Weisman, 1992 for a comparison with field results). Operant techniques have also been used to investigate relative pitch in a number of species of song birds (Hulse and Cynx, 1985; 1986; Page et al. 1989; Weary and Weisman, 1991). I think these experiments are interesting, in terms of what they show about relative pitch perception, and perhaps more importantly, about absolute pitch perception.

Evidence for Relative Pitch Perception: Chickadee Operant Data

I will describe an experiment on black-capped chickadees (Weary and Weisman, 1991). As illustrated in Figure 7.1, these birds have a simple two note song with a major change in frequency between the notes. In this experiment, we used a procedure very similar to that described for the great tits. Male captive subjects were trained to discriminate between certain songs, and then tested with other songs. We used three types of song as training stimuli: ones in which the ratio between the notes was much smaller than normal, ones where the ratio was normal, and ones in which the ratio was larger than normal. When subjects visited the feeder following playback of the normal song, the feeder opened allowing them access to food. Feeder visits following playback of either type of song with unnatural ratios were not rewarded. In this way, after a number of sessions, the chickadees learned to discriminate between the two classes of songs.

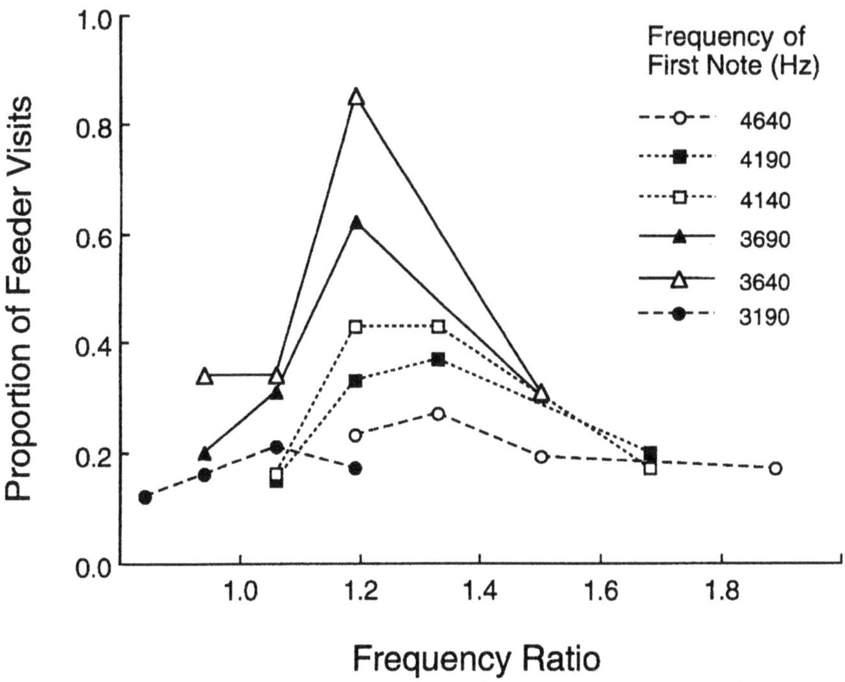

FIGURE 7.6 The proportion of visits to the feeder in response to 36 synthetic songs forming a broad range of note frequencies and frequency ratios. Response varies as a function of the frequencies of both the first and second notes and their interaction (i.e. the frequency ratio). Responses to each song are scaled, separately for each subject, as a proportion of responses to the song with the highest number of feeder visits. Data were averaged across the four chickadees used in the experiment.

The birds were then presented with test songs so as to determine the basis of this discrimination. In one set of test songs the frequency of both notes was varied independently on either side of the mean. These manipulations allowed us to test the effects of the frequencies of both notes, and the effect of the ratio, on song perception.

Figure 7.6 illustrates how the chickadees responded to these new test songs. The birds' response varied with the frequency of both notes. Response tended to be strongest at intermediate frequencies of both the first and second note. Thus, the frequencies of both notes are important to the birds. In addition, there is an interaction between these effects. This interaction results from an especially high rate of response to songs with the appropriate combination of frequencies, or in other words, songs with an appropriate frequency ratio.

The results from this experiment indicate that the chickadees used both the frequencies of the notes independently, as well as the relationship between them, to discriminate between the songs. Thus, this chickadee operant experiment again indicates the importance of relative pitch perception in song recognition. However, it also suggests a role for absolute pitch, because the frequency of each note was independently important.

The other operant studies of relative pitch in song birds have not shown strong evidence for relative pitch (Hulse and Cynx, 1985; 1986; Page et al. 1989). This difference may be due to the use of different species or different methodologies. In each of the studies showing relative pitch that I have described, songs or song-like stimuli were used. This was not the case for the three operant studies cited above. However, these other operant studies have also illustrated what may be an absolute pitch effect.

Absolute Pitch in Birds?

Other than the operant experiments that I have just described, what evidence do we have that birds use absolute pitch perception to recognize their songs? Because of the nature of the relative pitch studies, where notes are presented together in close succession, they are not well suited for directly examining absolute pitch. To study this perceptual mechanism, we need to present sounds in isolation, with long periods of time between successive presentations of relevant stimuli. With this in mind we can perhaps go back to the results of the experiment I described on great tits. Recall that notes were presented individually, and intervals between notes were typically greater than a minute. Despite these constraints, the birds still relied on the frequency of the notes to discriminate between them. Thus the results from this experiment can perhaps be considered as evidence for absolute pitch perception in song birds. Much better evidence would come from experiments where subjects are trained to identify a number of frequencies.

Others have argued that birds could be using some form of absolute pitch perception in a number of song recognition tasks. For example, there is some evidence from neighbor recognition studies that birds might use absolute pitch to recognize songs. In a study on white-throated sparrows, when the frequency of the entire song or just the frequency of the first note was lowered, recognition seemed to disappear (Brooks and Falls, 1975). Similar experiments were also performed on field sparrows (*Spizella pusilla*; Nelson, 1989a). Although these studies do not allow us to rule out a relative pitch interpretation, they at least allow room for one based on absolute pitch.

Summary

The work that I have reviewed shows some of what is known about how birds use frequency in song recognition. The first playback study on veeries, and the operant study on great tits, illustrated that frequency is an important feature in their recognition. This work, and the other studies cited also indicate that this reliance on frequency may be general to a variety of species of song birds.

The subsequent field work on veeries, as well as the playback studies cited on black-capped chickadees and white-throated sparrows, show more than just that frequency is important. They also indicate that a specific mechanism of pitch recognition, relative pitch, is being used by several species. Thus, not only does the use of frequency by birds seem general, but the specific mechanism by which they use frequency to recognize their songs also may be shared. There may also be exceptions, Carolina chickadees being one. The operant tests of relative pitch perception also shed some doubt on the generality of this process. One study on black-capped chickadees has shown that subjects can use relative pitch to discriminate songs, but the evidence from other species is less convincing. However, these operant studies do seem to implicate another pitch recognition mechanism, absolute pitch. Of course different song recognition mechanisms may be involved in different tasks: relative cues may be particularly important in species recognition, but birds might find absolute cues more useful in recognizing specific songs or individuals.

More than three decades ago, Marler (1960) argued that those features which are sung with the least variation should be those that are used by others for recognition, and since that time this idea has been a guiding principle for the field (e.g. Nelson, 1989b). The relative pitch results appear to support this idea, as the frequency ratio is a feature which shows little variation in comparison to other features of the song (e.g. Weisman et al. 1990, Hurly et al. 1991). I agree that there may be a relationship between the variation with which features are produced and their importance in recognition, but I suggest a different interpretation for this relationship. As a result of general selection on the avian auditory system, birds are better at discriminating differences in certain features, like frequency, than others, like duration (see Dooling, 1982). Those features which birds are best at discriminating should also be those which can be used most reliably for the purposes of recognition, and thus will tend to be favored in this task (see Weary, 1990). In addition, the features which birds are best at discriminating should be those which they can learn to reproduce most accurately, hence resulting in the relatively low degree of variation for these features when they are sung. Thus the relationship between the variability with which a feature is produced and the importance of that feature in recognition, may arise because both production and recognition are affected by the bird's ability to perceive differences in the feature. The fact that birds are particularly apt at perceiving changes in frequency may explain why the studies reviewed here have found this feature to be so important in recognition.

This reasoning may also be able to account for the importance of relative pitch in song recognition. When psychologists test the perceptual abilities of animals they tend to set relative tasks. For example, they train subjects to respond only when two tones are different, and then play them tones which vary in their degree of similarity to one another. It is likely that the subject's performance at this task would deteriorate as the temporal separation between two tones increased. In other words, as the task became less relative and more absolute the ability of the subject to discriminate differences in the feature would probably worsen.

I have argued that birds use frequency to recognize their songs because they are particularly good at perceiving differences in this feature. But why are birds (and indeed many other animals) better at detecting differences in frequency than other acoustic features? I believe the answer to this question can be extracted from studies of how sounds transmit during broadcast through natural habitat. These studies have shown that frequency information is relatively impervious to degradation during transmission relative to the amplitude and duration characteristics of the sound (see Wiley and Richards, 1982). It makes sense that animals should be best at detecting differences in those features which are transmitted most reliably. It is difficult to imagine the selective advantage in being able to detect differences in a feature which are smaller than can be reliably transmitted over the distances and through the habitat necessary for vocal communication.

Many of the chapters in this book, including the two others which concentrate on the perception of bird vocalizations (Cynx, Chapter 4, Dooling et al., Chapter 6), emphasize the specializations shown by different species. I began this chapter by addressing the differences between species in song recognition, and indeed the history of research on this subject seems almost to show as many specializations as there have been studies (see Becker, 1982). However, I think that the exciting aspect about this work on frequency perception in song birds is that it suggests generalities across species. Each species is specially adapted in certain ways to best exploit its own specific habitat, but it is likely that a great many adaptations are shared, and in some respects it may be these general processes that deserve our attention.

Acknowledgments

My thanks to John Krebs, Bob Lemon and Ron Weisman for their many important contributions to the work described. I am grateful to Jeff Cynx, Andy Horn, Johanne Mongrain, Scott Shackleton, Andrea Simmons, Katherine Weary and the Editors of this volume for their comments on earlier versions of this chapter. While performing the work described I was supported by doctoral and post-doctoral scholarships from the Natural Sciences and Engineering Research Council of Canada and Fonds pour la Formation de Chercheurs et L'Aide à la Recherche du Québec.

References

Becker, P.H. (1982) The coding of species-specific characteristics in bird sounds. In D.E. Kroodsma and E.H. Miller (Eds.) *Acoustic Communication in Birds, Vol. 1*. New York: Academic Press, pp. 213-252.

Brooks, R.J. and Falls, J.B. (1975) Individual recognition by song in white-throated sparrows. III. Song features used in individual recognition. *Can. J. Zool. 53*: 1749-1761.

Catchpole, C.K. (1973) The functions of advertising song in the reed warbler. *Behav. 46*: 300-320.

Deutsch, D. (1982) The processing of pitch combinations. In D. Deutsch (Ed.) *The Psychology of Music*. New York: Academic Press, pp. 271-312.

Dooling, R.J. (1982) Auditory perception in birds. In D.E. Kroodsma and E.H. Miller (Eds.) *Acoustic Communication in Birds, Vol. 1*. New York: Academic Press, pp. 95-130.

Dooling, R. J., Park, T. J., Brown, S. D., Okanoya, K. and Soli, S. D. (1987) Perceptual organization of acoustic stimuli by budgerigars (*Melopsittacus undulatus*): II. Vocal signals. *J. Comp. Psych. 101*: 367-381.

Eriksson, D. and Wallin, L. (1986) Male bird song attracts females - a field experiment. *Behav. Ecol. Sociobiol. 19*: 297-299.

Falls, J.B. (1982) Individual recognition by sounds in birds. In D.E. Kroodsma and E.H. Miller (Eds.) *Acoustic Communication in Birds, Vol. 2*. New York: Academic Press, pp. 237-278.

Falls, J.B. (1992) Playback: a historical perspective. In P.K McGregor (Ed.) *Playback and studies of animal communication*. New York and London: Plenum Press, pp. 11-34.

Horn, A.G. (1992) Field experiments on the perception of song types by birds. In P.K McGregor (Ed.) *Playback and studies of animal communication*. New York and London: Plenum Press, pp. 191-200.

Hulse, S. H. and Cynx, J. (1985) Relative pitch perception is constrained by absolute pitch in songbirds (*Mimus, Moluthrus*, and *Sturnus*). *J. Comp. Psych. 99*: 176-196.

Hulse, S. H. and Cynx, J. (1986) Interval and contour in serial pitch perception by a songbird *Sturnus vulgaris*. *J. Comp. Psych. 100*: 215-228.

Hurly, T.A., Ratcliffe, L. and Weisman, R. (1990) Relative pitch recognition in white-throated sparrows, *Zonotrichia albicollis*. *Anim.Behav. 40*: 176-181.

Hurly, A., Ratcliffe, L., Weary, D.M. and Weisman, R.G. 1993. White-throated sparrows can perceive pitch change using the frequency ratio independent of the frequency difference. *J. Comp. Psych. 106*: 388-391.

Hurly, T.A., Weisman, R.G., Ratcliffe, L. and Johnsrude, I.S. 1991. Absolute and relative pitch production in the songs of the white-throated sparrow (*Zonotrichia albicollis*). *Bioacoustics 3*: 81-91.

Johnsrude, I.S., Weary, D.M., Ratcliffe, L.R. and Weisman, R.G. 1994. Tests of song perception by brown-headed cowbirds. *J. Comp. Psych. 108*: 172-178.

Krebs, J., Ashcroft, R. and Webber, M. (1978) Song repertoires and territory defence in the great tit. *Nature 271*: 539-542.

Kroodsma, D.E. and Miller, E.H. (Eds) (1982) *Acoustic Communication in Birds*. New York: Academic Press.

Lambrechts, M. (1992) Male quality and playback in the great tit. In P.K McGregor (Ed.) *Playback and studies of animal communication*. New York and London: Plenum Press, pp. 135-152.

Lohr, B., Nowicki, S. and Weisman, R. (1991) Pitch production in Carolina chickadee songs. *Condor 93*: 197-199.

Marler, P. (1960) Bird songs and mate selection. In W.E. Lanyon and W.N. Tavolga (Eds.) *Animal sounds and communication*. Washington: American Institute of Biological Sciences, pp. 348-367.

McGregor, P.K. and Krebs, J.R. (1982) Song types in a population of great tits (*Parus major*): their distribution, abundance and acquisition by individuals. *Behav. 79*: 126-151.

Miyazaki, K. (1989) Absolute pitch identification: Effects of timbre and pitch region. *Music Percep. 7:* 1-14.

Mountjoy, D. J. and Lemon R. E. (1991) Song as an attractant for male and female European starlings, and the influence of song complexity on their response. *Behav. Ecol. Sociobiol. 28*: 97-100.

Nelson, D.A. (1989 a) Song frequency as a cue for recognition of species and individuals in the field sparrow (*Spizella pusilla*). *J. Comp. Psych. 103*: 171-176.

Nelson, D.A. (1989 b) The importance of invariant and distinctive features in species recognition of bird song. *Condor 91*: 120-130.

Page, S. C., Hulse, S. H. and Cynx, J. (1989) Relative pitch perception in the European starling (*Sturnus vulgaris*); further evidence for an elusive phenomenon. *J. Exp. Psychol.: Anim.Behav. Proc. 15*: 137-146.

Ratcliffe, L. and Weisman, R.G. (1992) Pitch processing strategies in birds: a comparison of laboratory and field studies. In P.K McGregor (Ed.) *Playback and studies of animal communication*. New York and London: Plenum Press, pp. 211-224.

Searcy, W.A. (1992) Measuring responses of female birds to male song. In P.K McGregor (Ed.) *Playback and studies of animal communication*. New York and London: Plenum Press, pp. 175-190.

Shackleton, S.A., Ratcliffe, L. and Weary, D.M. (1992) Relative frequency parameters and song recognition in black-capped chickadees. *Condor 94*: 782-785.

Ward, W. D., and Burns, E. M. (1982) Absolute pitch. In D. Deutsch (Ed.) *The Psychology of Music*. New York: Academic Press, pp. 431-449.

Weary, D. M. (1990) Categorization of song notes by great tits: which acoustic features are used and why? *Anim. Behav. 39*: 450-457.

Weary, D.M. (1992) Bird song and operant experiments: a new tool to investigate song perception. In P.K McGregor (Ed.) *Playback and studies of animal communication*. New York and London: Plenum Press, pp. 201-211.

Weary, D.M., Lemon, R.E. and Date, E.M. (1986) Acoustic features used in song discrimination by the Veery. *Ethology 72*: 199-203.

Weary, D.M. and Weisman, R.G. (1991) Operant discrimination of frequency and frequency ratio in the black-capped chickadee (*Parus atricapillus*). *J. Comp. Psychol. 105*: 253-259.

Weary, D.M, Weisman, R.G., Lemon, R.E., Chin, T. and Mongrain, J. (1991) Use of the relative frequency of notes by veeries in song recognition and production. *Auk 108*: 977-981.

Weisman, R.G. and Ratcliffe, L. (1989) Absolute and relative pitch processing in black-capped chickadees, *Parus atricapillus*. *Anim. Behav. 38*: 685-692.

Weisman, R.G., Ratcliffe, L., Johnsrude, I. and Hurly, T.A. (1990) Absolute and relative pitch production in the song of the black-capped chickadee (*Parus atricapillus*). *Condor 92*: 118-124.

Wiley, R.H. and Richards, D.G. (1982) Adaptations for acoustic communication in birds: sound transmission and signal detection. In D.E. Kroodsma and E.H. Miller (Eds.) *Acoustic Communication in Birds, Vol. 1*. New York: Academic Press, pp. 132-181.

8

Development of Perceptual Mechanisms in Birds: Predispositions and Imprinting

Johan J. Bolhuis

Abstract

The newly hatched young of a number of precocial bird species rapidly form a social preference for a conspicuous object to which they are exposed. In the natural situation the preference is formed for the mother, but in the laboratory the young animals can develop a preference for a wide range of inanimate mother surrogates. In domestic chicks, Gallus gallus domesticus, this process involves the development of at least two different perceptual mechanisms. The first of these mechanisms is involved in learning about stimuli to which the animals are exposed, a process known as filial imprinting. The mechanisms underlying this learning process are compared with those of conventional associative learning paradigms. The second perceptual mechanism involves a predisposition to approach stimuli resembling conspecifics. The predisposition emerges in dark-reared chicks given some non-specific experience during a sensitive period, and is expressed as a relatively general preference for stimuli with a head and neck region. During the development of fililal preferences the two perceptual mechanisms appear to interact at the behavioral level; the predisposition does not constrain the range of stimuli for imprinting. In the natural situation, the animal's response may be biased by the predisposition, and through the learning process it then comes to prefer individuals.

Introduction

The development of a social preference of a young animal for its parent has intrigued students of animal behavior ever since Konrad

Lorenz's vivid descriptions of avian imprinting (Lorenz, 1935, 1937; see Spalding, 1873, for an early discussion). This chapter is concerned with the development of perceptual mechanisms (Hogan, 1988) involved in such filial behavior in precocial birds. I will briefly describe some general characteristics of the formation of early filial preferences (see Bolhuis, 1991, for a more detailed review), and then discuss the underlying perceptual mechanisms. There is increasing evidence that at least two kinds of perceptual mechanisms play an important role in the development of filial behavior. The first type of perceptual mechanism is essentially a representation or memory of stimuli to which the animal is exposed. The formation of this mechanism is usually known as filial imprinting (Bateson, 1966; Sluckin, 1972; Bolhuis, 1991). The mechanisms of learning in imprinting have been the subject of considerable study (see Hoffman and Ratner, 1973; Bolhuis et al. 1990; Van Kampen, 1993, 1995 for reviews), and will be discussed briefly here.

The second type of perceptual mechanism involved in the development of filial behavior was already indicated by Lorenz (1935), when he pointed out that imprinting is not a universal phenomenon that applies to all species of birds. For example, he noted that newly hatched chicks of the curlew (*Numenius arquata*) could not be made to follow their human keeper, while chicks of the greylag goose (*Anser anser*) could. In Lorenz's terms, the curlew has an 'innate perceptory pattern' (Lorenz, 1937, p. 247) of the appropriate 'companion' ('Kumpan'), whereas in the greylag goose, the perceptory pattern is formed by experience with a particular object. The latter is equivalent to imprinting, while the former is roughly equivalent to a perceptual mechanism which we have called a predisposition (e.g. Horn and McCabe, 1984). Recent evidence suggests that more generally there is an important role for predispositions in the development of early filial behavior in certain precocial bird species. In this context, predispositions refer to perceptual preferences that develop in young animals without experience with the particular stimuli involved (for reviews, see Gottlieb, 1971; Horn, 1985; Bolhuis, 1991; Johnson and Bolhuis, 1991; Lickliter, 1993; ten Cate, 1994). In this chapter, the nature of these predispositions in the domestic chick (*Gallus gallus domesticus*) will be discussed. In contemporary ethology there is a tendency to avoid the use of Lorenzian concepts such as 'Innate Releasing Mechanism'. However, as we will see, much of Lorenz's intuition about the development of filial preferences is supported by the evidence that is now available.

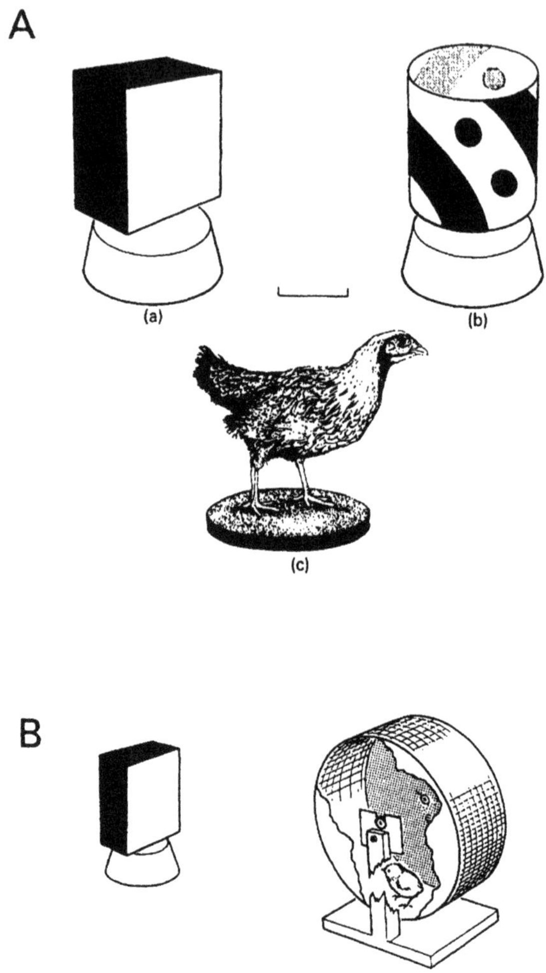

FIGURE 8.1 A. Examples of stimuli used in laboratory experiments investigating the development of filial preferences in chicks. All of the stimuli are mounted on motors with a central axle that can rotate. The box (a) and the cylinder (b) are made of perspex. They have colored filters where they are not black, and they are lit from within. Stimulus (c) is a stuffed Burmese red jungle fowl hen (*Gallus gallus spadiceus*). Scale bar is 10 cm. B. Diagram of a chick in a running wheel. The floor of the running wheel is made of wire mesh, and the sides are made of wood, painted matt black. The diameter of the wheel is 30 cm. For the purposes of illustration one of the opaque sides of the wheel is shown partly removed. In this example the chick can view an imprinting stimulus (in this case a rotating box as in Figure 1A) through the wire mesh floor. The stimulus is approximately 50 cm away from the center of the running wheel. When attempting to approach the stimulus or to move away, the chick rotates the wheel. In many of the experiments discussed in this chapter the chicks were placed in running wheels such as this one in darkness, without an imprinting stimulus present. Adapted from Horn (1985), by permission of the Oxford University Press.

The Development of Filial Preferences in Precocial Birds

In research into the formation of social preferences a distinction is made between filial and sexual imprinting. Filial imprinting is involved in the formation, in young animals, of an attachment to, and a preference for the parent, parent-surrogate or siblings. Sexual imprinting is involved in the formation of mating preferences that are expressed in later life; it will not be discussed in detail in this chapter (see ten Cate, 1989b; Bischof, 1994, for recent reviews). The formation of filial preferences has been studied mainly in precocial birds such as ducklings and chicks. These birds can move about shortly after hatching and will approach and follow a conspicuous object to which they are exposed. In a natural situation this object is likely to be the bird's mother or sibling, but in the absence of these, a wide range of inanimate mother-surrogates can also elicit filial behavior (see Bateson, 1966; Sluckin, 1972; Bolhuis, 1991, for reviews). When e.g. a chick is close to an appropriate object it will approach it and attempt to snuggle up to it. When the object is removed the bird becomes restless and emits shrill calls. Initially the young bird approaches a wide range of conspicuous objects to which it is exposed, but after it has had a certain amount of experience with a particular object the bird remains close to it and avoids novel ones. When given a choice between the familiar stimulus and a novel one the bird preferentially approaches the familiar stimulus. Filial imprinting is defined as the process through which social behavior of the young animal becomes limited to a particular object or class of objects, as a result of exposure to an object (Bateson, 1966; Bolhuis, 1991). This contemporary definition actually differs from Lorenz's (1937) original concepts, an issue to which we will return later.

To study imprinting in the laboratory, chicks or ducklings may be hatched and reared in darkness, and exposed to a conspicuous object (see Figure 8.1A) when they are about 24 h old (see Horn, 1985). Chicks may be exposed to a stimulus (training) while they are in individual running wheels, as illustrated in Figure 8.1B. The chick can see the imprinting object through the wire mesh floor of the running wheel. After imprinting training the animals are returned to a holding incubator and kept in darkness, until their preferences are measured in a choice test involving sequential or simultaneous exposure to the training stimulus and a novel stimulus. Although other measures have also been used, filial preferences are often expressed as the relative approach (or attempted approach) to the training stimulus, compared to the approach to the training stimulus and a novel stimulus. Bateson (1966) concluded that, in general, the more conspicuous an object is to the human eye, the more effective it will be as an imprinting stimulus. Movement of the object increases its attractiveness (e.g. ten Cate, 1989a) as does the presentation of an auditory stimulus with a visual stimulus (e.g. Van Kampen and Bolhuis, 1993; see below). Imprinting is not an instantaneous process: it takes ≥ 1 hour of exposure for a chick to achieve a significant preference for a simple artificial object, depending on the object. However, a significant preference can be achieved with a considerably shorter period of exposure when the imprinting stimulus is presented with a variable duration / variable interval schedule (Bolhuis and Johnson, 1988).

Learning Mechanisms in Filial Imprinting

Lorenz (1935) initially suggested that the process of acquisition of knowledge about the object to which social behavior becomes directed 'has nothing to do with learning' (p. 378: " [...] der mit Lernen nichts zu tun hat"), mainly because it was thought to occur within a sensitive period and to be irreversible. However, in the later English version of his paper he suggested that "Of course, it is a matter of personal opinion how much or how little importance one should attribute to these differences from associative learning" (Lorenz, 1937, p. 266); and we shall see that when the label 'imprinting' is removed, Lorenz's views on the learning processes involved appear to be similar to current interpretations. There are two main reasons for this apparent reconciliation of Lorenz's views with contemporary evidence. First, it has become clear that the criteria of a sensitive period and irreversibility are not as strict as Lorenz suggested (see Bateson, 1966; Bolhuis, 1991, for reviews), although they may still play a role (e.g. Cherfas and Scott, 1981; Bolhuis and Bateson, 1990; Cook, 1993; Johnson et al. 1989). Thus, these criteria do not necessarily distinguish imprinting from other forms of learning (Bolhuis et al, 1990). Second, Lorenz (1937, p. 262) actually distinguished between imprinting-- which he saw as the acquisition of a 'perceptual pattern' of the species-- on the one hand, and "an acquiring process, that is [..] conditioning in a broad sense of the word", on the other. Lorenz went on to say that the latter process is "true associative learning which unites the different sets of releasing stimuli emanating from an individual companion" (ibid.). Thus, Lorenz clearly acknowledged that there was a process of acquisition of knowledge of the individual characteristics of stimuli. In fact, it is this process which is called 'filial imprinting' by contemporary researchers (e.g. Bateson, 1990; Horn, 1985; Bolhuis, 1991).

The question remains whether filial and sexual imprinting involve the acquisition of a preference for individuals, or for the species (as Lorenz maintained), and whether the latter involves a special process.

As regards filial preferences it has been shown that chicks can acquire preferences for individual stuffed fowl (Johnson and Horn, 1987) and ducklings and chicks can learn to prefer maternal calls from particular individuals (Gottlieb, 1988; Van Kampen and Bolhuis, 1991; see Bolhuis and van Kampen, 1992, for a review). Thus, imprinting is now considered to be concerned with the acquisition of preferences for individuals (cf. Bateson, 1990), as well as for artificial objects (Horn, 1985). Further, birds from a number of precocial species have been shown to have naive preferences or predispositions that appear to be species-specific, both in the auditory (Gottlieb, 1971) and in the visual modality (Lickliter, 1993; Bolhuis and Hampton, 1994). Thus, imprinting is not necessary to achieve a filial preference for the animal's own species. It is often suggested that sexual imprinting involves the acquisition of a preference for the species (e.g. ten Cate, 1989b) although in the sexual context too, birds can acquire preferences for individuals (Bateson, 1982; Bolhuis et al. 1989b), and an apparent preference for the species could be a secondary consequence of a preference for individuals (Bolhuis, 1991).

It has become clear that filial imprinting does not involve simple stimulus-response associations, but that a representation of the training stimulus is formed, that can be used flexibly. For instance, Honey et al. (1993) found that following filial imprinting, chicks showed significantly faster rates of discrimination learning if one of the discriminanda was the imprinting stimulus. This is despite the fact that approach to the imprinting stimulus was not rewarded in the discrimination task. Some indication of the nature of the representation of the imprinting stimulus was provided by Bolhuis and Horn (1992), who demonstrated that there is considerable generalization of learned preferences in imprinting.

Imprinting and Associative Learning

Bolhuis et al. (1990) reviewed some of the evidence for an associative learning account of imprinting, and formulated the requirements for such an interpretation. The most important assumption is that, since filial imprinting proceeds in the absence of a conventional reinforcer, the imprinting object itself must provide the unconditioned stimulus (US). However, the nature of this US is not apparent. Hoffman and his collaborators have formulated a theory of imprinting as Pavlovian conditioning (Hoffman and Ratner, 1973), in which movement is thought to act as the US. During exposure, the bird associates initially neutral aspects of the object (e.g. color and shape) with the US (movement).

Bolhuis et al. (1990) reviewed evidence showing that, if it is assumed that movement of an imprinting object is the US, certain predictions from animal learning theory, such as latent inhibition and extinction are not confirmed by the data available. Furthermore, several studies have shown that movement of the imprinting object is not necessary for imprinting to occur. Exposure to a conspicuous object (CS) may lead to the occurrence of an internal motivationally significant event, which could serve as a US. Experiments in which presentation of a conspicuous object is used as a US in an operant learning task (e.g. Bateson and Reese, 1968) show that presentation of an imprinting object can be reinforcing in itself. Assuming then that a US occurs at presentation of an imprinting object, certain predictions can be made with regard to the occurrence of associative learning phenomena such as blocking and overshadowing during imprinting. Basically, these involve impairment of learning with one stimulus as a result of simultaneous presentation with another stimulus. Experimental results (de Vos and Bolhuis, 1990; Van Kampen, 1993) are consistent with these predictions, but further research is needed to establish the nature of the underlying mechanisms.

Under natural conditions, the chick is exposed to an object that has both visual and auditory characteristics; the mother hen has a particular size, shape and coloration and emits a specific call. Similarly, during imprinting training in the laboratory the young birds are typically exposed to a visual stimulus accompanied by a recording of the maternal call of a hen. The presence of a call during visual imprinting training can enhance the preference acquired for the visual stimulus. Van Kampen and Bolhuis (1993) exposed chicks to either a visual stimulus together with a maternal call, or to the visual stimulus alone. Subsequently, both groups received a preference test in which the visual training stimulus and a novel visual stimulus were presented without a maternal call. Chicks given the maternal call during training showed a significantly greater preference for the training stimulus than those that had not. In a similar way, the acquisition of a preference for a specific maternal call (Gottlieb, 1988) can be enhanced if that call is accompanied by a visual stimulus during training (Van Kampen and Bolhuis, 1991; see Van Kampen and de Vos, 1991, 1994, for similar results within the visual modality). An interesting interpretation of these findings is that a similar mechanism might underlie the facilitation of imprinting and the potentiation of learning observed when subjects receive conditioning with a simultaneous compound (for a review see LoLordo and Droungas, 1989). For instance, in rats, aversive conditioning of an odor is enhanced when that odor is paired with a taste during the conditioning session. Rescorla and Durlach (1981) have suggested that such potentiation may be a result of within-event learning. That is, during conditioning associations may form between the representations of the two elements of the compound stimulus. Similarly, during imprinting training with an audio-visual compound, representations of the auditory and visual components of the compound may become associated. The formation of such an association will mean that during a test in which only one of the components is presented there will be two potential sources of preferential responding-- that are acquired by the tested component itself and that are due to the associatively-activated representation of the other component of the compound (see Rescorla and Durlach, 1981; Bolhuis and Honey, 1994).

Bolhuis and Honey (1994) tested some of the predictions of such a within-event learning account of potentiation of imprinting. They exposed chicks to a compound stimulus containing an auditory element (a taped maternal call) and a visual element (a moving two-dimensional image). The potentiation effect was replicated in subsequent preference tests with the training visual stimulus and a novel visual stimulus. That is, chicks that had been exposed to the compound stimulus showed a significantly greater preference for the training (visual) stimulus than chicks that had been exposed to the visual stimulus only. The authors found that additional imprinting training with the call reduced the preference for the visual stimulus, whether exposure to the call occurred after or before exposure to the audio-visual compound. These results are analogous to findings in similar ('sensory preconditioning') paradigms in rats (Rescorla and Durlach, 1981). They support the suggestion that the potentiation of imprinting is a product of within-event learning that is undermined when one element of the compound is presented in isolation.

Configural Representations

Rescorla and Durlach (1981) and Bolhuis and Honey (1994) suggested an alternative interpretation of their respective findings, in which the representation of a compound stimulus is thought to be configurational. That is, under certain circumstances the elements of e.g. an audio-visual compound stimulus might come to address a single or unitary representation, and activity in this representation controls performance. Such an interpretation is accommodated by a neural net model of imprinting and object recognition that was developed by Bateson and Horn (1994; cf. O'Reilly and Johnson, 1994). Figure 8.2 illustrates the simplest version of the model, consisting of three layers, an Analysis, Recognition and an Executive system, respectively. A basic feature of the network is that links between modules are strengthened when there is simultaneous activity in those modules, and that links are weakened when activity in a Recognition module is not accompanied by activity in

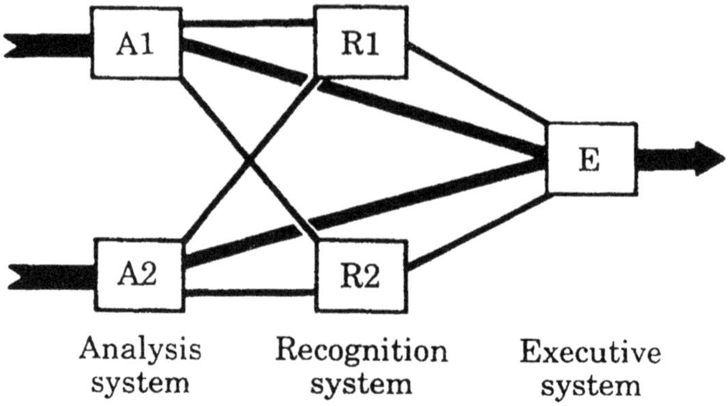

FIGURE 8.2 Flow diagram of the simplest version of the neural net model by Bateson and Horn (1994). The network consists of three systems (Analysis, Recognition and Execution) that contain modules. For simplicity only 2 Analysis modules and 2 Recognition modules are shown. All modules in the Analysis system are linked to all modules in the Recognition system which, in turn, are linked to all modules in the Executive system. All modules in the Analysis system are also linked directly to the module in the Executive system that controls filial approach behavior. The strength of the links connecting the modules is reflected in the thickness of the lines. From Bateson and Horn (1994), by permission of Academic Press.

an Analysis module. Suppose that the auditory and the visual stimulus activate analysis modules A1 and A2, respectively. If one assumes that the dual input provided by the elements of the compound stimulus enhances the subsequent likelihood for either element to activate a module in the Recognition system, say R1, then compound training would result in a potentiation effect. That is, the links between A1 and R1 and between A2 and R1 become stronger than they would have been when either of the stimuli had been presented alone. To account for the observation that separate exposure to an element of the compound reduces the potentiation effect (Bolhuis and Honey, 1994) one would need to assume that if activity in R1 is not accompanied by the presentation of one of the elements, then its capacity to activate R1 will be reduced. Given that presentations of the auditory stimulus alone will produce activity in R1 that is unaccompanied by input from the visual stimulus, it is to be expected that input from the visual stimulus will become less able to activate R1.

The results of Bolhuis and Honey (1994), van Kampen and de Vos (1991, 1993) and those of van Kampen and Bolhuis (1991, 1993) are remarkably consistent with Lorenz's suggestion, quoted above, concerning the integration of stimulus elements into a unified 'perceptual pattern' involved in imprinting on individual stimuli.

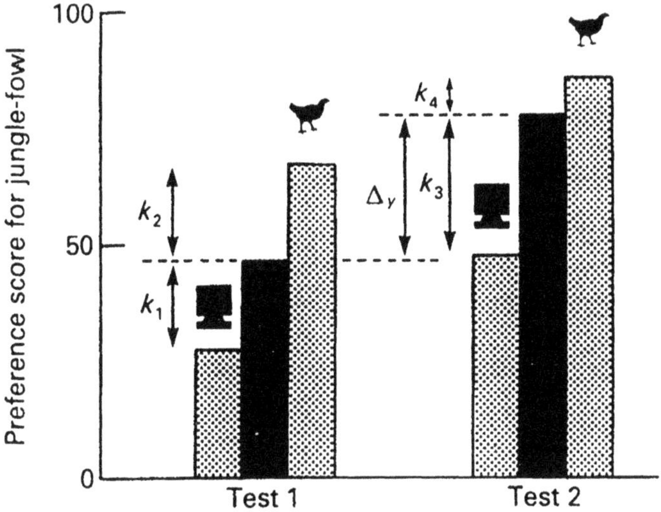

FIGURE 8.3 Mean preference scores, expressed as a preference for the stuffed fowl, of chicks previously trained by exposure to a rotating stuffed jungle fowl, a rotating red box, or exposed to white light. Preference scores are defined as: activity when attempting to approach the stuffed jungle fowl divided by total approach activity. Preferences were measured in a simultaneous test either 2 h (Test 1) or 24 h (Test 2) after the end of training. K1-K4 represent the differences between the preferences of the trained chicks and the controls; ΔY represents the difference in preference between the control chicks at Test 2 and at Test 1. See text for further explanation. (From Horn (1985), by permission of the Oxford University Press, After Johnson et al. 1985).

Predispositions

As outlined in the Introduction of this chapter, the development of filial behavior in the chick involves two perceptual mechanisms that are neurally and behaviorally dissociable (Horn and McCabe, 1984; Horn, 1985; Johnson et al. 1985, 1992; Bolhuis et al. 1985, 1989a). I have discussed some of the characteristics of the first mechanism, which underlies filial imprinting. Behavioral evidence for the second mechanism, that of a predisposition (Horn and McCabe, 1984), was provided by Johnson et al. (1985). These authors gave day-old dark-reared chicks imprinting training by exposing them to either a rotating red box or to a rotating stuffed jungle fowl hen (*Gallus gallus spadiceus;* see Figure 8.1). The approach preferences of the chicks were measured in a subsequent test where the two training stimuli were presented simultaneously. Preferences were tested at either 2 h (Test 1) or 24 h (Test 2) after the end of training (see Figure 8.3). At Test 1 the chicks preferred the object to which they had been exposed previously. At Test 2 there was a significantly greater preference for the fowl in both experimental groups. Thus, the preference for the jungle fowl increased from the 2 h to the 24 h test, and did so regardless of the stimulus with which the chicks had been trained. Chicks in a control group that had been placed in the training apparatus with no stimulus except for an overhead light also showed an increasing preference for the jungle fowl with time in preference tests conducted at the same time as in the experimental groups.

Johnson et al. (1985) proposed that the preferences of trained chicks are influenced by at least two different systems. On the one hand, there is an effect of experience with particular stimuli (reflected in the differences k1-k4 in Figure 8.3), that is, filial imprinting. On the other hand, there is an emerging predisposition to approach stimuli resembling conspecifics (reflected in Figure 8.3 as Δy for the control group). The existence of these two interacting mechanisms was confirmed in a series of behavioral studies (e.g. Bolhuis and Trooster, 1988; Bolhuis et al. 1989a; Johnson et al. 1992). The results in Figure 8.3 suggest that training with a particular stimulus is not necessary for the predisposition to emerge. It was found that the predisposition can in fact emerge in dark-reared chicks, provided that they receive a certain amount of non-specific stimulation within a certain period in development (e.g. Bolhuis et al. 1985; Johnson et al. 1989; see below).

Induction and Development of a Filial Predisposition

Bolhuis et al. (1985) found that visual experience was not necessary to 'trigger' or induce (Gottlieb, 1976) the predisposition. Chicks that were placed in wire mesh running wheels (see Figure 8.1B) in total darkness for 2 periods of 1 h preferred the jungle fowl to the red box in a simultaneous choice test 24 h after the end of the period in the wheels. Placement in the running wheels, in darkness, for 2 periods of 1 h (with 1 h in between during which the chicks were kept in a dark incubator) is a standard procedure that is used in many of the experiments discussed here, and that is sufficient to induce a predisposition. However, Bolhuis et al. (1985) found that chicks that had remained in their dark incubator from the time of hatching to the time of testing did not develop a preference. Therefore, some experience associated with being placed in the running wheels is sufficient to trigger the emergence of the preference for the jungle fowl.

In the study by Bolhuis et al. (1985) the period in which the chicks were in the running wheels, in darkness, began approximately 24 h after the birds had hatched. A significant preference for the stuffed jungle fowl was found 24 h, but not 2 h after the end of the period in running wheels. Hampton et al. (1995) investigated whether 24 hours need to pass in order for the underlying mechanisms to develop or to exert their influence on the chicks' behavior. It was found that the predisposition becomes manifest at 10 h and 5 h either after placement in running wheels, or after the chicks had been trained by exposure to a rotating red box. Further, these authors showed that placement in running wheels was not necessary for the predisposition to emerge. Chicks that were taken out of their incubator and handled at the same times that the wheel-treated chicks would normally be, showed a significant preference for the fowl over the red box at 5 h, but not at 2 h after the end of this experience.

Lickliter and his colleagues provided evidence to suggest that in the development of filial preferences in Bobwhite quail (*Colinus virginianus*), the expression of auditory predispositions is affected by visual experience, and *vice versa* (see Lickliter 1993, for a review). Hampton et al (1995) investigated whether, similarly, auditory stimulation is sufficient for the induction of the visual predisposition in the chick. Exposing 24-old dark-reared chicks to a maternal call of a chicken for 2 h was found to be sufficient to induce the development of a visual predisposition 24 h later. Thus, the evidence from the study by Hampton et al. (1995) suggests that different kinds of experience, in different modalities, are sufficient for the induction of the predisposition.

Hampton et al.'s findings may be similar to those in studies of the development of filial preferences in Bobwhite quail by Lickliter and his colleagues (e.g. Lickliter, 1993). These authors reported that both visual (Lickliter, 1990) and auditory (Lickliter and Stoumbos, 1991) stimulation given before hatching affected the time of expression of a species-specific visual preference. Lickliter and co-workers have reported either a facilitation (Gottlieb, 1976, see below) of, or interference with, perceptual development after auditory or visual stimulation (e.g. Banker and Lickliter, 1993; Lickliter, 1990; Lickliter and Stoumbos, 1991; McBride and Lickliter, 1994). In Bobwhite quail chicks that received standard rearing conditions, without additional auditory or visual experience, a species-specific visual preference developed normally (Banker and Lickliter, 1993). In the studies by Lickliter et al. standard rearing conditions involved placement in social groups immediately after hatching, and visual exposure to siblings in the incubator. Socially reared quail chicks that were visually deprived until testing did not show a species-specific visual preference (Banker and Lickliter, 1993), suggesting that the non-visual experience involved in social rearing is not sufficient to induce a visual predisposition in Bobwhite quail. McBride and Lickliter (1993) found that exposure to the visual features of siblings, but not to those of scaled quail chicks (*Callipepla squamata*) was sufficient for a species-specific visual preference to be expressed. The authors note that Bobwhite quail chicks are marked and colored much like adult Bobwhites. Thus, in this case, very specific visual experience is required for a species-specific preference to be expressed. Hampton et al. (1995) suggested that the Bobwhite quail's preferences may be a result of generalization of learning of the visual characteristics of siblings, rather than of a predisposition, which may develop in the absence of specific experience.

Sensitive Periods

As mentioned earlier in this chapter, Lorenz (1935, 1937) suggested that one of the characteristics that made imprinting unique was that it occurred within a 'critical period'. Bateson and Hinde (1987) reviewed some of the relevant evidence and suggested that rather than 'critical period', which implies a fixed period, the term 'sensitive period' should be used. In their words "The sensitive period concept implies a phase of great susceptibility preceded and followed by lower sensitivity, with relatively gradual transitions" (p. 20). These authors concluded that external experience played an important role in changes in 'imprintibility', and that the end of an apparent sensitive period was largely due to the imprinting experience itself (cf. Sluckin and Salzen, 1961), and not so much a consequence of endogenous changes.

In dark-reared chicks there is a sensitive period for the induction of the filial predisposition. Johnson et al. (1989) found that dark-reared chicks which were placed in the running wheels (for 2 h) at 12 h, 42 h or 48 h after hatching did not show a significant preference in a preference test 24 h later, but chicks that received a 2 h period in the running wheels at 24 h or 36 h after hatching significantly preferred the stuffed jungle fowl to the red box. It was not clear from these results whether the crucial factor was the time at which the chicks were placed in the wheels, or when they were tested. For example, chicks placed in the wheels 48 h after hatching were 72 h old when tested; chicks placed in the wheels 24 h after hatching were 48 h old when tested. The former did not show a preference for the jungle fowl whereas the latter did. To clarify this issue, chicks in a control group were placed in the wheels at 24 h after hatching and given a preference test 72 h after hatching (i.e. at the same time as the chicks in the 48 h wheel placement group). These chicks showed a significant preference for the jungle fowl, demonstrating that the crucial factor is the time at which the chicks are placed in the wheels, not the time of testing. Thus, in dark-reared chicks there is a sensitive period approximately between 20-40 h after hatching during which placement in the wheels may induce the emergence of the predisposition. There is a parallel of this finding in work on auditory predispositions in ducklings. Gottlieb (1985) found that duck embryos needed to be exposed to embryonic calls within a certain period of time to ensure maintenance of species-specific auditory preferences after hatching.

The results of a study by Bolhuis et al. (1989a) were consistent with the existence of a sensitive period for the induction of a predisposition. When chicks were placed in running wheels for 2 x 1 h, in darkness, at approximately 30 h after hatching, exposure to a rotating red box at approximately 70 h after hatching did not result in a significant preference in a simultaneous test with the red box and stuffed jungle fowl. However, when the chicks were placed in the wheels at 45 h after hatching, subsequent box-training at 70 h after hatching resulted in a significant preference for the red box over the stuffed fowl. Thus, in chicks that received non-specific stimulation within the sensitive period, the learned preference had to compete with the effects of the predisposition; when the non-specific experience was given after the end of the sensitive period, when preferences are presumably not influenced by the predisposition, the chicks' preferences were determined by the birds' training experience.

The sensitive period for the induction of the predisposition that was demonstrated in dark-reared chicks (Johnson et al, 1989) is by no means absolute. Visual experience may alter the sensitive period (Hampton, 1995), while manipulation of catecholamine levels in the brain (Davies et al. 1992) or anesthesia (Bolhuis and Horn, 1995) may delay its onset (see below).

Interaction Between Perceptual Mechanisms

The results of Johnson et al. (1985) are interpreted as reflecting an interaction between two perceptual mechanisms. The chance-level performance of the box-trained chicks at the second preference test (Figure 8.3) is thought to result from an interaction between the effects of training to the red box and an emerging predisposition for the stuffed fowl. There are, however, other explanations for these (Figure 8.3) results. For example, box-trained chicks may have learned less, or processed the acquired information about the red box less deeply, or have forgotten more quickly, than was the case for chicks exposed to the stuffed fowl. Thus, for one or all of the preceding reasons, by the time the second preference test was given, the red box-trained birds failed to show a preference for the red box, whereas the fowl-trained birds still had a strong preference for their training object. Johnson et al. (1992) addressed this hypothesis in a study where they repeated the training procedure with a red box. However, rather than the preference test consisting of a choice between the red box and a stuffed fowl as before, the chicks were given two choice tests: one between the box to which they had been exposed and a box of a novel color, and one between the familiar box and a novel stuffed fowl. There was a decline in preference for the red box over the fowl from the first to the second test, as before. However, the acquired preference for the red box over a blue box was robust even in the second preference test. Thus, these results replicated previous findings and suggested that they cannot be interpreted in terms of differential forgetting.

Bolhuis et al. (1989a) investigated whether the predisposition served as a filter or 'template' (Marler, 1976), restricting the information that the chick can store in memory, or, alternatively, whether the predisposition acts independently of the effects of learning during development. Dark-reared chicks were allowed to develop the predisposition, manifest as a significant preference for the stuffed fowl over the red box, by placing them in running wheels in darkness for 2 h when they were approximately 24 h old. When the chicks were approximately 50 h old, they were trained by exposing them to the rotating red box. In subsequent preference tests, it was found that these chicks, after the predisposition had developed, were still capable of learning the characteristics of conspicuous objects. These results suggest that the two mechanisms interact at the behavioral level rather than the predisposition constraining the information that can be stored. That is, the activated predisposition will bias the chick's approach response to certain stimuli, and the animal may learn the characteristics of those stimuli; but the chick is still capable of learning the characteristics of other objects.

Stimulus Selection in Filial Predispositions

Baerends and Kruijt (1973) defined stimulus selection as "(..) the phenomenon whereby the various physical stimuli activating the receptors of an organism are not quantitatively equivalent with regard to their controlling effect on a particular behavior pattern" (p. 23). From an extensive series of experiments on the egg-retrieving response of the herring gull (*Larus argentatus*), they concluded that the response to stimulus eggs obeyed the principle of heterogeneous summation, whereby different stimulus characteristics (color, speckling, size) contribute to the induction of the response independently (cf. Baerends et al. 1982; see Baerends, 1993 for similar findings in filial behavior of cichlid fish).

The stimulus characteristics that are important for the filial predisposition to be expressed were first examined by Johnson and Horn (1988). These authors attempted to 'titrate' the visual complexity of the stuffed jungle fowl against its 'fowl-like' characteristics. To this end, the preferences of dark-reared chicks, which had been placed in running wheels for 2 h when they were approximately 24 h old, were tested either 2 or 24 h later. Stimuli used during the tests were an intact stuffed jungle fowl versus one in a series of increasingly degraded versions of a stuffed jungle fowl. The degraded versions ranged from one where different

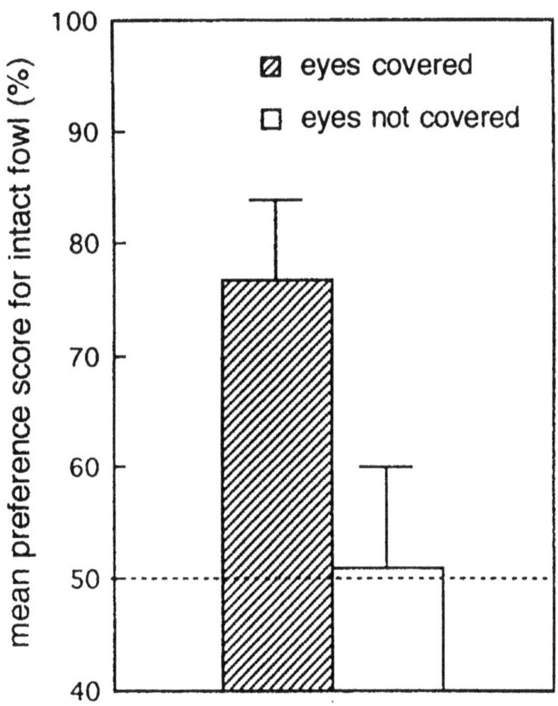

FIGURE 8.4 Mean preference scores (see Figure 3) in one of the experiments by Hampton (1995). Dark-reared chicks were allowed to develop a predisposition, then at approximately 50 h after hatching approach preferences were examined in a test involving the simultaneous presentation of two stuffed jungle fowls. Scores are expressed as preferences for the intact stuffed jungle fowl over the alternative stuffed jungle fowl. In the case of the 'eyes covered' group, the alternative stimulus was another stuffed jungle fowl which had its eyes covered with patches of jungle fowl feathers. In the 'eyes not covered' group, the alternative stimulus was another intact stuffed jungle fowl. The results of chicks that were active during the preference test are shown only. See text for further explanation. Adapted from Hampton (1995), with permission.

parts of the model (wings, head, torso, legs) were reassembled in an unnatural way, to one in which the pelt of a jungle fowl had been cut into small pieces that were stuck onto a rotating box. It was found that the intact model was preferred at the second test, when the chicks were approximately 52 h old, only when the degraded object possessed no distinguishable jungle fowl features. In addition, Johnson and Horn (1988) demonstrated that chicks did not prefer an intact jungle fowl model over an alternative object that contained only the head and neck of a stuffed fowl. Thus, it seems likely that the head and neck region contains stimuli that are relevant for the predisposition. In subsequent experiments, Johnson and Horn (1988) found that the chicks did not prefer a stuffed jungle fowl hen over a stuffed Gadwall duck (*Anas strepera*) or even a stuffed polecat (*Mustela putorius*). Thus, the predisposition is not species- or even class specific.

Hampton (1995) provided dark-reared chicks with non-specific stimulation (placement in running wheels) that would induce the emergence of the predisposition, and 24 h later their preferences were tested in simultaneous tests involving different pairs of stimuli. One of the test stimuli used was a stuffed jungle fowl hen, of which the eyes had been covered with a patch of color-matched jungle fowl feathers attached to plasticine. Chicks showed an equally strong preference for a stuffed fowl with its eyes covered with patches of feathers and for an intact stuffed fowl, compared to a red box. These findings indicate that areas outside the eyes provide stimulus elements that are sufficient for the predisposition to be expressed. However, in two separate experiments chicks showed a significant preference for an intact stuffed fowl over a stuffed fowl with covered eyes (see Figure 8.4). These results suggest that eyes are an important stimulus, but that other aspects of the stimulus are also sufficient for the expression of a predisposition.

Thus, it appears that the eyes are not a sign stimulus that is uniquely important for the expression of the predisposition. It may be that stimuli outside the head also play a role in the expression of the predisposition. It is possible that, as in the work of Baerends and collaborators (Baerends and Kruijt, 1973; Baerends et al. 1982; Baerends, 1993), the stimulus selection involved in the predisposition conforms to the rule of heterogeneous summation, with the eyes forming an important stimulus, as well as other stimuli. Alternatively, it is conceivable that the chicks' response is governed by a combination of a configurational stimulus (Tinbergen, 1951), that needs to have a head and neck as well as a body, with added to that certain stimulus elements (such as the eyes) that are separately important. The bird's response is then governed by the rule of heterogeneous summation. This interpretation is consistent with

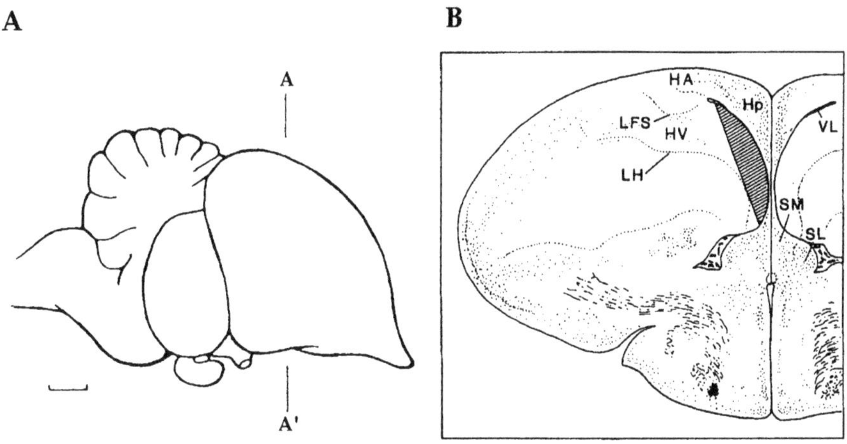

FIGURE 8.5 Schematic drawings of the brain of the domestic chick. A. Lateral aspect of the brain. The vertical lines aa′ indicate the approximate plane of the coronal section outline of the brain in B. Scale bar: 2 mm. Adapted from Horn (1991b). According to Horn (1991a), the rostro-caudal extent of the IMHV is roughly from A 6.4 to A 9.6 in the stereotaxic atlas by Kuenzel and Masson (1988). B. Simplified diagram of a coronal section of the chick brain at the level of the IMHV (Approximately A 7.6 in the atlas by Kuenzel and Masson (1988)), adapted from Ambalavanar et al. (1993). The extent in the coronal plane of the left IMHV, as removed in biochemical studies (see Horn, 1985) is indicated by the hatched area. From Bolhuis (1994), by permission of Cambridge University Press.

Tinbergen's views, who suggested that configurational sign stimuli may 'cooperate according to the rule of heterogeneous summation' (Tinbergen, 1951, p. 101).

Neural Mechanisms of Imprinting and Predispositions

The neural mechanisms of learning and memory processes involved in filial imprinting have been investigated by Horn and his collaborators (Horn, 1985). This work has been reviewed elsewhere (e.g. Horn, 1985, 1992), and will be discussed only briefly here, mainly with regard to the distinction between the mechanisms involved in learning and in predispositions (see also Bolhuis, 1994).

Imprinting is associated with a number of neural changes in a restricted region of the chick forebrain referred to as the intermediate and medial hyperstriatum ventrale (IMHV; see Figure 8.5). These changes include an increase in RNA synthesis, an increase in the size of postsynaptic densities of a certain class of synapses, an increase in N-methyl-D-aspartate receptor binding and an increase in phosphorylation of MARCKS protein, a protein kinase C substrate (Horn, 1985, 1992; Sheu et al. 1993). Imprinting also leads to a learning-related increase in immunoreactivity of the immediate early gene product Fos in the IMHV (McCabe and Horn, 1994; cf. Ambalavanar et al. 1993). Bilateral lesions to the chick IMHV impair both acquisition and retention of filial preferences. However, Horn and McCabe (1984) reported that the effects of lesions to the IMHV were dependent on the stimulus used for imprinting training. IMHV lesions greatly impaired filial preferences of chicks trained with a red box, but had a relatively small (but significant) effect on the preferences of chicks trained by exposure to the stuffed fowl. These findings led to the suggestion that the chicks' preferences are influenced not only by learning, but also by a predisposition to approach certain stimuli such as the stuffed fowl.

Although lesion studies led to the original discovery in the chick of the emerging predisposition (Horn and McCabe, 1984), little is known about the underlying neural and physiological mechanisms. Johnson and Horn (1986) found that the development of the predisposition was not affected by early bilateral lesions to the IMHV. Thus, it is likely that the neural substrate involved in the predisposition is located outside the IMHV. However, recent electrophysiological findings (Brown and Horn, 1994; cf. Nicol et al. 1995) suggest that this neural substrate may influence neuronal activity in the IMHV. Brown and Horn (1994) trained dark-reared chicks by exposing them to a rotating red box or a rotating blue cylinder (see Figure 8.1). One day after the end of imprinting training, recordings were made from groups of neurons in the left IMHV. There was an effect of training on the response to presentation of a stuffed jungle fowl. A significantly higher proportion of responses to the fowl were excitatory in trained chicks (68%), irrespective of the training object, than in dark-reared chicks (7%). The chicks had received imprinting training with a compound stimulus of the visual imprinting object and the taped call of a hen. It is known that such compound training can lead to potentiation of learning about the elements of the compound (Van Kampen and Bolhuis, 1991; Bolhuis and Honey, 1994). It is also possible that exposure to the call had an inducing effect on predispositions (Bolhuis and Hampton, 1994), thereby influencing the response of neurones in the IMHV to presentation of the fowl during test.

Davies et al. (1992) found that injections of the catecholaminergic neurotoxin DSP4 delayed the onset of the sensitive period for the induction of the predisposition in dark-reared chicks. Chicks were given intraperitoneal injections of DSP4 or vehicle shortly after hatching. The chicks were then placed in running wheels for 2 x 1 h, in darkness, at a particular time after hatching, and their filial preferences were tested 24 h later. Unlike vehicle-injected controls, DSP4-treated chicks did not show a significant preference for the jungle fowl over the red box when they had been placed in the wheels at 24 h or 36 h after hatching. However, DSP4-treated chicks showed a significant preference for the fowl when they had been placed in the wheels at 42 h or 48 h after hatching, i.e. at times when this wheel placement did not induce a significant preference in the controls. Preliminary results indicate that a similar delay in the onset of the sensitive period is achieved after equithesin anesthesia (Bolhuis and Horn, 1995). After the preference tests, Davies et al (1992) determined the levels of dopamine and noradrenaline in forebrain samples containing predominantly IMHV. In control chicks there was a significant negative correlation between forebrain dopamine concentration and preference score for the fowl; no such correlation was found for noradrenaline. As the authors note, if dopamine is involved in the expression of the predisposition, it is likely to involve brain regions outside the IMHV, as lesions to this structure do not affect the predisposition (Johnson and Horn, 1986).

It is not known whether catecholamines are involved in the induction of the predisposition. It has been suggested that androgens may have such a role (Bolhuis et al. 1985, 1986; Horn, 1985). Bolhuis et al. (1986) reported that in chicks trained by exposure to a stuffed jungle fowl, there was a significant positive correlation between plasma levels of testosterone and preference score for the stuffed fowl. No such relationship was found in chicks trained by exposure to a rotating red box. Further, subcutaneous injections of a long-lasting testosterone ester shortly after hatching significantly increased the mean preference score of chicks trained on a stuffed fowl, not that of red box-trained chicks. These findings suggest that the preference for the stuffed fowl is limited by the concentration of androgens in the plasma, and they led to the suggestion (Bolhuis et al. 1985; Horn, 1985) that testosterone may be involved in the development of the predisposition. Horn (1985) suggested that increased levels of testosterone or other hormones could render the systems involved in the predisposition operational.

The experience of handling, placement in running wheels, or exposure to a call, might lead to an increase in plasma levels of certain hormones, including testosterone, or catecholamines (Davies et al. 1992). It is known that under certain circumstances, handling leads to an increase in the plasma levels of corticosteroids in laying hens (Beuving and Vonder, 1978) and androgens in Japanese quail, *Coturnix coturnix japonica* (Ramenofsky, 1985). Within 0.5 min. of being placed into an arena, male Japanese quail showed an increase in plasma testosterone concentration which lasted for approximately 7 min. Further experiments are needed to investigate the role of androgens in the development of the predisposition.

Predispositions, Imprinting, and the Development of Behavior

From a functional perspective, Bolhuis et al. (1985) suggested that a filial predisposition may ensure that the young animal directs its attention to a particular class of objects (i.e. stimuli with a head and neck (Johnson and Horn, 1988) or indeed to conspecifics (Gottlieb, 1971; Lickliter, 1993; Bolhuis and Hampton, 1994). After the predisposition has fully developed, learning the visual and auditory characteristics of a particular object may lead to the recognition of individuals, in filial as well as in sexual imprinting. It is important to realize how the development of filial preferences might proceed in the natural situation. The dynamics of the process as they have been reported may be peculiar to the laboratory conditions that were used (e.g. rearing in visual and tactile isolation, in the absence of maternal vocalizations), and do not necessarily reflect the normal course of development (cf. Rogers, 1995). In this chapter I have reviewed evidence that suggests that the complex visual and auditory experience that the young bird receives before and after hatching would secure a rapid induction and development of a predisposition. For instance, Hampton et al. (1995) demonstrated that in the laboratory, the predisposition in the chick is manifest between 2 and 5 h after the inducing experience. Furthermore, Bolhuis et al (1985) found that the predisposition is expressed as early as 2 h after complex visual experience. Also, it is likely that in the natural situation, the necessary experience for the induction of the predisposition is available to the young bird at the appropriate time in the 'sensitive period', which, as we have seen, is not a strict and absolute constraint.

The terms maintenance, facilitation, and induction that were used earlier, were introduced by Gottlieb (1976, 1980), to distinguish some of the ways in which experience can affect the development of 'species-typical perception'. All these three possibilities have been shown to occur in the development of filial preferences (Bolhuis, 1991). Hampton et al. (1995) concluded that, within the time course used in their own and related studies, the effects of experience on the development of the predisposition may be categorized as a case of induction; the predisposition is not expressed initially, and only emerges after non-specific experience. Gottlieb's terms are of course descriptive, and do not elucidate the underlying mechanisms (Hampton et al. 1995). Shettleworth (1994) noted that Gottlieb' s classification was introduced as "an alternative to viewing early development in terms of conventional categories of learning" (p. 370). She pointed out the difference between Gottlieb's descriptive concepts and e.g. Rescorla's (1988) theoretical distinctions of learning processes. However, as Shettleworth asserts, Gottlieb's classification may be a useful way of categorizing effects of experience on behavioral development.

The finding that exposure to different kinds of stimuli may lead to the emergence of a predisposition (Hampton et al. 1995) is an example of induction of a perceptual mechanism by non-specific experience (cf. Hogan, 1988, 1994). Similar results have been reported for instance in the development of food recognition in chicks (Hogan, 1977). The work by Lickliter and his colleagues suggests that the same principle applies to the facilitation of the development of perceptual mechanisms. Oetting et al. (1995) provided evidence for maintenance (termed 'consolidation' Kruijt and Meeuwissen, 1991, 1993; Bischof, 1994) or 'stabilization' (Bischof and Clayton, 1991; Oetting et al, 1995) of sexual preferences in the zebra finch through non-specific experience (cf. Bischof, 1994, Kruijt and Meeuwissen, 1993). As Hogan's (1988, 1994) reviews of developmental paradigms suggest, the (inducing, maintaining or facilitating) effect of non-specific, seemingly unrelated experience on specific perceptual mechanisms appears to be a fundamental principle of behavioral development.

Acknowledgments

The research of mine reported in this chapter was supported by a Postdoctoral Research Fellowship from the Biotechnology and Biological Sciences Research Council (UK). I am grateful to Gabriel Horn, Sara Shettleworth, Jerry Hogan and Cynthia Moss for their comments on the manuscript.

References

Ambalavanar, R., Van der Zee, E.A., Bolhuis, J.J., McCabe, B.J. and Horn, G. (1993) Co-expression of Fos immunoreactivity in protein kinase (PKCγ)-positive neurons: quantitative analysis of a brain region involved in learning. *Brain Res. 606:* 315-318.

Baerends, G.P. (1993) A comparative study of stimulus selection in the filial following response of fry of substrate spawning ciclid fish. *Behavior 125:* 79-155.

Baerends, G.P., Blokzijl, G.J. and Kruijt, J.P. (1982) The processing of heterogenous information of the egg features inducing the retrieval response. *Behavior 82:* 212-214.

Baerends, G.P. and Kruijt, J.P. (1973) Stimulus selection. In R.A. Hinde and J. Stevenson-Hinde (Eds.) *Constraints on learning.* New York: McGraw-Hill, pp. 23-50.

Banker, H. and Lickliter, R. (1993) Effects of early and delayed visual experience on intersensory development in bobwhite quail chicks. *Devel. Psychobiol. 26:* 155-170.

Bateson, P.P.G. (1966) The characteristics and context of imprinting. *Biol. Rev. 41:* 177-220.

Bateson, P.P.G. (1982) Preferences for cousins in Japanese quail. *Nature 295:* 236-237.

Bateson, P. (1990) Is imprinting such a special case? *Phil. Trans. Roy. Soc. Lond. B. 329:* 125-131.

Bateson, P. and Hinde, R.A. (1987) Developmental changes in sensitivity to experience. In M.H. Bornstein (Ed.) *Sensitive periods in development.* Hillsdale, NJ. : Lawrence Erlbaum Associates, pp. 19-34.

Bateson, P. and Horn, G. (1994) Imprinting and recognition memory: a neural net model. *Anim. Behav. 48:* 695-715.

Bateson, P.P.G. and Reese, E.P. (1968) Reinforcing properties of conspicuous objects before imprinting has occurred. *Psychon. Sci. 10:* 379-380.

Beuving, G. and Vonder, G.M.A. (1978) Effect of stressing factors on corticosterone levels in the plasma of laying hens. *Gen. Comp. Endocrinol. 35:* 153-159.

Bischof, H.-J. (1994) Sexual imprinting as a two-stage process. In J.A. Hogan and J.J. Bolhuis (Eds.) *Causal mechanisms of behavioural development.* Cambridge: Cambridge University Press, pp. 82-97.

Bischof, H.-J. and Clayton, N.S. (1991) Stabilization of sexual preferences by sexual experience in male zebra finches *Taeniopygia guttata castanotis. Behaviour 118:* 144-155.

Bolhuis, J.J. (1991) Mechanisms of avian imprinting: A review. *Biol. Rev. 66:* 303-345.

Bolhuis, J.J. (1994) Neurobiological analyses of behavioural mechanisms in development. In J.A. Hogan and J.J. Bolhuis (Eds.) *Causal mechanisms of behavioral development.* Cambridge: Cambridge University Press, pp. 16-46.

Bolhuis, J.J. and Bateson, P.P.G. (1990) The importance of being first: A primacy effect in filial imprinting. *Anim. Behav. 40:* 472-483.

Bolhuis, J.J. De Vos, G.J. and Kruijt, J.P. (1990) Filial imprinting and associative learning. *Quart. J. Exp. Psychol. 42B:* 313-329.

Bolhuis, J.J. and Hampton, N.G. (1994) A species-specific predisposition in the domestic chick. *Europ. J. Neurosci. suppl. 7:* 175.

Bolhuis, J.J. and Honey, R.C. (1994) Within-event learning during filial imprinting. *J. Exp. Psychol.: Anim. Behav. Proc. 20:* 240-248.

Bolhuis, J.J. and Horn, G. (1992) Generalization of learned preferences in filial imprinting. *Anim. Behav. 44:* 185-187.

Bolhuis, J.J. and Horn, G. (1995) Effects of anaesthesia on the development of a filial predisposition in the chick. In M. Burrows, T. Matheson, P.L. Newland and H. Schuppe (Eds.) *Nervous systems and behaviour. Proceedings of the 4th International Congress of Neuroethology.* Stuttgart: Georg Thieme Verlag, p. 192.

Bolhuis, J.J. and Johnson, M.H. (1988). Effects of response-contingency and stimulus presentation schedule on imprinting in the chick. *J. Comp. Psychol. 102: 61-65.*

Bolhuis, J.J., Johnson, M.H. and Horn, G. (1985) Effects of early experience on the development of filial preferences in the domestic chick. *Devl. Psychobiol. 18:* 299-308.

Bolhuis, J.J., Johnson, M.H. and Horn, G. (1989a) Interacting mechanisms during the formation of filial preferences: The development of a predisposition does not prevent learning. *J. Exp. Psychol.: Anim. Behav. Proc. 15:* 376-382.

Bolhuis, J.J., Johnson, M.H., Horn, G. and Bateson, P. (1989b) Long-lasting effects of IMHV lesions on social preferences in domestic fowl. *Behav. Neurosci. 103:* 438-441.

Bolhuis, J.J., McCabe, B.J. and Horn, G. (1986) Androgens and imprinting: Differential effects of testosterone on filial preference in the domestic chick. *Behav. Neurosci. 100:* 51-56.

Bolhuis, J.J. and Trooster, W.J. (1988) Reversibility revisited: Stimulus-dependent stability of filial preference in the chick. Anim. Behav. 36: 668-674.

Bolhuis, J.J. and Van Kampen, H.S. (1992) An evaluation of auditory learning in filial imprinting. *Behaviour 122:* 195-230.

Brown, M.W. and Horn, G. (1994) Learning-related alterations in the visual responsiveness of neurons in a memory system of the chick brain. *Europ. J. Neurosci. 6:* 1479-1490.

Cherfas, J. and Scott, A. (1981) Impermanent reversal of imprinting. *Anim. Behav. 30:* 301.

Cook, S.E. (1993) Retention of primary preferences after secondary filial imprinting. *Anim. Behav. 46:* 405-407.

Davies, D.C., Johnson, M.H. and Horn, G. (1992) The effect of the neurotoxin DSP4 on the development of a predisposition in the domestic chick. *Devl. Psychobiol. 25:* 251-259.

De Vos, G.J. and Bolhuis, J.J. (1990) An investigation into blocking of filial imprinting in chicks during exposure to a compound stimulus. *Q. J. Exp. Psychol. 42B:* 289-312.

Gottlieb, G. (1971) *Development of species identification in birds.* Chicago: University of Chicago Press.

Gottlieb, G. (1976) The roles of experience in the development of behavior and the nervous system. In G. Gottlieb (Ed) *Neural and behavioral specificity: studies in the development of behavior and the nervous system.* New York: Academic Press, pp. 237-280.

Gottlieb, G. (1980) Development of species identification in ducklings: VI. Specific embryonic experience required to maintain species-typical perception in Peking ducklings. *J. comp. physiol. Psychol. 94:* 579-587.

Gottlieb, G. (1985) Development of species identification in ducklings: XI. Embryonic critical period for species-typical perception in the hatchling. *Anim. Behav. 33:* 225-233.

Gottlieb, G. (1988) Development of species identification in ducklings: XV. Individual auditory recognition. *Devl. Psychobiol. 21:* 509-522.

Hampton, N.G. (1995) *An investigation of filial predispositions in the domestic chick.* Unpublished dissertation, University of Cambridge.

Hampton, N.G., Bolhuis, J.J. and Horn, G. (1995) Induction and development of a filial predisposition in the chick. *Behaviour 132*: 451-477.

Hoffman, H.S. and Ratner, A.M. (1973) A reinforcement model of imprinting: Implications for socialization in monkeys and men. *Psychol. Rev. 80:* 527-544.

Hogan, J.A. (1977) The ontogeny of food preferences in chicks and other animals. In L.M. Barker, M. Best and M. Domjan (Eds.) *Learning mechanisms in food selection.* Waco, Texas: Baylor University Press, pp. 71-97.
Hogan, J.A. (1988) Cause and function in the development of behavior systems. In E.M. Blass (Ed.) Hand*book of behavioral neurobiology, vol. 9* New York: Plenum Press, pp. 63-106.
Hogan, J.A. (1994) Development of behavior systems. In J.A. Hogan and J.J. Bolhuis (Eds.) *Causal mechanisms of behavioural development.* Cambridge: Cambridge University Press, pp. 242-264.
Honey, R.C., Horn, G. and Bateson, P. (1993) Perceptual learning during filial imprinting: Evidence from transfer of training studies. *Q. J. Exp. Psychol. 46B:* 253-269.
Horn, G. (1985) *Memory, imprinting, and the brain.* Oxford: Clarendon Press.
Horn, G. (1991a) Technique for removing IMHV from the chick brain. In R.J. Andrew (Ed.) *Neural and behavioural plasticity: the use of the domestic chick as a model.* Oxford: Oxford University Press, pp. 44-48.
Horn, G. (1991b) Imprinting and recognition memory; a review of neural mechanisms. In R.J. Andrew (Ed.) *Neural and Behavioural Plasticity: The Use of the Domestic Chick as a Model.* Oxford: Oxford University Press, pp. 219-261.
Horn, G. (1992) Brain mechanisms of memory and predispositions: Interactive studies of cerebral function and behavior. In M.H. Johnson (Ed.) *Brain development and cognition: a reader.* Oxford: Blackwell, pp. 485-513.
Horn, G. and McCabe, B.J. (1984) Predispositions and preferences. Effects on imprinting of lesions to the chick brain. *Anim. Behav. 32:* 288-292.
Johnson, M.H. and Bolhuis, J.J. (1991) Imprinting, predispositions and filial preference in the chick. In R.J. Andrew (Ed.) *Neural and behavioural plasticity.* Oxford: Oxford University Press, pp. 133-156.
Johnson, M.H., Bolhuis, J.J. and Horn, G. (1985) Interaction between acquired preferences and developing predispositions during imprinting. *Anim. Behav. 33:* 1000-1006.
Johnson, M.H., Bolhuis. J.J. and Horn, G. (1992) Predispositions and learning: Behavioral dissociations in the chick. *Anim. Behav. 44:* 943-948.
Johnson, M.H., Davies, D.C. and Horn, G. (1989) A critical period for the development of a predisposition in the chick. *Anim. Behav. 37:* 1044-1046.
Johnson, M.H. and Horn, G. (1986) Dissociation between recognition memory and associative learning by a restricted lesion to the chick forebrain. *Neuropsychologia 24:* 329-340.
Johnson, M.H. and Horn, G. (1987) The role of a restricted region of the chick forebrain in the recognition of conspecifics. *Behav. Brain Res. 23:* 269-275.
Johnson, M.H. and Horn, G. (1988) Development of filial preferences in dark-reared chicks. *Anim. Behav. 36:* 675-683.
Kruijt, J.P. and Meeuwissen, G. (1991) Sexual preferences of male zebra finches: effects of early and adult experience. *Anim. Behav. 42:* 91-102.
Kruijt, J.P. and Meeuwissen, G. (1993) Consolidation and modification of sexual preferences in adult male zebra finches. *Neth. J. Zool. 43:* 68-79.
Kuenzel, W.J. and Masson, M. (1988) *A stereotaxic atlas of the brain of the chick (Gallus domesticus).* Baltimore: Johns Hopkins University Press.

Lickliter, R. (1990) Premature visual experience facilitates visual responsiveness in bobwhite quail neonates. *Inf. Behav. Dev. 13:* 487-496.

Lickliter, R. (1993) Timing and the development of perinatal perceptual organization. In G. Turkewitz and D. A. Devenny (Eds.) *Developmental time and timing.* Hillsdale, NJ: Lawrence Erlbaum Associates, pp. 105-123.

Lickliter, R. and Stoumbos, J. (1991) Enhanced prenatal auditory experience facilitates postnatal visual responsiveness in bobwhite quail chicks. *J. Comp. Psychol. 105:* 89-94.

LoLordo, V.M. and Droungas, A. (1989) Selective associations and adaptive specializations: Taste aversion and phobias. In S.B. Klein and R.R. Mowrer (Eds.) *Contemporary learning theories. II. Instrumental conditioning theory and the impact of biological constraints on learning.* Hillsdale, NJ: Lawrence Erlbaum, pp. 145-179.

Lorenz, K. (1935) Der Kumpan in der Umwelt des Vogels. *J. f. Ornithol. 83:* 137-213, 289-413.

Lorenz, K. (1937) The companion in the bird's world. *Auk 54:* 245-273.

Marler, P. (1976). Sensory templates in species specific behavior. In J.C. Fentress (Ed.) *Simpler Networks and Behavior.* Sunderland, Mass.: Sinauer, pp. 314-329.

McBride, T.C. and Lickliter, R. (1993) Social experience with siblings fosters species-specific responsiveness to maternal visual cues in bobwhite quail chicks (*Colinus virginianus*). *J. Comp. Psychol. 107:* 320-327.

McBride, T. and Lickliter, R. (1994) Specific postnatal auditory stimulation interferes with species-typical visual responsiveness in Bobwhite quail chicks. *Devl. Psychobiol. 27:* 169-183.

McCabe, B.J. and Horn, G. (1994) Learning-related changes in Fos-like immunoreactivity in the chick forebrain after imprinting. *Proc. Natl. Acad. Sci. U.S.A. 91:* 11417-11421.

Nicol, A.U., Brown, M.W. and Horn, G. (1995) Neurophysiological investigations of a recognition memory system for imprinting in the domestic chick. *Europ. J. Neurosci. 7*: 766-776.

Oetting, S., Pröve, E. and Bischof, H.-J. (1995) Sexual imprinting as a two stage process: Mechanisms of information storage and stabilisation. *Anim. Behav. 50: 393-403.*

O'Reilly, R.C. and Johnson, M.H. (1994) Object recognition and sensitive periods: a computational analysis of visual imprinting. *Neural Comput. 6:* 357-389.

Ramenofsky, M. (1985) Acute changes in plasma steroids and agonistic behavior in male Japanese quail. *Gen. Comp. Endocrinol. 60:* 116-128.

Rescorla, R. A. (1988) Pavlovian conditioning: It's not what you think it is. *American Psychologist, 43:* 151-160.

Rescorla, R. A., and Durlach, P. J. (1981) Within-event learning in Pavlovian conditioning. In N.E. Spear and R. R. Miller (Eds.) *Information processing in Animals: Memory Mechanisms.* Hillsdale, N.J.: Lawrence Erlbaum, pp. 81-111.

Rogers, L.J. (1995). *The development of brain and behaviour in the chicken.* Wallingford: CAB International.

Shettleworth, S.J. (1994) The varieties of learning in development: toward a common framework. In J.A. Hogan and J.J. Bolhuis (Eds.) *Causal mechanisms of behavioural development.* Cambridge: Cambridge University Press, pp. 358-376.

Sheu, F-S., McCabe, B.J., Horn, G. and Routtenberg, A. (1993) Learning selectively increases protein kinase C substrate phosphorylation in specific regions of the chick brain. *Proc. Natl. Acad. Sci. (U.S.A). 90:* 2705-2709.

Sluckin, W. (1972) Imprinting and Early Learning. London: Methuen.

Sluckin, W. and Salzen, E.A. (1961) Imprinting and perceptual learning. *Quart. J. Exp. Psychol. 8:* 65-77.

Spalding, D.A. (1873) Instinct, with original observations on young animals. *Macmillan's Magazine 27*: 282-293. Reprinted in 1954 in: *Brit. J. Anim. Behav. 2:* 2-11.

ten Cate, C. (1989a) Stimulus movement, hen behaviour and filial imprinting in Japanese quail (*Coturnix coturnix japonica*). *Ethology 82:* 287-306.

ten Cate, C. (1989b) Behavioral development: Toward understanding processes. In P.P.G. Bateson and P. Klopfer (Eds.) *Perspectives in Ethology, vol. 8.* New York: Plenum Press, pp. 243-269.

ten Cate, C. (1994) Perceptual mechanisms in imprinting and song learning. In J.A. Hogan and J.J. Bolhuis (Eds.) *Causal mechanisms of behavioural development.* Cambridge: Cambridge University Press, pp. 116-146.

Tinbergen, N. (1951) *The study of instinct.* Oxford: Oxford University Press.

Van Kampen, H.S. (1993) Filial imprinting and associative learning: similar mechanisms? *Neth. J. Zool. 43:* 143-154.

Van Kampen, H.S. (1995) A framework for the study of filial imprinting and the development of attachment. *Psychon. Bull. Rev.,* in press.

Van Kampen, H.S. and Bolhuis, J.J. (1991) Auditory learning and filial imprinting in the chick. *Behaviour 117:* 303-319.

Van Kampen, H.S. and Bolhuis, J.J. (1993) Interaction between auditory and visual learning during imprinting. *Anim. Behav. 45:* 623-625.

Van Kampen, H.S. and de Vos, G.J. (1991) Learning about the shape of an imprinting object varies with its colour. *Anim. Behav. 42:* 328-329.

Van Kampen, H.S. and de Vos, G.J. (1994) Potentiation in learning about the visual features of an imprinting stimulus. *Anim. Behav. 47:* 1468-1470.

9

Neural Codes for "Pitch" Processing in a Unique Vertebrate Auditory System

Andrea M. Simmons and Rebecca C. Buxbaum

Abstract

Pitch is a critical dimension of human auditory perception, yet its neural bases are not well-understood, due to the multiplicity of neural codes available for extracting acoustic features relevant for pitch. One strategy for attacking this problem is to examine pitch extraction algorithms in an animal species where the contributions of different neural codes can be easily separated. This paper describes a possible mechanism for periodicity extraction in such a model system, the auditory system of anuran amphibians. Behavioral studies conducted with several anuran species show that these animals are sensitive to features of sounds that mediate pitch perception in humans. The anuran peripheral auditory system extracts periodicity cues from multiple-harmonic sounds primarily by time-domain processing. Eighth nerve fibers are sensitive to the amplitude envelope, harmonic structure, and phase structure of complex sounds, and transmit this information via arrays of phase-locked activity to the central auditory system, which may convert this timing information into a place representation.

Introduction

For human listeners, one of the most salient characteristics of the sounds we hear is their pitch. Pitch is important in the identification and segregation of many kinds of sounds, including speech and music (Hartmann, 1988). It is often defined in terms of musical pitch, the attribute by which sounds can be ordered on a musical scale (Moore,

1989). How we recognize pitch is a complicated problem in auditory research, because pitch is a subjective, psychological feature of a sound, rather than an objective physical feature such as frequency or intensity. The physiological basis for pitch perception has long been one of the dominant themes in research on hearing. It has proven difficult to define unambiguously the features of the acoustic waveform that determine its pitch, and how these features are encoded by the ear and interpreted by the brain to create pitch sensations. These difficulties motivate the comparative analysis of pitch-like phenomena offered here.

Tonal Pitch and Complex Pitch

The significance of pitch as a theoretical issue derives from the fact that most kinds of sounds evoke a sensation of pitch. Sounds which have different physical characteristics and which are easily discriminable can nonetheless be judged to have the same pitch. Human listeners are able to assign pitch values to sounds as physically different as simple sine waves (pure tones), tone complexes made up of multiple frequencies, interrupted noise bursts, amplitude modulated (AM) tones and noise, and trains of clicks or impulses (Bilsen, 1966; Burns and Viemeister, 1976; Flanagan and Guttman, 1960; Moore, 1989; Thurlow and Small, 1953). The auditory system can match a pitch of a single tone (simple or tonal pitch) to the pitch of a more complex sound (complex or periodicity pitch), indicating that it treats these different sounds as having some features in common.

Pitch has acoustic correlates. Tonal pitch is typically defined along a psychological scale whose values are correlated with frequency (and vary slightly with other acoustic parameters such as intensity and duration; Stevens, 1935). Although humans can discriminate single tones across the entire range of hearing (20 Hz to 20,000 Hz), it is only those tones with frequencies below about 5000 Hz that evoke a clear musical pitch (Attneave and Olson, 1971; Semal and Demany, 1990).

Complex sounds containing more than one frequency also can evoke distinct sensations of pitch. Harmonic sounds contain frequency components that are integer multiples of a base or fundamental frequency. The pitch of a harmonically-structured sound, periodicity pitch, is given by this fundamental frequency. The fundamental frequency of a harmonic sound is the reciprocal of the period of the sound. The period of a sound can be extracted from the sound's envelope, the changes in amplitude of the sound over time. Two sounds with the same envelope period can be identified as having the same pitch, even in cases where the specific harmonics making up these sounds differ. The specific harmonic structure (the amplitudes and frequencies of the individual components) of a complex sound is called its fine-structure. Psychologically, fine-structure is a cue for timbre. The differences between pitch and timbre are easily explained in musical terms. The same note played by two different instruments can be identified as having the same pitch, even though the notes sound different in timbre (Wightman, 1973).

Periodicity pitch in humans is evoked most strongly by sounds that contain frequency components (harmonics) below about 5000 Hz, with fundamental frequencies (frequency spacing between harmonics) below about 800 Hz (Ritsma, 1962, 1963). The fundamental frequency need not actually be physically present in the complex sound in order to evoke a pitch; that is, a sound consisting of only high harmonics of a low fundamental frequency will still be perceived as having a pitch described by the fundamental frequency (Fletcher, 1934). The low pitch that can be extracted from a series of high frequency harmonics, when the fundamental frequency is not physically present in the sound, is called residue pitch (Schouten, 1940; Schouten et al. 1962). For these stimuli with missing fundamentals, the perceived pitch corresponds approximately to the rate of repetition of the signal envelope, or the frequency spacing between successive harmonics, both of which can serve as a cue for the fundamental frequency.

The observation that many different types of sounds can produce the same percept of pitch implies that pitch is created by several different neural processes that encode more than one aspect of sounds in parallel pathways and then somehow transform these parallel codes into a single neural representation of pitch. These parallel representational modes are identified with the frequency domain, exemplified by the tuning of auditory neurons to different frequencies in sounds, and the time domain, exemplified by the synchronization of neural discharges to particular phases of the stimulus waveform. The early history of pitch theories is largely that of attempts to account for the coding of tonal pitch in the inner ear in terms of either frequency or time coding of fundamental frequency and periodicity (de Boer, 1976). The same mechanisms proposed to account for coding of tonal pitch were also invoked to account for extraction of pitch from more complex stimuli.

Theories of pitch processing in the frequency domain are based on the known tonotopic organization of the basilar membrane in the mammalian cochlea. An incoming sinusoid gives rise to a travelling wave which propagates along the basilar membrane, and which reaches its maximal amplitude at that place on the membrane tuned to the particular frequency of the sinusoid (Bekesy, 1960). In this way, the basilar membrane effectively carries out a Fourier transformation of the sound, by which the original waveform is decomposed into its different constituent frequencies. These different frequencies are represented by the height of the travelling wave at different places along the basilar membrane and then in the amount of activity (number of action potentials) evoked in different auditory nerve fibers. Auditory nerve fibers reflect the pattern of basilar membrane vibration by changes in their rate of firing. The pitch of a simple sinusoid would correspond to the position of maximal excitation along the basilar membrane as reflected in high firing rates of nerve fibers innervating that place (Siebert, 1970; Stevens et al. 1935; Zwislocki, 1965). Such a place coding scheme accounts well for the ability of humans to discriminate high frequency tones (above about 5000 Hz), where there are sharp, well-defined peaks at the appropriate place along the basilar membrane (Moore, 1989). It is not clear how well a simple place scheme can account for discrimination of pure tones of lower frequency (Siebert, 1970; Wier et al. 1977). Low frequency tones produce more diffuse patterns of vibration along the basilar membrane, and how well the perceived pitch of these tones can be accounted for by a simple place mechanism is unclear (Goldstein and Srulovicz, 1977; Siebert, 1970). It is also unclear how effective a simple place coding scheme is for perception of periodicity pitch. Different frequencies in complex, multiple harmonic sounds are mapped onto different places along the basilar membrane; however, this mapping does not occur with equal precision at all places, and some frequencies are "resolved" or separated better than others (Schouten, 1940). Whether resolved frequencies contribute more to the percept of pitch than those frequencies whose basilar membrane vibration patterns overlap is still a matter of debate (Moore, 1989; Ritsma, 1967). Moreover, the perceived pitch of stimuli with missing fundamental frequencies cannot be accounted for by stimulation of the place on the basilar membrane corresponding to the low perceived pitch (Licklider, 1956). Low frequency residue pitches must therefore be extracted from nerve fibers tuned to some higher harmonics of the complex sound (Licklider, 1956; Schouten, 1940; Schouten et al. 1962).

Alternative models of peripheral processing of pitch are based on a mechanism to measure the rate of repetition (the time interval between peaks in the envelope or fine-structure) of the stimulus waveform. These periodicity theories are concerned with transforming the time or period between successive neural discharges into the physiological parameter defining pitch (Meddis and Hewitt, 1991; Patterson, 1987; Wever, 1949). Auditory nerve fibers can discharge in a synchronized fashion or phase lock to particular phases of sound waves (Rose et al. 1967; Wever, 1949). The time intervals between successive action potentials tend to cluster near multiples of the stimulus period. In response to sound frequencies from about 20 to 1000 Hz, synchronized discharges can occur on every stimulus cycle. In the frequency range of about 1000 to about 5000 Hz, an individual auditory nerve fiber does not necessarily discharge to each cycle of the stimulus waveform; however, volleying of discharges occurring collectively across all the fibers in one auditory channel will represent the periodicity of the stimulus (Wever, 1949). For harmonically-structured sounds, the pattern of neural synchronization would measure the frequency spacing between successive harmonics by phase-locking to peaks in the envelope of the stimulus (Evans, 1978; Javel, 1980). The auditory system could then measure the time intervals between major peaks in the envelope of the stimulus, and extract a pitch from these time intervals (Schouten, 1940).

The dilemma for understanding the coding of pitch in the mammalian auditory system is that both frequency and time cues are plausibly available for most sounds that evoke a sensation of pitch. The frequency region for which a place representation is demonstrated to exist encompasses those frequencies for which a periodicity code is demonstrated to exist, too. For example, the 5000 Hz upper limit of phase-locking of mammalian auditory nerve fibers (Rose et al. 1967) corresponds well to the upper limit of component frequencies in multiple-harmonic sounds which produce distinct sensations of periodicity pitch in humans (Ritsma, 1962, 1963). This correspondence does not necessarily imply that pitch is coded solely by periodicity because a place code also operates in this frequency range (Evans, 1978; Javel and Mott, 1988). Moreover, some psychoacoustical phenomena, such as acuity in discriminating change in tonal pitch and the pitch of inharmonic signals, are not readily explained by either type of code alone (Moore, 1989). How place and periodicity cues are balanced against each other for any particular stimulus, and how this balance changes as, for example, the frequency of the sound changes, are difficult questions. Duplex theories of pitch acknowledge the importance of both neural frequency tuning and neural discharge timing in coding pitch (Licklider, 1951; Wever, 1949). Duplex theories typically allocate these cues according to the predominant frequency of the sound, or according to the particular region of frequencies within the overall spectrum of the sound. High frequency tonal pitch is ascribed to a place mechanism, and the low pitch of periodic complexes is ascribed to a periodicity mechanism (Wever, 1949).

More recent developments in pitch theories have also been concerned with the central process that converts the peripheral code, whether based on place alone, time alone, or a combined place-time mechanism, into a central neural representation of pitch (Goldstein, 1973; Goldstein and Srulovicz, 1977; Langner, 1981; Licklider, 1951; Loeb et al. 1983; Wightman, 1973). These theories devote attention to different ways in which peripheral cues are integrated at a higher level into a uniform dimension of pitch. If, for example, complex pitch is coded differently than tonal pitch, or if different kinds of complex sounds are coded differently, then the central auditory process that creates the dimension of pitch must receive inputs from different peripheral coding systems that are then somehow interleaved or merged into a single perceptual scale. Assuming that tonal and complex pitch differ in their underlying representation but still produce a sensation of pitch as a common attribute, one might speculate that the auditory system converts one of the two codes into the format used by the other code prior to being assigned a pitch value, for example. That is, an initial place code might be transformed centrally into something more nearly like period than frequency (Licklider, 1951).

Spectral models of pitch formation in the central auditory system postulate that the waveform of a complex signal undergoes an initial spectral or place analysis in the cochlea. Different frequencies in a complex sound are coded by peaks in neural firing rates at specific locations along the basilar membrane, forming an array of neural activity along the tonotopic axis. This stage of processing is then followed by a central stage of pitch synthesis, which acts to transform this tonotopic representation to extract pitch information. This central mechanism examines the distributed patterns of peripheral maxima and minima of neural excitation across the tonotopic distribution, and assigns a unique pitch to different patterns of excitation. This central processing could be based on either spectral or temporal processing (Goldstein, 1973; Patterson, 1987; Wightman, 1973). For example, cells in the central auditory system might fire to specific time intervals between peaks in the neural activity along the tonotopic array. Alternatively, cells might be specifically tuned to particular rates of periodicity, but not by firing in a phase-locked fashion.

Theories emphasizing periodicity mechanisms in the auditory periphery postulate that the composition of the stimulus is revealed by the patterns of phase-locked activity that occur at different places on the basilar membrane corresponding to the various frequencies in the stimulus (Licklider, 1951; Lazzaro and Mead, 1989; Meddis and Hewitt, 1991; Patterson, 1987). In these models, the initial frequency-place analysis in the cochlea is followed by an analysis of the phase-locked activity occurring at these different places. The central processing of periodicity extracts these local periodicities from the incoming stream of synchronized neural discharges and compares them to detect consistent patterns. In this way, the auditory system effectively carries out a transformation of the time-series waveform of sounds into a new time-series representation of nerve discharges, and pitch arises from estimating the period of this new time series. One way in which this can be accomplished is by a series of coincidence neurons that detect the temporal relationships between incoming neural discharges (Licklider, 1951; Lazzaro and Mead, 1989; Loeb et al. 1983). Such a mechanism might involve a neural delay scheme that stores the timing of discharges making up one volley and measures the time interval that elapses until the next volley of discharges occurs (Licklider, 1951; Langner, 1981; Langner and Schreiner, 1988). The size of the time interval or period is represented by which neurons detect a coincidence between the delayed and incoming discharges, thus deriving a place representation of periodicity as a second stage of auditory coding. A central coincidence model that performs a correlation between the fine-structure and the envelope of the stimulus has also been proposed (Langner, 1981, 1992). This model incorporates neurons with intrinsic oscillations, which can become entrained to the periodicity of the sound. Such neurons have been identified in the auditory midbrains of cats and guinea fowl (Langner, 1981; Langner and Schreiner, 1988).

The very sophistication of the mammalian auditory system that normally is an obstacle to isolating the role of both time-domain and frequency-domain cues underlying complex pitch perception has led in some other vertebrate species to a divergence in the way these cues are used for biological purposes. For example, for some animals, biological pressures have led to emphases on perception of different frequency ranges than for humans. The comparison of auditory processing in other species can exploit evolutionary differences to alter the balance between the use of time- and frequency-domain codes, thus isolating the contributions of one representation or the other to perception. Knowledge of the biological significance of different types of sounds in the animal's life is a crucial part of this comparative approach. Here, we discuss complex sound processing in the auditory system of anuran amphibians (frogs and toads). This is an example of a system where place cues are less well-developed than in mammals, and time-domain (periodicity) cues might be paramount in extraction of acoustic features relevant to a sensation of pitch.

Auditory Processing by Frogs

Acoustic Behavior of Anurans

Frogs are anuran (tailless) amphibians that have well-known acoustic behaviors associated with communication. Vocal communication plays a paramount role in regulating important aspects of anuran social behavior, including territorial interaction and mating (Gerhardt, 1988; Wells, 1988). In many different anuran species, males gather at ponds and emit species-specific advertisement calls to attract females, and to announce occupation of a calling site to neighboring males. More than one species can often be found calling together at the same site; in some locations, males of different species might be separated by only a few centimeters (Schwartz and Wells, 1985). To choose a mate from these choruses, the female frog faces the task of identifying and localizing conspecific males while avoiding heterospecific males, while in the presence of background noise of both biotic and abiotic origin. The male's advertisement call thus plays an essential role in aiding species, and perhaps individual, recognition, and is crucial for maintaining reproductive isolation of sympatric species. Both spectral and temporal properties of this vocal signal play important roles in aiding the female frog in perceptually isolating species-specific vocalizations under noisy acoustic conditions (Gerhardt, 1988).

Figure 9.1 shows spectrograms of segments of natural choruses of male bullfrogs (*Rana catesbeiana*) and male green treefrogs (*Hyla cinerea*).

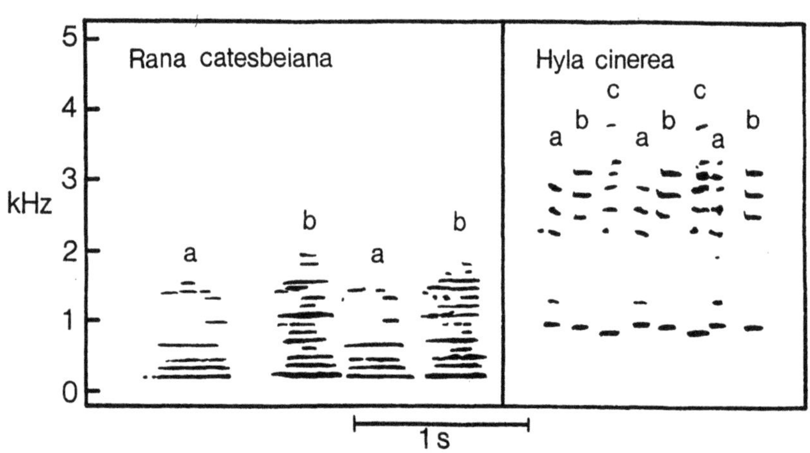

FIGURE 9.1 Spectrograms of segments of natural choruses of 2 interacting bullfrogs (a,b; *Rana catesbeiana*, left) and 3 interacting green treefrogs (a,b,c; *Hyla cinerea*, right). Recordings of the vocalizations were made at natural ponds near Sterling, Massachusetts (bullfrogs) and near Savannah, Georgia (green treefrogs; these recordings were made by Susan E. Allan). Vocalizations were recorded using an Azden microphone onto a Marantz (PMD-430) stereo tape deck. Spectrograms were made using a Unigon II spectrograph. The spectrogram decomposes the sound into its individual Fourier components, and displays the identity of the individual frequencies (y axis) in the vocalizations of each animal over time (x axis). The darkness of the line corresponding to each frequency represents the relative amplitude of that frequency component. There is a consistent frequency spacing between the individual components of each animal's call. This indicates that the calls are harmonic in nature, and the frequency spacing represents the fundamental frequency of the sound.

These spectrograms identify the specific component frequencies in the complex advertisement calls. The vocalizations of both species are complex, multiple-harmonic sounds that humans hear as having a prominent pitch related to the first harmonic (fundamental frequency) of the complex, which often is missing from the spectrum of the signals. In both chorus segments, the vocalizations of several individuals occur in alternation (a, b for bullfrogs; a, b, c for treefrogs). The bullfrog calls have several prominent low-frequency components between about 200 to 600 Hz and several prominent high-frequency components between about 1000 to 1500 Hz, with a spacing of about 100 Hz between adjacent components. The fundamental frequency of 100 Hz is present in the vocalizations of most males, but it is typically reduced in amplitude (by about 20-30 dB) from the amplitude of the dominant low-frequency peak at 200 or 300 Hz. The green treefrogs (Hyla) emit sounds with energy concentrated in a low-frequency range between 900 and 1200 Hz and again between 2500 to 3500 Hz, with a spacing of about 300 Hz between adjacent components. The fundamental frequency of 300 Hz is not present in the sounds. The vocalizations of the chorusing individuals in both species represented here are shifted with respect to the actual place (frequency) of the harmonics in the calls of each individual, possibly to avoid acoustic interference between neighboring chorusing males. Yet the overall frequency spacing between harmonics remains relatively constant, and the periodic structure of the vocalizations remains intact.

In general, the exact number and frequency of harmonics that appear for individual frogs within a chorus will differ because the locations of the individuals differ with respect to the distance and orientation of the recording microphone (and therefore of the approaching female frog). Although it might appear from Figure 9.1 that one individual male could be distinguished from another by the specific harmonic frequencies actually present in the vocalizations of this male, this identifying clue can become attenuated or obscured when the recording microphone is moved. Moreover, the environment itself might alter the relative strengths of the harmonics reaching the female's ears. Depending on such factors as vegetation and temperature, for example, different frequencies — both the actual spectral peaks in the vocalization and how these peaks vary in amplitude over time — will be selectively attenuated and degraded (Richards and Wiley, 1980). This means that individual males of any species cannot be reliably located or isolated from each other by the specific frequencies and amplitudes of the various harmonics in their advertisement calls alone. As the female moves to different locations in the course of finding a mate, the relative strength of the harmonics received from each male changes, but the combined harmonic composition of the sound — the presence of a particular band-limited spectrum or a particular waveform periodicity — might remain intact as a reliable index for isolating one male from another. This overall harmonic composition is represented by the envelope of the stimulus, the spacing of adjacent harmonics, or the frequency of the missing first harmonic of the sounds. The regular, periodic repetition of the vocalization might serve as another identifying feature, and aid in segregation of the chorus from aperiodic background noise. Field recordings of mixed-species choruses of hylid frogs show that chorus sounds recorded as far as 160 m from the chorusing site retain their overall envelope structure, even though the fine-temporal properties of the calls of individuals are obscured (Gerhardt and Klump, 1988). Sonagrams of these chorus sounds resemble modulated or shaped noise.

Another example of this effect from our own work is presented in Figure 9.2. Here, we show a time-series waveform of the envelope and fine-structure of an advertisement call of a male bullfrog recorded in the field at different distances and microphone orientations from the male. The fine-structure of the sound, which is prominently displayed at some distances and orientations (0.5 m, 45°), is attenuated and changed in its details both with distance (6.5 m) and even with microphone orientation at the same recording distance (6.5 m, 30° and 180°). The periodic nature of the vocalization, given by the envelope and by peaks in the fine-structure, remains intact. These observations suggest that the periodic nature of the envelope of the vocalization is available as an acoustic cue to identify the vocalization at large distances from the calling male, while the detailed fine-structure of the signal might only be available at relatively short distances from the caller.

Sensitivity to Periodicity Cues

Many species of frogs are sensitive to temporal properties of complex sounds (Gerhardt, 1988). The temporal features that are most relevant to discussion of periodicity extraction include the repetition rate of individual pulses or clicks within a complex call, waveform periodicity, amplitude modulation (AM), harmonic structure, and phase structure. All of these acoustic features influence perception of periodicity pitch in humans (Moore, 1989; Plomp, 1976). In anurans, several of these cues differ between the advertisement calls of sympatric species, and they also vary between different functional categories of vocalizations within a particular species (for example, the aggressive and the advertisement calls of one species might differ in their rates of periodicity; Gerhardt, 1988). These cues thus play crucial roles in both species and call identification (Allan and Simmons, 1994; Capranica, 1965; Gerhardt, 1978, 1981, 1982, 1988; Loftus-Hills and Littlejohn, 1971; Schwartz, 1987; Schwartz and Wells, 1984, 1985; Straughan, 1975).

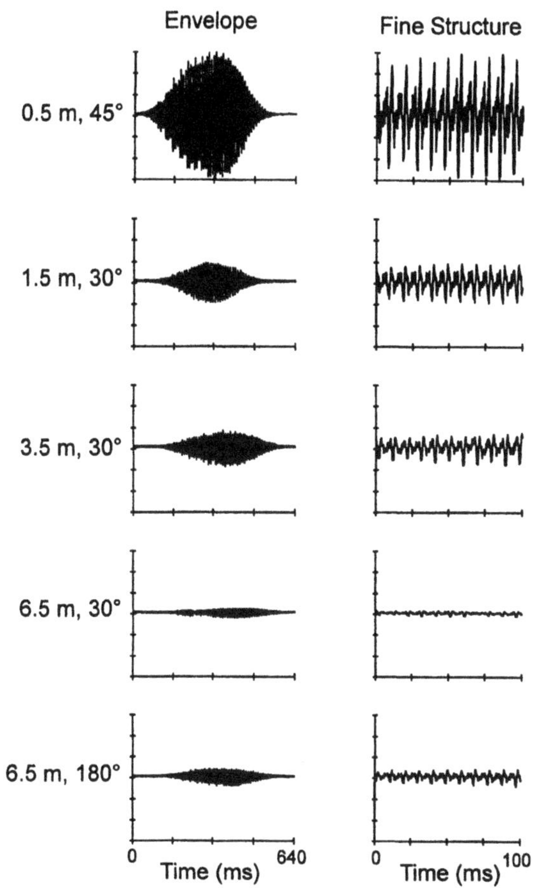

FIGURE 9.2 Time-series waveforms of bullfrog advertisement calls recorded from one individual in a natural chorus. The overall envelope of the entire vocalization is shown in the left column. In the right column, a segment of the entire vocalization is chosen and magnified to display the details of the fine-structure of the sound. There are different time scales for the 2 representations of the vocalization. The y axis is the same for all envelope representations (first column), but differs from that used for the fine-structure representations (second column). The displays in the left column show that the overall shape of the stimulus envelope remains relatively consistent and identifiable over distances up to 6.5 m under these recording conditions. The displays in the right column show that the peaks in the fine-structure of the waveform also remain relatively consistent and identifiable over distance, even though the specific individual frequencies making up the fine-structure become obscured. The time interval between the peaks in the fine-structure are the reciprocal of the fundamental frequency of the waveform. Both the overall shape of the stimulus envelope and the time interval between peaks in the fine-structure can serve as cues for periodicity. Recordings were made using an Azden microphone placed at the listed distances and orientations from the midline of the vocal sac of the calling male. The vocalizations were digitized with an RC Electronics isc-67 interface board and accompanying software on an IBM PC/AT computer at a sampling rate of 20 kHz.

Envelope periodicity is an important cue mediating communication in the green treefrog. The two major vocalizations in the male green treefrog's repertoire, the advertisement call (Figure 9.1) and the aggressive call, contain similar spectral energy, but differ in waveform periodicity. The aggressive call is produced by pulsing the advertisement call, resulting in a sound with an amplitude modulation of about 50 Hz superimposed on the original waveform periodicity of 300 Hz. Moreover, these two conspecific vocalizations differ in periodicity from the advertisement call of a sympatric species of treefrog, *H. squirella* (Gerhardt, 1978). The acoustic differences between such signals are perceived by human listeners as differences in the pitch produced by the process of modulation. Green treefrogs also distinguish between these two signals, and are probably not using the spectral changes associated with the process of modulation to guide their discrimination (Gerhardt, 1978).

Figure 9.3 shows how the evoked calling behavior of male green treefrogs, studied in a natural chorus, is affected by the envelope periodicity of synthetic signals. In these experiments, males are presented with playbacks of various signals; if the males recognize the signal as having some feature in common with a species-specific vocalization, they will vocalize in response to these playbacks (Capranica, 1965). In the experiments described here (Allan and Simmons, 1994), male green treefrogs were presented with playbacks of harmonically-structured complex sounds whose spectral peaks (900, 2700, and 3000 Hz) match those in the conspecific vocalizations (Figure 9.1). These signals (AM calls) were then amplitude modulated at various rates including those of the conspecific advertisement call (unmodulated and 300 Hz) and aggressive call (50 Hz), and of the advertisement call of *H. squirella* (100 Hz). Males were also presented with playbacks of AM noise, which were white noise bursts modulated at particular envelope rates. Unlike AM calls, AM noise contains no spectral cues (sidebands) produced by the process of modulation. This means that discrimination of noise modulated at various rates must be achieved by discrimination of the modulation rate (envelope periodicity) itself, rather than on the basis of some frequency component in the sound. For human listeners, AM noise has a pitch associated with the frequency (envelope) of the modulation (Burns and Viemeister, 1976). Examples of some of these different stimuli, to show how modulation rate affects the stimulus envelope, are also shown in Figure 9.3. The males' calling response is clearly affected by the rate of amplitude modulation of the stimuli (Figure 9.3). Significantly more responses were evoked by playbacks of AM calls modulated at high rates (225 to 450 Hz) than at low rates (50 to 100 Hz). Responses to unmodulated stimuli were intermediate. The relatively higher rates of response to 300 Hz modulated stimuli than to unmodulated stimuli may

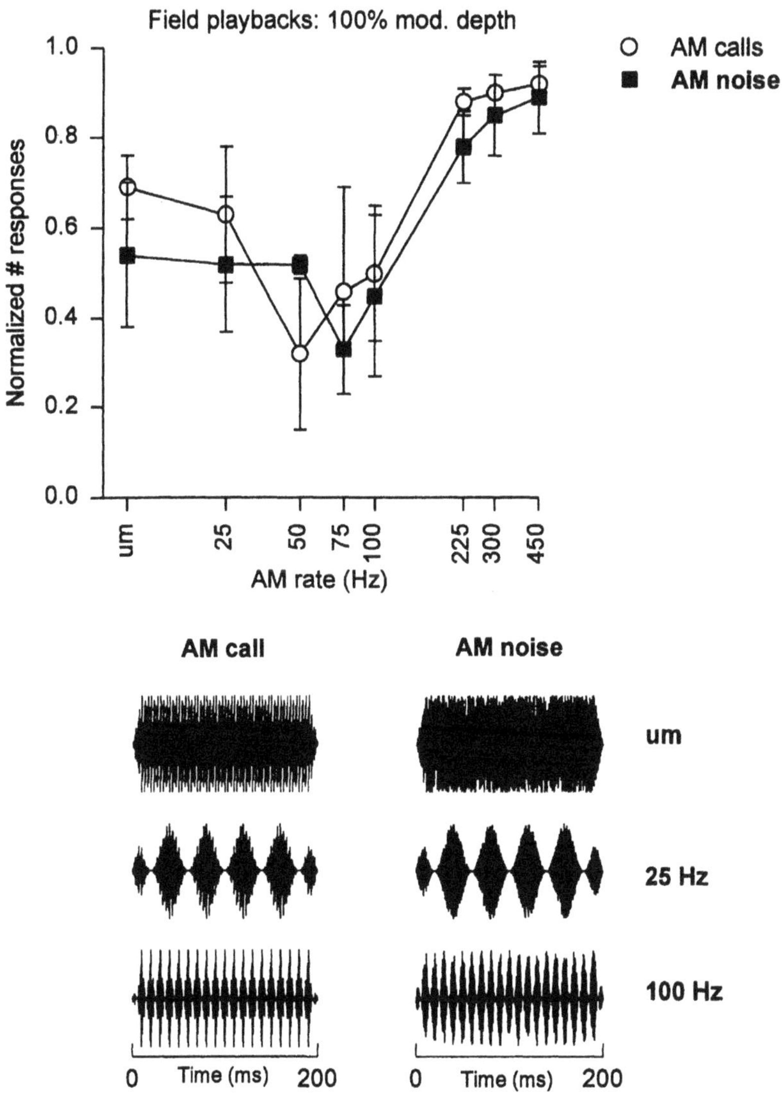

FIGURE 9.3 Evoked vocal responses of male green treefrogs to playbacks of AM calls and AM noise at different rates of modulation, at modulation depths of 100%. Experiments were conducted in a natural calling site near Savannah, Georgia. Signals were digitally generated using a Data Translation DT2801A board with a sampling rate of 20 kHz and an IBM PC/AT 286 computer. They were filtered and recorded onto cassette tapes using a Marantz stereo cassette recorder for later playback in the field. Examples of the stimuli are shown in the bottom of the figure. AM calls consisted of 3 equal-amplitude frequency components of 900, 2700, and 3000 Hz. Shown are a display of an unmodulated call (um), a call modulated at a rate of 25 Hz, and a call modulated at a rate of 100 Hz. The modulation rate imposes changes in the overall envelope of the call, and introduces extra frequency components in the form of sidebands. AM noise consisted of broad-

be due to the extra salience of the periodicity in the former stimuli produced by the modulation process. Males also respond more to playbacks of AM noise modulated at high than at low rates (Figure 9.3). The similarity in response to playbacks of AM calls and AM noise confirms that the temporal cue associated with the envelope (AM rate) is guiding the males' behavior.

The stimuli used to collect the data in Figure 9.3 were all modulated at a depth of 100%; this depth of modulation produces prominent fluctuations in the stimulus envelope. AM signals can also be modulated at lower depths to produce less prominent changes in the stimulus envelope. Figure 9.4 shows results of playback experiments in which males were presented with stimuli modulated at different rates but at a depth of 50% (Allan and Simmons, 1994). At this lower depth, the frogs no longer showed any evidence of recognizing the differences between the various stimuli. These data indicate that the prominence of the waveform envelope is a significant cue mediating vocal communication in frogs.

The harmonic structure of a sound is a salient perceptual cue for pitch to humans. If the components in a complex sound are moved to frequency values that do not have an integer common divisor, then the resulting sound is inharmonic. The degree of harmonicity or inharmonicity can be thought of as the extent to which the frequency components match or fail to match harmonic frequency values, and the sharpness of the perceptual quality of periodicity pitch tracks this match (de Boer, 1956; Schouten et al. 1962). Whether frogs can detect differences in the harmonic structure of complex signals has been studied in

---

band noise (frequency range 20-6000 Hz) upon which the same modulation rates were imposed. Shown are a display of unmodulated noise (um), noise modulated at 25 Hz, and noise modulated at 100 Hz. The modulation rate again imposes changes in the envelope of the stimulus, but without extra frequency sidebands. All stimuli were 160 ms in duration, with rise/fall times of 20 ms, and were presented to the animal at a repetition rate of 1/800 ms. Each different modulation rate was presented separately. Playback intensity was 80 dB SPL at the frog's ear. Data are presented as means and standard errors of the normalized number of answering vocal responses from 8 individual males. Responses were normalized for each animal individually, by dividing the actual number of responses to a particular stimulus by the maximum number of responses evoked by any one stimulus in the same playback session. There are significant differences in responding due to modulation rate for both AM calls [$X^2 = 18.45$, df = 7, $p < .01$] and AM noise [$X^2 = 14.6$, df = 7, $p < .05$]. The frogs vocalized more in response to stimuli modulated at higher rates (225-450 Hz) than at lower rates (50-100 Hz). The similarity in the pattern of responding to AM calls and AM noise suggests that the frogs were using the differences in the stimulus envelope to guide their vocal responses, and were not responding to the extra frequency sidebands in the AM calls. Data are replotted from Allan and Simmons (1994). Reprinted by permission.

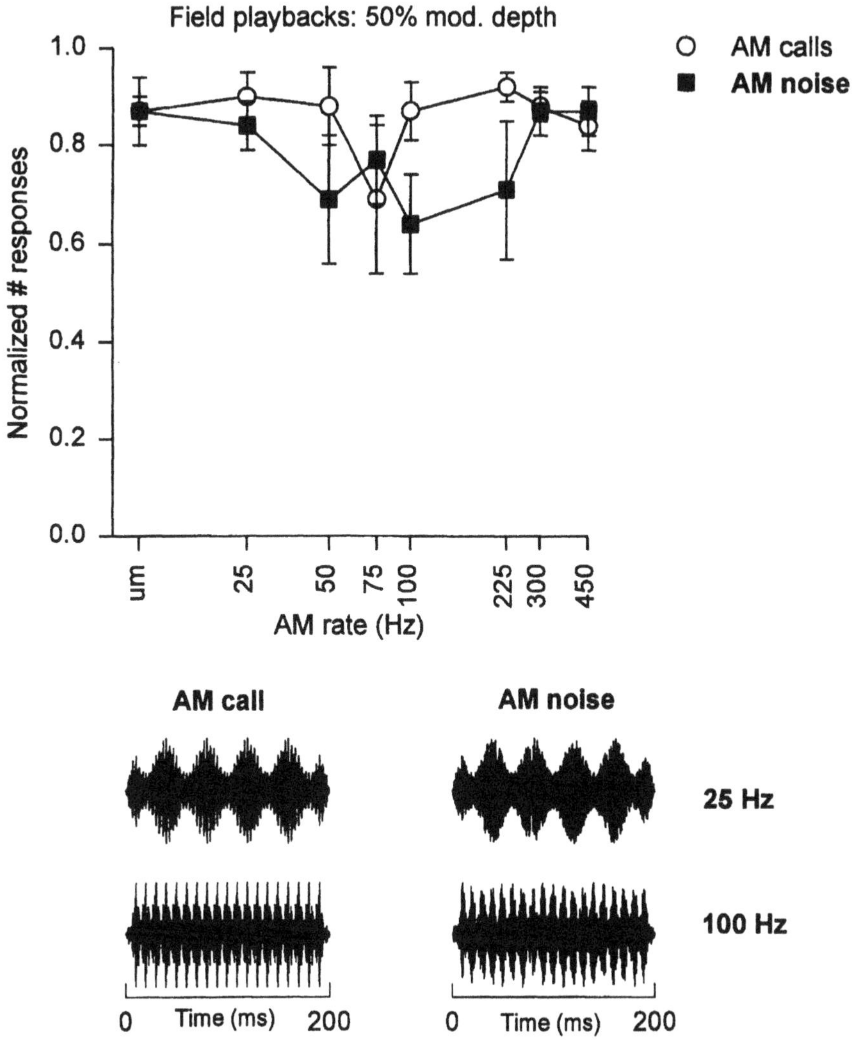

FIGURE 9.4 Evoked vocal responses of male green treefrogs to stimuli modulated at depths of 50%. Examples of the stimuli are shown in the bottom of the figure. As compared to the stimuli in Figure 9.3, these stimuli have less pronounced fluctuations in their envelope due to the differences in modulation rate. There were no statistically significant differences in the evoked calling response due to the differences in modulation rate at this modulation depth. This suggests that the prominence of the waveform envelope is crucial in mediating the frogs' perception of periodicity. Data are replotted from Allan and Simmons (1994). Reprinted by permisssion.

several different kinds of experiments (Gerhardt et al. 1990; Simmons, 1988a; Simmons et al. 1993b). Psychophysical experiments estimate the frog's absolute threshold to sounds using a laboratory behavior (reflex modification) not contingent upon a natural communication response of the animal, and under conditions where the acoustic environment can be more strictly controlled than is possible under field conditions. Data from such experiments show that green treefrogs have a lower threshold by 5-10 dB for detecting in noise harmonically-structured sounds than inharmonic sounds. These lower thresholds result in a more favorable signal-to-noise ratio for extraction of the sound from the background noise (Simmons, 1988a). Other experiments testing sensitivity to harmonic structure using an ethologically-based behavioral response such as selective approach to a sound source (phonotaxis) or vocal response to sound playback (evoked calling) found no reliable differences in responding due to differences in harmonic structure of the stimuli (Gerhardt et al. 1990; Simmons et al. 1993b). These experiments suggest that green treefrogs probably do not employ the cue of harmonic structure as a means of separating sound sources in the natural environment, where other identifying acoustic cues are available. These differences in experimental results may be related to methodological differences. In the psychophysical experiments, animals were tested with stimuli at near-threshold levels presented individually in lengthy, aperiodic series. In the evoked calling and selective phonotaxis experiments, animals were presented with stimuli at higher intensities sufficient to evoke the behavioral response, and the stimuli were presented at periodic rates matching the natural calling rates. Under these conditions, the slight differences in harmonic structure in the stimuli may not have been a salient enough perceptual cue to override the periodicity produced by the repetition rate of the stimuli themselves.

Another factor influencing the salience of periodicity pitch in human listeners is the phase relations between the harmonic components of a complex sound. Changing the starting phase of individual harmonics affects the abruptness or impulsive nature of the envelope of the signal, in much the same way as changing the depth of amplitude modulation. As the envelope waveform becomes abrupter, with more prominent peaks, the perceived pitch is more salient (Hartmann, 1988; Licklider, 1956; Lundeen and Small, 1984). If the phase changes make the envelope or peaks in the fine-structure less prominent, and less easily distinguishable from background noise, then the perceived pitch is less obvious. Experiments using the evoked calling technique have shown that changes in the internal phase spectrum of a synthetic advertisement call are not readily perceptible by green treefrogs, as long as prominent peaks remain in the envelope of the sound (Simmons et al. 1993b). Male bullfrogs, on the other hand, can discriminate complex acoustic signals differing in their internal phase spectrum (Hainfeld et al. 1995). In these experiments, vocalizing male bullfrogs were presented with playbacks of synthetic advertisement calls (spectrum similar to that of bullfrog *b* in Figure 9.1). The starting phases of the harmonics in these signals were varied, but the overall harmonic composition of the sound remained the same. The ability of the animals to discriminate these changes in phase spectra was examined in a habituation design. Frogs were presented with repeated playbacks of one signal; after calling to this signal declined in response to repeated presentations (Megela and Capranica, 1983), a signal with a different phase spectrum was presented. Figure 9.5 shows an example of how these changes in phase spectrum affect the evoked calling response. In this example, latency of calling declined in response to repeated presentations of one stimulus (the alternating phase stimulus). In the alternating phase stimulus, each harmonic component is shifted in its starting phase by 90° from the previous component. This results in a signal with an envelope periodicity of twice the fundamental frequency. In contrast, a sine phase stimulus has all starting phases at 0 degrees, and the envelope periodicity is at the fundamental. Changing the phase spectrum of the stimulus from alternating phase to sine phase resulted in a significant increase in response latency. These changes in latency with changes in the stimulus suggest that the animal can discriminate between the stimuli. The differences in results between the green treefrog experiments and the bullfrog experiments may be due to differences in the sensitivity of the behavioral measure (recognition as opposed to discrimination), or to differences in the frequencies of the harmonics comprising the complex signals. The signals used in the green treefrog experiments contained higher frequency components (maximum of 4000 Hz) than in the bullfrog experiments (maximum of 2000 Hz); phase changes may be more distinguishable in the lower-frequency stimuli.

Neural Mechanisms of Periodicity Extraction

Structure of the Auditory Periphery

Anurans are the only vertebrates which possess three different organs in their inner ear specialized for the processing of airborne sounds. These are the saccule, the amphibian papilla (AP), and the basilar papilla (BP; Figure 9.6). The saccule is innervated by the anterior branch of the eighth cranial nerve, and responds best to low-frequency sounds and vibrations (Yu et al. 1991). The AP and the BP are both innervated

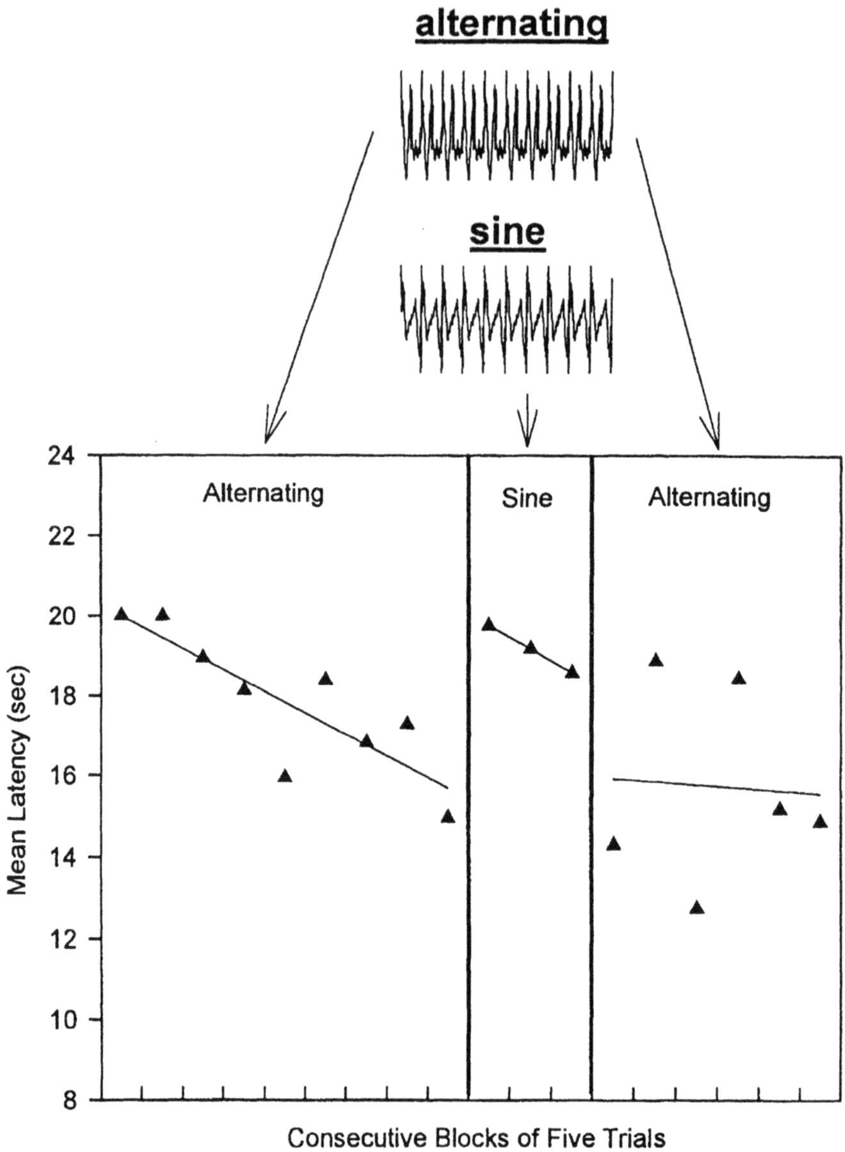

FIGURE 9.5. Habituation and dishabituation of evoked vocal responses of male bullfrogs to complex stimuli differing in their internal phase spectra. Examples of stimuli (50 ms segments) are shown at the top of the figure. The stimuli all consisted of 21 harmonics of 100 Hz, chosen in accord with the natural bullfrog advertisement call (Figure 9.1). The individual harmonics were summed together in different starting phases, which produced changes in the details of the fine-structure of the waveform. In the sine phase stimulus, all starting phases were set at 0°. In the alternating phase stimulus, starting phases alternated between 0° and 90°. The alternating phase stimulus has a waveform periodicity

by the posterior branch of the eighth cranial nerve, but there is no peripheral crosstalk between these two organs. None of these inner ear organs contains a basilar membrane. The absence of this particular morphological basis for Fourier analysis of sounds suggests that place coding mechanisms are less well-developed in anurans than in mammals or birds.

In derived (modern) frogs like the ranids, the AP is an S-shaped organ with a club-like head region and an elongated tail region. It contains hair cells which rest on a layer of stationary supporting cells, with a tectorial membrane overlying them (Wever, 1985). Even though it contains no basilar membrane, the AP is tonotopically-organized; low frequencies below about 400 Hz are represented in the rostral (head) patch, and higher frequencies up to about 1000 Hz are represented in the caudal (tail) patch (Lewis et al. 1982). The tonotopy of the AP may be imparted by some sort of traveling wave mechanism, perhaps propagated along the tectorial membrane (Hillery and Narins, 1984; Lewis, 1992). The variations in the mass of the tectorial membrane along the extent of the AP could provide a basis for this kind of effect (Shofner and Feng, 1983). Fibers innervating the AP are tuned to frequencies in the range of 100-1000 Hz, a range that encompasses the low frequency spectral energy present in anuran vocalizations (Capranica and Moffat, 1983).

The BP is a smaller, cup-shaped organ. It also contains hair cells resting on supporting cells and an overlying tectorial membrane which does not appear to be graded in mass (Wever, 1985). The BP is not tonotopically-organized; instead, it is a simple resonant structure, in which all auditory nerve fibers in an individual animal are tuned to approximately the same resonant frequency (which could differ slightly

---

of double the fundamental frequency of the sine phase stimulus. Stimuli were digitally synthesized using R/C Electronics series 200 hardware and software (sampling rate 20 kHz) and an IBM PC/AT computer. Each individual stimulus ("croak") was 800 msec in duration, with rise-fall times of 10 msec. Stimuli were presented in series of 6 "croaks" (silent interval of 200 msec) at rates of 1 series of croaks every 20 sec. This mimics the natural calling sequence of male bullfrogs. The sounds were filtered and recorded on a Marantz stereo cassette recorder, and presented to individual bullfrogs at a playback intensity of 80 dB SPL. The synthetic calls were presented in a habituation paradigm, in which one series of croaks was presented repeatedly for 45 trials. At trial 46, a call with a different phase spectrum was presented for 15 trials. At trial 61, the original call was presented again. Data are averaged over 5 consecutive trials. The latency of response to the playbacks declined with repeated presentations of one stimulus (the alternating phase stimulus). When the stimulus was changed to the sine phase stimulus, the latency to response increased. These changes are statistically significant [$F(1,43) = 4.4$, $p < .05$]. These data suggest that differences in the internal phase spectrum of complex signals can be discriminated by bullfrogs. Data are from Hainfeld et al. (1995).

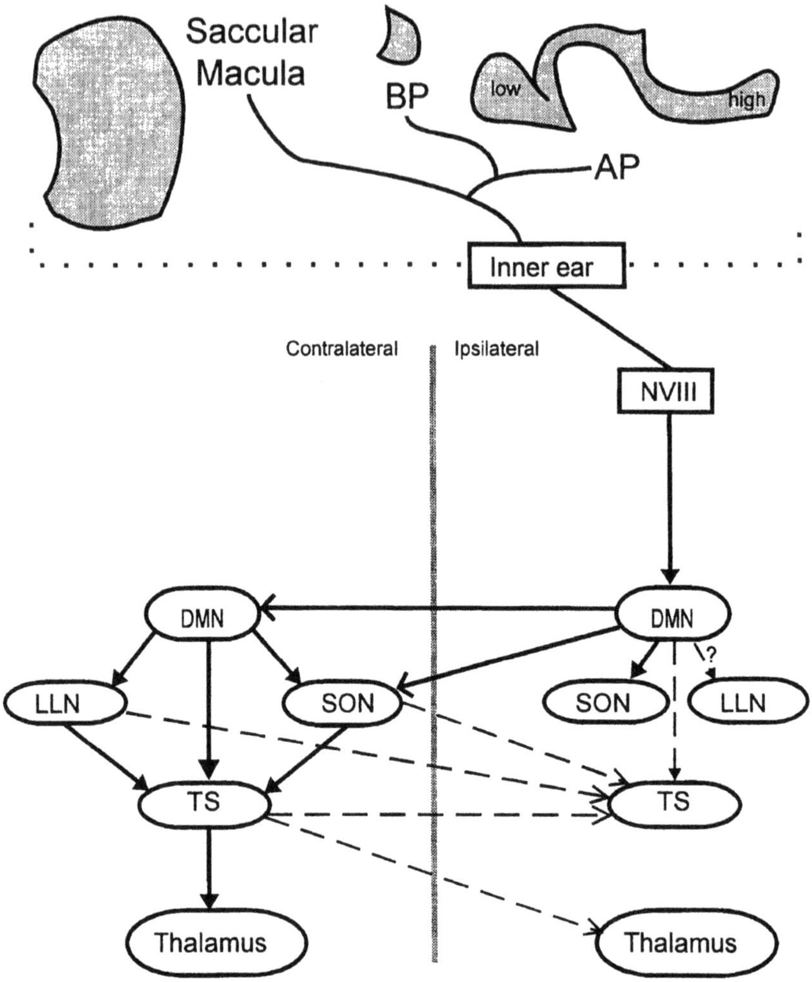

FIGURE 9.6 Schematic of the peripheral sensory epithelia and the ascending and commissural auditory pathways of anuran amphibians, as adapted from Capranica (1976); Feng and Lin (1991); Hillman (1976); Lewis et al. (1982); and Shofner and Feng (1983). Solid lines indicate well-established, strong connections; dashed lines indicate weaker connections. The three sensory surfaces in the inner ear (saccular macula, BP, AP) are innervated by the eighth cranial nerve (NVIII). The frequency sensitivities of the different patches of the AP are noted. NVIII projects to the dorsal medullary nucleus (DMN) in the caudal brainstem. The DMN is considered to be the homolog of the mammalian cochlear nucleus (Feng and Lin, 1994). Efferents from the DMN project to the nucleus of the lateral lemniscus (LLN), superior olivary nucleus (SON), and the torus semicircularis (TS) on the contralateral side. The DMN also projects to the ipsilateral DMN and SON. The TS, the homolog of the mammalian inferior colliculus, receives afferent information from the LLN and the SON as well as the DMN, and projects to the auditory thalamus (Feng and Lin, 1991). Cells with response properties relevant for periodicity extraction have been recorded from both the DMN and the TS.

between different animals of the same species; Zakon and Wilczynski, 1988). The BP is tuned to a narrow, species-specific range which is higher than the frequency response range of the AP and which typically matches a dominant high frequency component in the species vocalizations. In the bullfrog, for example, the BP is tuned to frequencies between 1000-2000 Hz (Feng et al. 1975; Freedman et al. 1988), while in the green treefrog, the BP is tuned to frequencies between 3000-4000 Hz (Capranica and Moffat, 1983). Males of both of these species have advertisement calls with energy in both a low-frequency and a high-frequency region (Figure 9.1), and the high-frequency energy in these calls falls within the frequency response range of the BP. For those species of frogs without such bimodal spectra in their calls, the dominant frequency component in the vocalization falls within the frequency-response region of the BP (Rose and Capranica, 1984). The upper limit of frequency response in the basilar papilla corresponds to the upper limit of hearing in frogs measured psychophysically (Megela-Simmons et al. 1985). This upper limit extends up to about 5000-6000 Hz in the species studied thus far.

Nerve fibers innervating the AP and the BP have V-shaped excitatory tuning curves that define the range of frequencies and intensities to which the fibers are sensitive. Examples of such tuning curves recorded from the bullfrog eighth nerve are shown in Figure 9.7. The frequency at which the fiber best responds at the lowest intensity (threshold) is called the best excitatory or characteristic frequency (CF). As the intensity of the stimulus increases, the fiber can respond to more frequencies. Tuning curves of anuran fibers tend to be broad in shape, implying poor frequency selectivity. Because frequency and period are reciprocal, broad spectral tuning implies good temporal tuning. The sharpness of tuning can be quantified by dividing CF by the frequency response bandwidth at 10 dB above threshold to form a measure called $Q_{10\,dB}$. The $Q_{10\,dB}$ of individual fibers typically varies between 0.5 and 6 (Zakon and Wilczynski, 1988). This is comparable to the tuning curve sharpness of mammalian eighth nerve fibers tuned to a comparable range of low frequencies (Kiang, 1965). But tuning sharpens in mammals for higher frequencies above about 1000 Hz, while it remains poor in the frog, up to the upper limit of BP tuning. A plot of the $Q_{10\,dB}$ values for a sample of 379 bullfrog eighth nerve fibers is shown in Figure 9.8. At most CFs, there is a range of $Q_{10\,dB}$ values, and there is no correlation between sharpness of tuning and fiber CF. In the cat auditory nerve, on the other hand, sharpness of tuning is significantly correlated with CF over a comparable frequency range (Kiang, 1965).

The broad frequency tuning of frog peripheral fibers is also reflected in physiological estimates of the ability of individual fibers to detect sinusoidal signals in broadband noise. Critical ratios (signal-to-noise ratios) for detection of tones in noise measured in the bullfrog's eighth nerve vary over a 30 dB range from 6-35 dB, and are broader for BP than for AP fibers (Freedman et al. 1988). These relatively high critical ratio values imply the presence of wide filters for signal detection and thus poor ability to resolve or separate out individual frequencies from the background noise. It is unlikely that this broad spectral tuning could support frequency analysis and discrimination of harmonic structure to the same degree as in mammals (Simmons, 1988a,b). It reinforces the idea that frequency tuning and place coding, although they exist in the

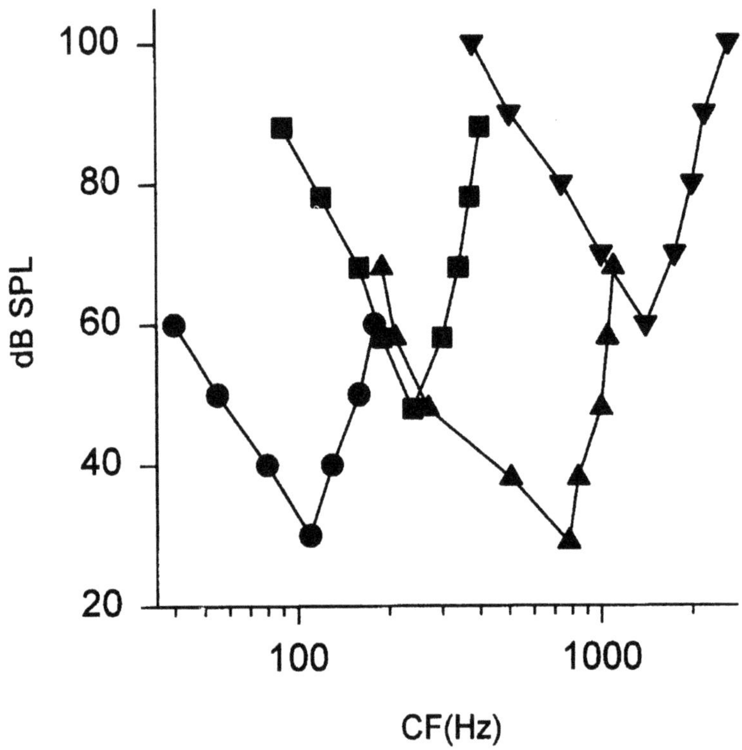

FIGURE 9.7 Tuning curves (plots of frequency sensitivity at different stimulus amplitudes) of 4 fibers recorded from the bullfrog's eighth nerve. Three of these fibers (CF 110 Hz, 245 Hz, and 790 Hz) innervate the AP, and one fiber (CF 1400 Hz) innervates the BP. Thresholds at CF for the 3 AP fibers (30 dB SPL, 48 dB SPL, and 29 dB SPL) are lower than that for the BP fiber (60 dB SPL). These differences in thresholds between AP and BP fibers are typical (Zakon and Wilczynski, 1988). The tuning curves of all 4 fibers are broad in shape, with Q_{10dB} values (a measure of the sharpness of tuning) of 2.2, 2.2, 2.5, and 1.9.

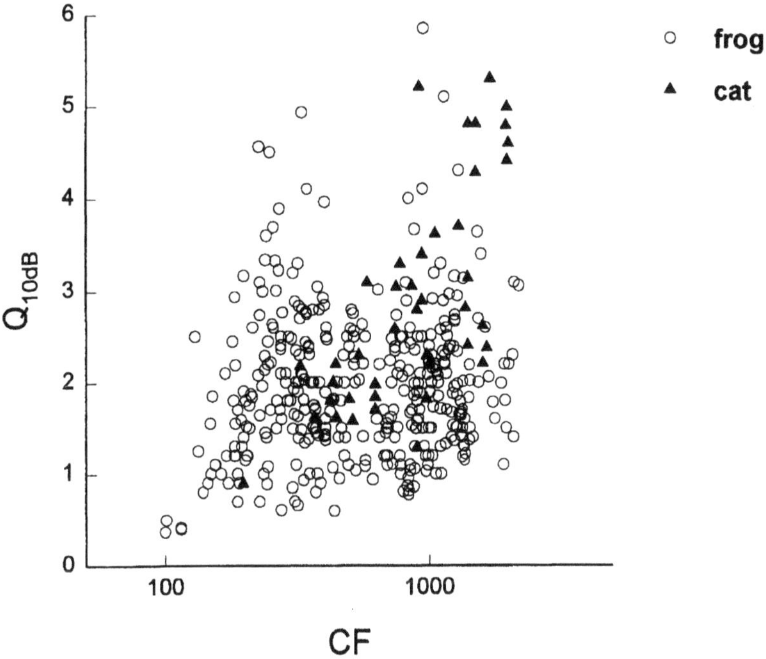

FIGURE 9.8 Plots of $Q_{10\ dB}$ values against CF for a sample of 379 single fibers from the bullfrog eighth nerve. Q_{10dB} is not significantly correlated with fiber CF [Q = 0.00132×(CF) + 1.86; r^2 = .006]. Q_{10dB} values from cat auditory nerve fibers (Kiang, 1965) are plotted for comparison. These data represent averages from 45 cat fibers, over 100 Hz bins. For these fibers, $Q_{10\ dB}$ is significantly correlated with CF [Q = 0.0019×(CF) + 1.006; r^2 = 0.57, $p < .01$].

frog auditory periphery, may not be paramount in underlying processing of complex sounds. This idea is examined more fully below.

Peripheral Coding of Periodicity

Both AP and BP fibers phase lock to the envelope waveform of simple sinusoids, but only when the frequencies of these sinusoids are below about 1000 Hz (Freedman et al. 1988; Hillery and Narins, 1987; Narins and Hillery, 1983). BP fibers thus do not synchronize to pure tones at CF, but will phase lock to lower frequency tones that fall in their tuning curves (Narins and Hillery, 1983; Narins and Wagner, 1989). The 1000 Hz upper limit of synchronization to pure tones is well below the 5000 Hz upper limit for synchronization in the mammalian auditory nerve (Kiang, 1965; Rose et al. 1967).

Both AP and BP fibers synchronize to the envelope of amplitude modulated tones and noise (Dunia and Narins, 1989; Feng et al. 1991; Rose and Capranica, 1985). Figure 9.9 shows plots of the 'existence region' of synchronization to the envelope of AM noise at 3 different modulation depths for fibers from the bullfrog eighth nerve. In these data, synchronization is quantified by the measure vector strength, which is computed from period histograms of fiber responses (Figure 9.10). Vector strength ranges from 0 (no synchronization) to 1 (perfect synchronization). At a modulation depth of 100%, vector strength and thus synchronization is high and statistically different from 0 up to modulation rates of 500 Hz. The plot of Figure 9.9 shows this limited existence region of modulation detection in the form of a modulation transfer function with a low-pass shape. At any given modulation rate, as the depth of modulation decreases below 50%, synchrony to the envelope

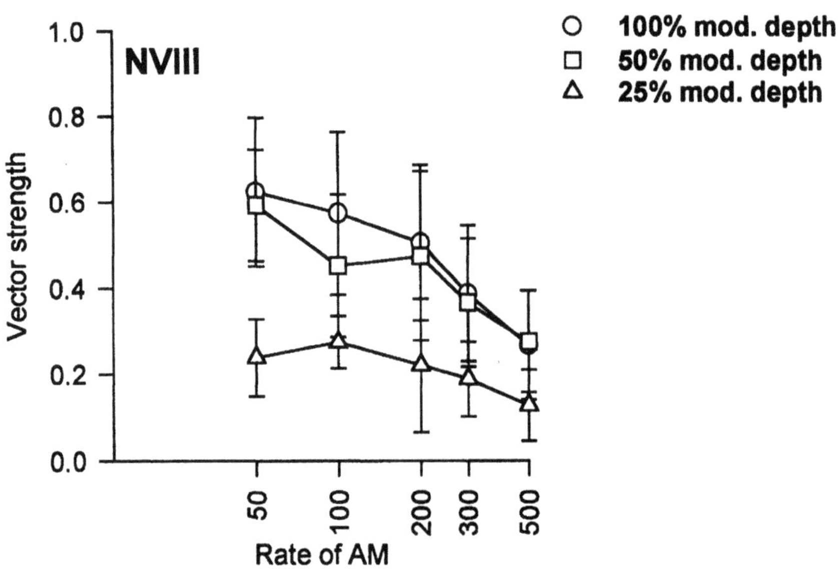

FIGURE 9.9 Vector strength of synchronized responses from 138 individual bullfrog eighth nerve fibers to the envelope of AM noise at different depths of modulation. Vector strength is calculated from period histograms, like those in Figure 9.10. It varies from 0, indicating no phase-locking, to 1.0, indicating perfect phase-locking. Values of 0.2 or below are not statistically different from 0 (Rayleigh test). Each data point represents mean vector strength, +/- standard deviation. At 100% modulation depth, mean vector strength is high at a AM rate of 50 Hz, and declines at higher rates. Values are still statistically significant at AM rates of 500 Hz. There is no significant difference in vector strength between responses at 100% and 50% modulation depths. Vector strength is significantly lower at depths of 25% than at rates of 50% or 100% at all rates tested.

also decreases. At modulation depths of 25%, synchronization to the waveform envelope is no longer statistically significant. These data indicate that the prominence of the waveform envelope influences the phase-locked responses of eighth nerve fibers. Note, however, that the eighth nerve seems to follow decreases in modulation depths better than the behavioral response, which shows a decrease in the ability to discriminate modulation depths at 50% depth (Figure 9.4), where the synchronization of eighth nerve fibers is still robust.

A recent series of experiments (Schwartz and Simmons, 1990; Simmons et al. 1992; Simmons et al. 1993a) analyzed how eighth nerve fibers in the bullfrog code the complex spectral and temporal structure of the species-specific advertisement call (Figures 9.1 and 9.2) over a biologically-realistic range of intensities. Figure 9.10 shows examples of period histograms of responses of 12 individual fibers to a synthetic bullfrog advertisement call presented at an intensity of 80 dB SPL. The 12 fibers have characteristic frequencies from 115 to 1636 Hz, and innervate either the AP (CFs 115-1010 Hz) or the BP (CFs 1168-1636 Hz). Period histograms mark the time-of-occurrence of single action potentials within each period of the stimulus. The period histograms in Figure 9.10 are compiled over the fundamental frequency (100 Hz; period 10 ms) of the signal. If the fiber is phase-locked to the period of the stimulus, then the period histogram has a peaked shape and the calculated vector strength is high. The number of discrete peaks in these histograms is related to the harmonic of the stimulus to which the fiber is synchronizing; for example, a single-peaked histogram compiled over the stimulus period indicates that the fiber is synchronizing to the first harmonic (fundamental frequency), a double-peaked histogram indicates synchronization to the second harmonic, and so on. The period histograms of responses of the fibers in Figure 9.10 all show similar single-peaked shapes, regardless of their CFs or presumed place of origin within the inner ear. The vector strengths to the fundamental frequency are high for all fibers, and statistically different from 0. This indicates that the fibers are synchronizing preferentially to the fundamental frequency of the synthetic call. This is true even though there is very little spectral energy in the signal at 100 Hz, and there are higher spectral peaks in the signal close to the CFs of these fibers.

Plotted in Figure 9.11 is the mean vector strength to the 100 Hz fundamental frequency of the synthetic call from a sample of 409 individual fibers. The stimulus was presented at 4 different intensities, from 70 to 100 dB SPL. Fibers were categorized into 3 groups based on their CFs (Zakon and Wilczynski, 1988). AP low fibers have CFs from about 100 to 499 Hz and presumably originate from the rostral portion of the AP; AP mid fibers have CFs from about 500 to about 1000 Hz and presum-

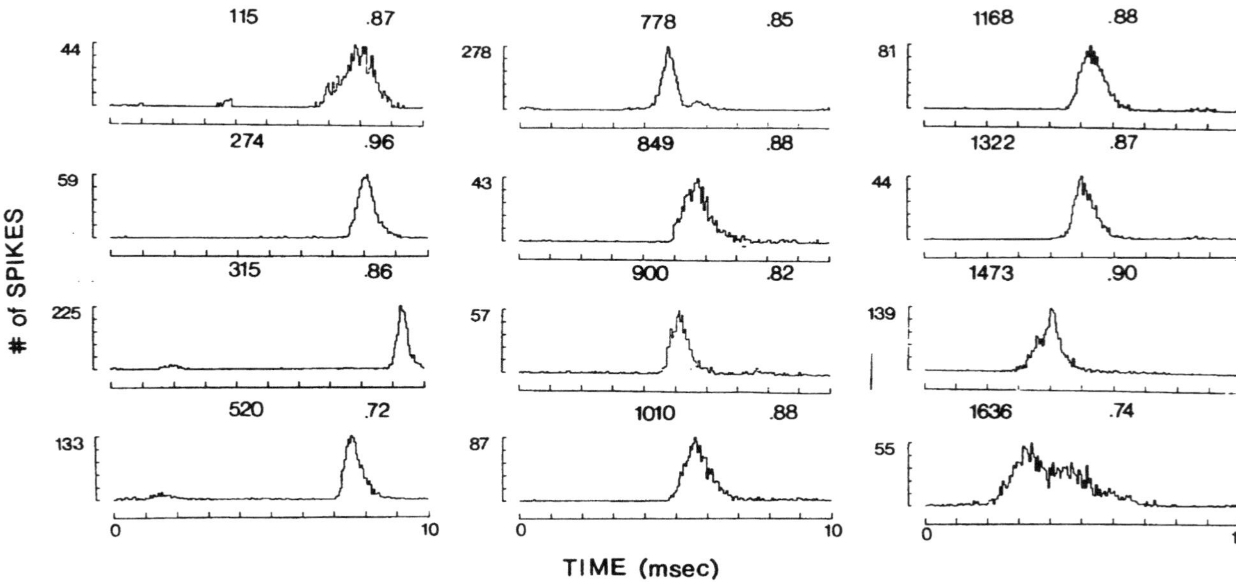

FIGURE 9.10. Period histograms of responses of 12 individual bullfrog eighth nerve fibers to a synthetic advertisement call, with a spectrum as in Figure 9.1. The synthetic call was synthesized with all harmonics in sine phase; stimulus duration was 800 msec, and intensity was 80 dB SPL. The period histograms are compiled over the 100 Hz fundamental frequency (10 ms period) of the stimulus. Fiber CF (from 115 to 1636 Hz) and the vector strength calculated from each period histogram is indicated at the top of each individual graph. All of the period histograms have a single peak, indicating that the fibers are phase-locking to the fundamental frequency of the stimulus. The vector strength reflects the sharpness of each peak. All vector strengths are high, and statistically different from 0. The fibers do not synchronize to the frequency in the stimulus closest to their CF, even though the stimulus contains dominant low frequency peaks within the existence region of phase-locking to single tones. These data indicate that all fibers, regardless of their CF or place of origin in the inner ear, are preferentially extracting the fundamental frequency (waveform periodicity) of the synthetic call in their patterns of phase-locked response. These data were collected by Schwartz and Simmons (1990).

ably originate from the caudal portion of this organ; and BP fibers have CFs from about 1000 to 2000 Hz (Lewis et al. 1982). Figure 9.11 shows that, at stimulus intensities of 70, 80, and 90 dB SPL, there are no significant differences in synchronization to the fundamental frequency of the signal within these groups of fibers. Both AP and BP fibers phase lock strongly to the waveform periodicity of the synthetic signal. In effect, the pitch of the signal is extracted from the phase-locked responses of the fibers. Figures 9.10 and 9.11 together show that these fibers do not typically synchronize to the harmonic in the signal closest to their CFs. If they did, then all of the period histograms in Figure 9.10 would

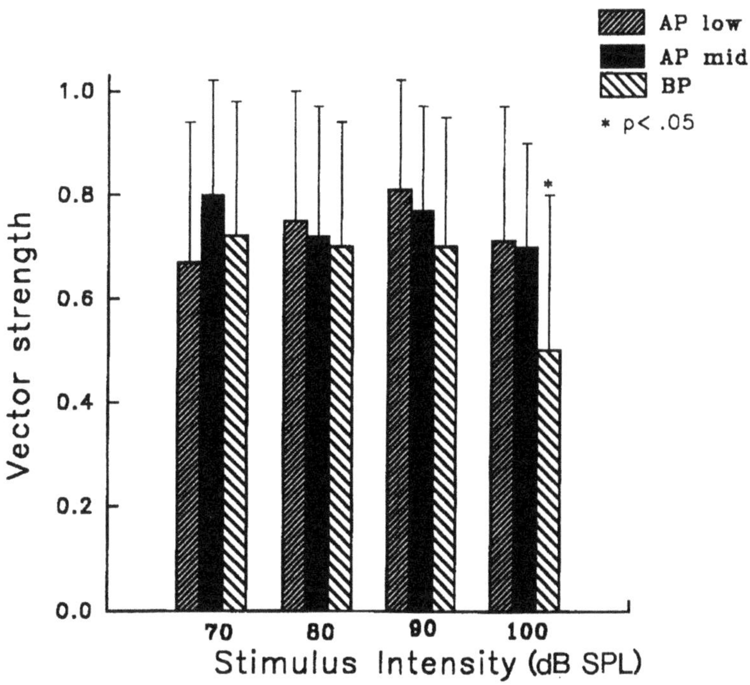

FIGURE 9.11. Mean vector strength (+/- standard deviation) to the fundamental frequency of a synthetic advertisement call recorded from a sample of 409 individual fibers from the bullfrog eighth nerve. Stimuli were presented at levels of 70, 80, 90 or 100 dB SPL. Fibers are divided into groups based on CF and presumed place of origin within the inner ear. AP low fibers have CFs between 100 and 499 Hz and presumably innervate the rostral patch of the AP; AP mid fibers have CF between 500 and 1000 Hz and presumably innervate the caudal extension of the AP; BP fibers have CFs between 1000 and 2000 Hz. There are no significant differences in vector strength between these three groups of fibers at either 70, 80 or 90 dB SPL. At 100 dB SPL, BP fibers show significantly less synchronization to the fundamental frequency than do either AP low or AP mid-fibers. Data were collected by Schwartz and Simmons (1990) and Simmons et al. (1992).

differ with fiber CF. This means that the spectral fine-structure (the distribution of harmonics) of the synthetic call is not accurately reflected in the timing of fiber firing, because firing is captured by the fundamental frequency.

This strong synchronization to the fundamental frequency remains robust under low to moderate levels of background noise. As noise level increases, the high-amplitude peaks in the stimulus envelope become relatively smaller as compared to the background; at very high noise levels, the envelope can become smeared and indistinguishable from the noise. Under these conditions (Figure 9.12), some fibers shift the timing of their firing from representing the fundamental frequency

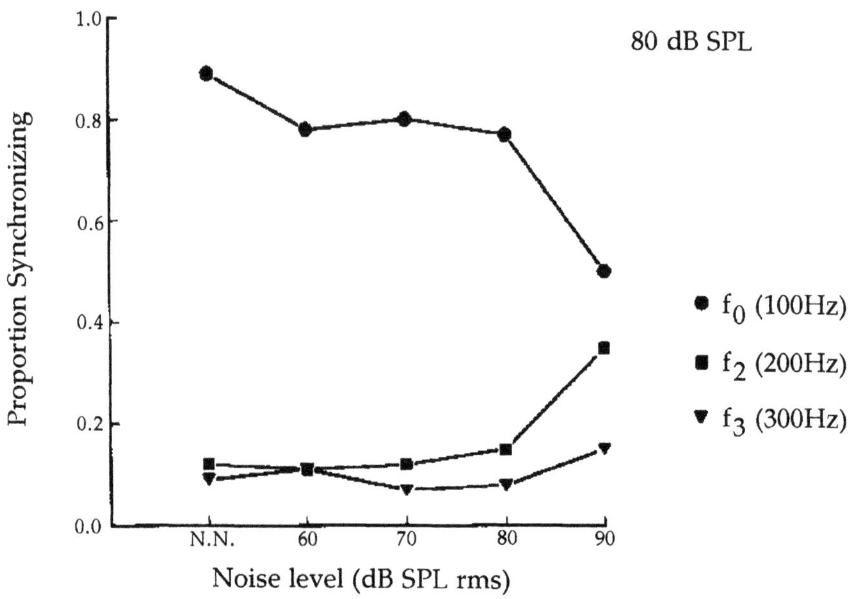

FIGURE 9.12. This figure plots the proportion of fibers synchronizing to the individual harmonics in the synthetic advertisement call presented against different levels of background masking noise. Stimulus intensity is 80 dB SPL, and noise varies from no noise (N.N.) to noise levels of 90 dB SPL rms. Data are from a sample of 125 individual fibers. Fibers consistently synchronize to only the first (100 Hz, fundamental frequency), second (200 Hz), or third (300 Hz) harmonic of the synthetic call. Under no noise conditions, about 90% of all fibers tested synchronize preferentially to the first harmonic (fundamental frequency). As background noise level increases, the proportion of fibers synchronizing to the fundamental begins to decline, but it is not until noise reaches a level of 90 dB SPL rms that the proportion of fibers synchronizing to the fundamental declines significantly. At this noise level, fibers either do not phase lock to any component in the stimulus, or they shift their synchronization to the second or third harmonic. Data are replotted from Simmons et al. (1992). Reprinted by permission.

to representing the second or third harmonic in the signal (Simmons et al. 1992). Some fibers continue to signal the fundamental frequency in their temporal response patterns, even at high noise levels, while others show masking of temporal response at these high noise levels and thus no synchronization to any harmonic. No fiber phase locks to a harmonic higher than the third. These effects are seen in the responses of both AP and BP fibers, and are unrelated to fiber CF. Fibers do not shift their synchronization to the stimulus component closest to CF as noise level increases. The robust synchronization of some fibers at high noise levels occurs in spite of the fact that rate responses (number of action potentials) no longer increase with stimulus intensity at these levels. This ensures that both the waveform periodicity (fundamental frequency) and the low frequency spectral fine-structure of the signal are being preserved under noisy acoustic conditions, and can operate to guide behavior in these situations. For example, the female frog approaching the chorus, when background noise levels are low, can be guided by the combined envelopes of the vocalizations of the calling males. In the midst of the chorus, where noise levels are high, discrimination of the spectral peaks themselves might be used to recognize individual males.

The influence of envelope shape on fundamental frequency extraction by bullfrog peripheral fibers was further examined by recording responses to a synthetic advertisement call whose individual frequency components were summed together in various starting phases (Simmons et al. 1993a). These phase configurations included a phase coherent version with all harmonics summed in sine phase and an alternating phase version, where the starting phases of consecutive harmonics vary between sine and cosine phases. Fibers typically respond in a periodic fashion to these different stimuli (Figure 9.13). These data are shown in the form of autocorrelation functions (ACFs), which are plots of the preferred time intervals between spike firing, and in the form of period histograms. The period histogram and the ACF both provide an estimate of the ability of a fiber to synchronize to a particular period in the stimulus. These two measures differ in that the period histogram shows only the time of occurrence of action potentials at one particular time interval or period (the period over which the histogram is compiled). In contrast, the ACF measures the time of occurrence of action potentials at all possible time intervals in the stimulus, out to the end of the ACF time axis. The power spectrum (Fourier transform) of the ACF can be used to identify the dominant period in the fiber's response, by giving a plot of the largest frequency component present in the ACF. As shown in Figure 9.13, capture of synchronized firing by the fundamental frequency is affected by the phase spectrum of the signal. In response to the sine phase stimulus, the dominant component in the power spec-

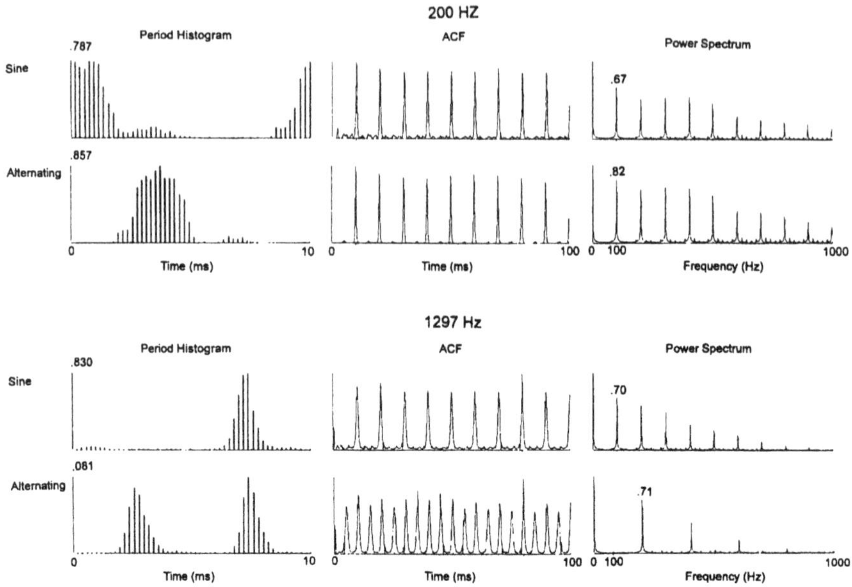

FIGURE 9.13 Period histograms (left column), autocorrelation functions (ACF, middle column), and the power spectrum of the ACF of responses of 2 individual eighth nerve fibers to synthetic advertisement calls with different phase spectra (sine phase or alternating phase). The fiber with CF 200 Hz (top) innervates the rostral patch of the AP, and the fiber with CF 1297 Hz (bottom) innervates the BP. Stimuli were presented at 10 dB above the fiber's threshold. The period histograms are compiled over the 100 Hz fundamental frequency of the stimulus. The calculated vector strength is shown at the top of each histogram. The ACF measures all time intervals between consecutive action potentials in the neural response. In response to the sine phase stimulus, the ACF of the AP fiber shows sharp peaks separated by 10 ms. The power spectrum (Fourier transform) of the ACF shows a dominant peak at 100 Hz, the fundamental frequency of the stimulus. The height of this peak is indicated by the number above each ACF; this number is equivalent to the square of the period histogram. The period histogram and the ACF to the alternating phase stimulus similarly indicate that the fundamental frequency is signalled in the fiber's phase-locked discharges. In contrast, the BP fiber shows a different pattern of response to the sine phase and alternating phase stimuli. The fiber extracts the 100 Hz fundamental frequency of the sine phase stimulus in its period histogram and ACF; however, the period histogram of the response to the alternating phase stimulus shows two peaks, indicating that the fiber is synchronizing to 200 Hz, the periodicity of this stimulus. Data are replotted from Simmons et al. (1993a). Reprinted by permission.

trum of the response of both the AP fiber (CF 200 Hz) and the BP fiber (CF 1297 Hz) is 100 Hz, the fundamental frequency of the stimulus. When the phase spectrum changes from sine phase to alternating phase, synchronized firing can either stay at the fundamental frequency (in the responses of the AP fiber), or shift to the second harmonic (in the responses of the BP fiber). These data show that BP fibers signal the changes in phase spectrum between a sine phase and an alternating phase signal. This is consistent with behavioral experiments (Hainfeld et al. 1995; see Figure 9.5) showing that bullfrogs can discriminate these two kinds of signals.

In the experiments described above, the signal used (Figure 9.1) was a multiple-harmonic signal in which the fundamental frequency was physically present (although of low amplitude). The coding of the envelope periodicity of stimuli with missing fundamentals was directly examined by analyzing fiber responses to three component signals with and without energy at the fundamental frequency (Simmons and Ferragamo, 1993). The harmonic structure of these stimuli was also shifted in a manner that paralleled psychoacoustic work on pitch-shift effects in humans (de Boer, 1956; Schouten et al. 1962). When the carrier frequency and the modulation frequency of AM tones are harmonically-related, then human listeners report a periodicity pitch associated with the modulation frequency. When the carrier is shifted in frequency so as not to be harmonically related to the modulation frequency, then listeners report a shift in the perceived pitch that is related to the change in the peaks in the fine-structure of the waveform (de Boer, 1956; Schouten et al. 1962). The pitch-shift effect has been interpreted by a temporal mechanism sensitive to high-amplitude peaks in the fine-structure of the waveform (Schouten et al. 1962).

Bullfrog eighth nerve fibers, regardless of their CFs, preferentially extract the fundamental frequency of the harmonic complexes in their patterns of phase-locked activity, whether or not the fundamental is physically present in the stimulus. This is shown in Figure 9.14 as a plot of the dominant component in the Fourier transform of the ACF of the responses of different fibers against fiber CF. The stimuli in this experiment had fundamental frequencies of 200 Hz, which were either present (Figure 9.14, top) or absent (Figure 9.14, bottom) from the signal. The data show that the 200 Hz fundamental was signalled in the phase-locked responses of most fibers, whether or not the fundamental was physically present in the stimulus. When the stimuli were made inharmonic, the overall pattern of synchronization altered. Some fibers continued to synchronize to the 200 Hz frequency spacing between harmonics, while others shifted their responses to signal another frequency in the stimulus. This means that, over all fibers, a pitch-shift between the harmonic and inharmonic complexes is signalled in fiber temporal response. These differences in the phase-locked activity of individual fibers in response to harmonic and inharmonic stimuli may serve as a basis for the animal's different behavioral responses to these kinds of stimuli (Simmons, 1988a).

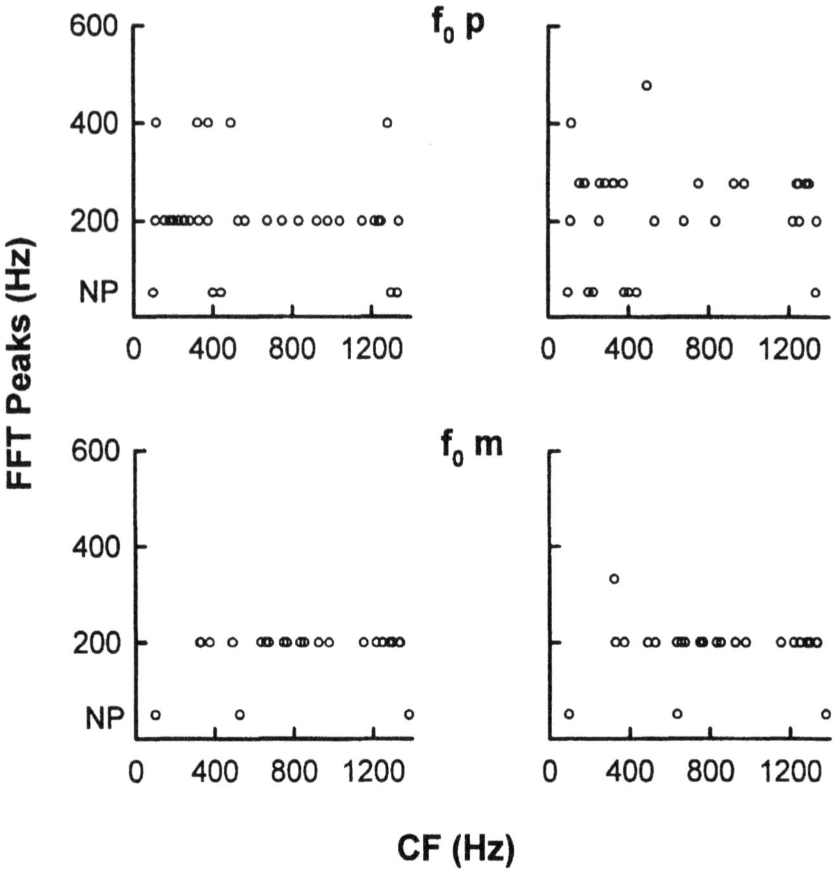

FIGURE 9.14 Plots of the dominant component in the power spectrum of the ACF of fiber response (FFT peaks, y axis) against fiber CF. Stimuli consisted of 3 frequencies with a fundamental frequency (frequency spacing) of 200 Hz. For the stimuli in the top row, the 200 Hz fundamental frequency was present in the stimulus (f_0 p). For the stimuli in the bottom row, the fundamental frequency was absent (f_0 m). The stimuli in the left column were harmonic, with all components being integer multiples of the fundamental frequency (e.g. 200, 400, and 600 Hz; 400, 800, and 1000 Hz). The stimuli in the right column were inharmonic, so that the individual components were not integer multiples of the fundamental (e.g. 277, 477, and 677 Hz; 477, 877, and 1077 Hz). Each data point represents the dominant peak in the power spectrum of the fiber's response to these different stimuli. NP = no significant peak in the fiber's response, indicating a lack of synchronization. Data are based on the responses of 47 individual fibers. In response to harmonic stimuli, most fibers signal the 200 Hz fundamental frequency in their phase-locked responses, whether or not the fundamental was present in the stimulus. In response to inharmonic stimuli, some fibers shift their responses away from the fundamental frequency when it was present in the stimulus. These data indicate that eighth nerve fibers can signal a pitch-shift between harmonic and inharmonic stimuli. Data are replotted from Simmons and Ferragamo (1993). Reprinted by permission.

All of these experiments demonstrate that the frog peripheral auditory system extracts the periodicity of complex signals by phase-locking to the peaks in the envelope of the stimulus. The representation of the periodicity of multiple-harmonic sounds in the frog's auditory nerve differs from the representation of periodicity in the response properties of mammalian auditory nerve fibers. The detailed spectral (formant) structure of synthetic vowel sounds is represented in the phase-locked activity of mammalian fibers with CFs below the 5000 Hz limit of synchronization. The fundamental frequency of these stimuli is signalled in the synchronized response of fibers with high CFs, and fibers with CFs between the dominant formant peaks in the vowel. Synchronized responses are not typically dominated by the fundamental frequency, because fibers also synchronize to the harmonic in the complex sound closest to their CF (Delgutte and Kiang, 1984; Horst et al. 1986; Javel, 1980; Palmer, 1990; Young and Sachs, 1979). That is, the relative homogeneity of response seen in bullfrog peripheral fibers to a biologically-relevant signal (Figure 9.10) is not typical of responses of mammalian fibers to synthetic speech sounds. Both place and timing mechanisms operate in the mammalian periphery, and the balance between these different coding strategies is affected by the type of acoustic stimulus used (its fine-structure and its intensity), and by the tuning characteristics (CF, threshold, spontaneous activity) of individual fibers (Delgutte and Kiang, 1984; Deng et al. 1987; Horst et al. 1986).

The strong similarity in the signalling of the fundamental frequency of the synthetic advertisement call by both AP and BP fibers shows that place mechanisms of frequency representation are not required for estimation of periodicity. The data suggest that the bullfrog's eighth nerve extracts periodicity through the timing of fiber firing. While the place code that might be used to determine first harmonic frequency is less useful due to the broad tuning of eighth nerve fibers, the periodicity code is explicitly enhanced by the presence of mechanisms that extract the first harmonic period of the stimulus.

Central Coding of Periodicity

The nature of the peripheral processing suggests that the central nervous system receives information about periodicity through arrays of phase-locked activity from the eighth nerve. But how is this periodicity information represented in the central auditory system? Several recent studies have explored this question, by recording neural responses to complex stimuli at different levels of the frog ascending auditory system (Figure 9.6).

The eighth nerve projects to the dorsal medullary nucleus (DMN) in the caudal brainstem. Some cells in the DMN phase-lock to single tones at frequencies up to 800 Hz, similar to the phase-locking responses of eighth nerve fibers. These cells also phase-lock to the envelope of AM tones, up to modulation rates of about 300 Hz (Feng and Lin, 1994). This cut-off is similar to, although slightly below, that seen in the eighth nerve (see Figure 9.9). These data suggest that cells in the DMN can encode the periodicity of complex acoustic signals by their phase-locking activity.

Several studies have examined the coding of AM sounds and repetitive trains of pulses in the auditory midbrain (torus semicircularis, TS) of several species of anurans (Diekamp and Schneider, 1988; Eggermont, 1990; Epping and Eggermont, 1986a,b; Feng et al. 1990; Gooler and Feng, 1992; Rose and Capranica, 1984, 1985; Rose et al. 1985; Walkowiak, 1984). The TS is the homolog of the mammalian inferior colliculus. It receives input from the DMN, as well as from the superior olivary nucleus and the nucleus of the lateral lemniscus (Figure 9.6). The TS contains 3 different nuclei that are responsive to sounds — the principal, magnocellular, and laminar nuclei. The principal nucleus is tonotopically organized, while the magnocellular and laminar nuclei are not (Feng and Lin, 1991). Cells in the principal and magnocellular nuclei of the TS synchronize to the envelope of AM tones and noise, but the existence region for synchronization is more restricted than that for auditory nerve fibers. In some species, synchronized responses of TS neurons extend only up to modulation rates of about 40 Hz (Gooler and Feng, 1992; Rose and Capranica, 1985). In the principal nucleus of bullfrog TS, synchronization to the envelope of AM noise, even at depths of 100%, is not significant at rates above 100 Hz (Figure 9.15). Moreover, synchronization declines at depths of 50% and lower, while remaining significant at depths of 50% in the eighth nerve. The effect of modulation depth on the temporal response properties of cells in the TS is more similar to the behavioral response (Figure 9.4) than are responses of eighth nerve fibers. Some auditory-responsive cells in the TS show no phase-locking at all, even in response to low modulation rates (Rose and Capranica, 1985). This restricted range of phase-locking suggests that cells in the TS do not distinguish between stimuli on the basis of this kind of temporal code.

Another way in which TS cells could distinguish between different rates of periodicity is not by changes in the pattern of phase-locked activity, but by changes in the rate of action potential (spike) firing to different stimuli. Although TS cells do not show a wide range of phase-locking to different AM rates, they can detect differences in modulation frequency by changes in their firing rates to these different stimuli. Rose and Capranica (1984, 1985) first identified what they called AM-tuned neurons in the TS, which are maximally sensitive (fire more strongly to,

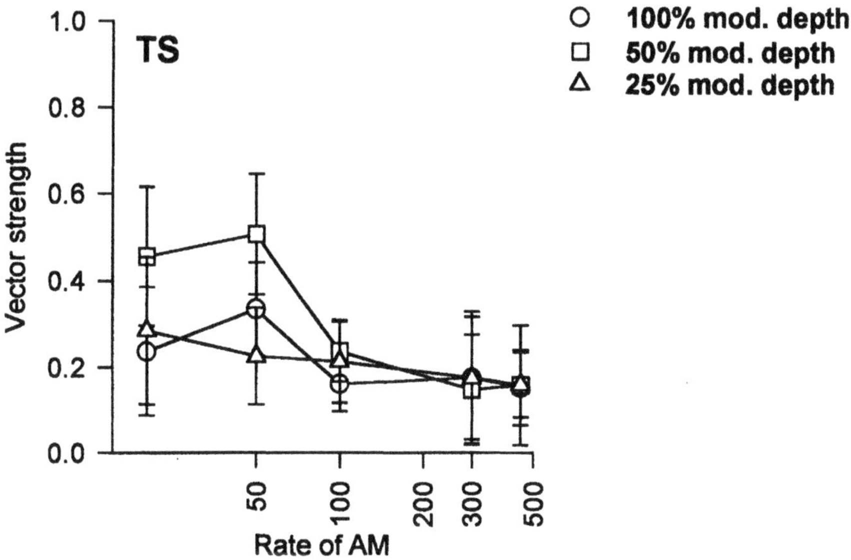

FIGURE 9.15 Vector strength (mean +/- standard deviation) of responses of cells in the TS to AM noise modulated at different rates and depths. Vector strength is calculated from period histograms. Data are based on the responses of 28 cell clusters in the principal nucleus of the TS. At a depth of 100%, vector strength is high at rates of 20 and 50 Hz, but reaches the limit of significance at rates of 100 Hz. Vector strength of responses to stimuli modulated at 50% and 25% depth is low at most AM rates. The phase-locking ability of cells in the TS falls below that of fibers in the eighth nerve.

although not in a synchronized manner) to a restricted range of AM rates. These neurons have tuning curves, not for CF as shown for eighth nerve fibers in Figure 9.7, but for AM rate. A particular nerve cell would then fire most strongly to a particular periodicity, but firing would not be phase-locked. Rose and Capranica (1984) suggested that AM-tuned neurons might form an array of what they called temporally-tuned band-pass filters for detecting biologically-relevant rates of periodicity. According to this argument, there should exist in the TS of a particular species cells specifically tuned to AM rates in that species' vocalizations. Rose and Capranica (1984) reported the existence of AM-tuned cells in two species of toads which had best modulation frequencies matching the modulation rate of the species advertisement call. Other experiments have shown a great deal of overlap in the distribution of best AM rates in the TS of different species (Eggermont, 1990; Rose and Capranica, 1984; Rose et al. 1985).

Figure 9.16 presents examples of AM tuning curves derived from responses of 4 different cell clusters in the principal nucleus of the bullfrog TS. Two of the cell clusters seem to show maximal responses at particular rates of AM, and so can be characterized by a best modulation frequency, the AM rate at which spike count is maximal. For one of these AM tuning curves, the best modulation frequency is at the 100 Hz periodicity of the bullfrog's advertisement call. The other tuning curves have best modulation frequencies unrelated to a periodicity in the spe-

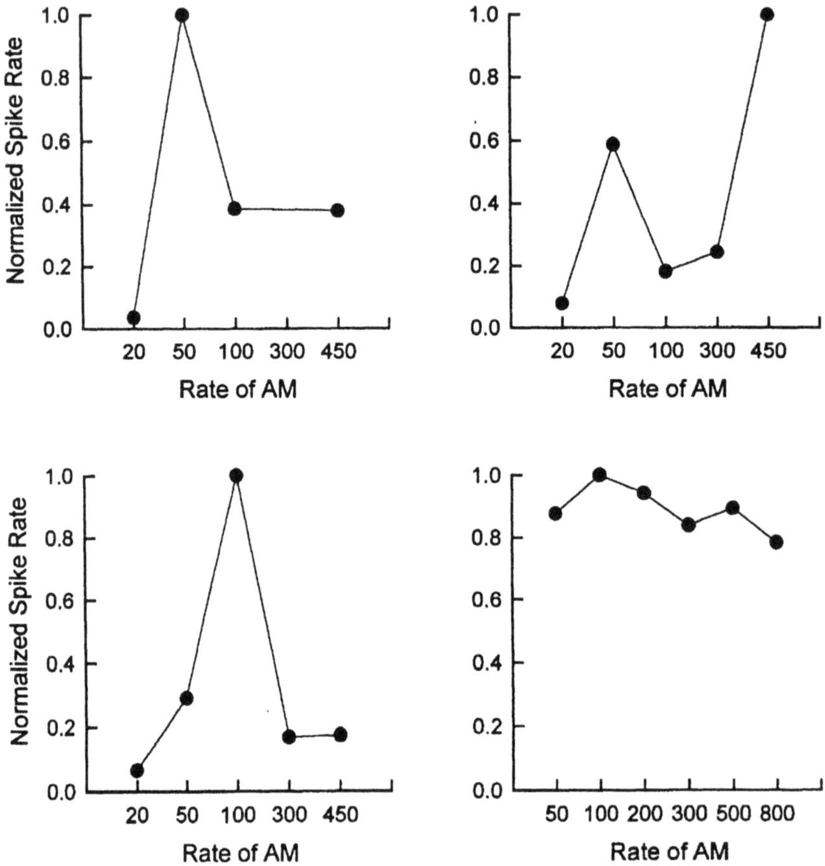

FIGURE 9.16 Tuning curves of 4 cells in the TS of the bullfrog in response to AM noise modulated at different rates. Data were analyzed as normalized spike (action potential) counts, so that the maximum spike rate to a given AM rate was set to 1.0, and all other spike rates were fractions of this value. All stimuli were modulated at a depth of 100%. The tuning curves in the left column show a neural response selective to a particular rate of modulation (50 Hz or 100 Hz). This means that the cells responded best to this particular periodicity. The tuning curves in the right column show less evidence of selective tuning. These cells responded similarly to all AM rates, or showed a more variable response pattern.

cies vocalizations, or do not show a clear preference for AM rate. All of these cell clusters, regardless of the shape of their tuning curves, show low-pass filter functions for synchronization (Figure 9.14). There is no clear relationship between the ability of a cluster to synchronize to stimuli and its AM tuning. Moreover, for some cell clusters, the best modulation frequency changes with stimulus intensity (Figure 9.17). These

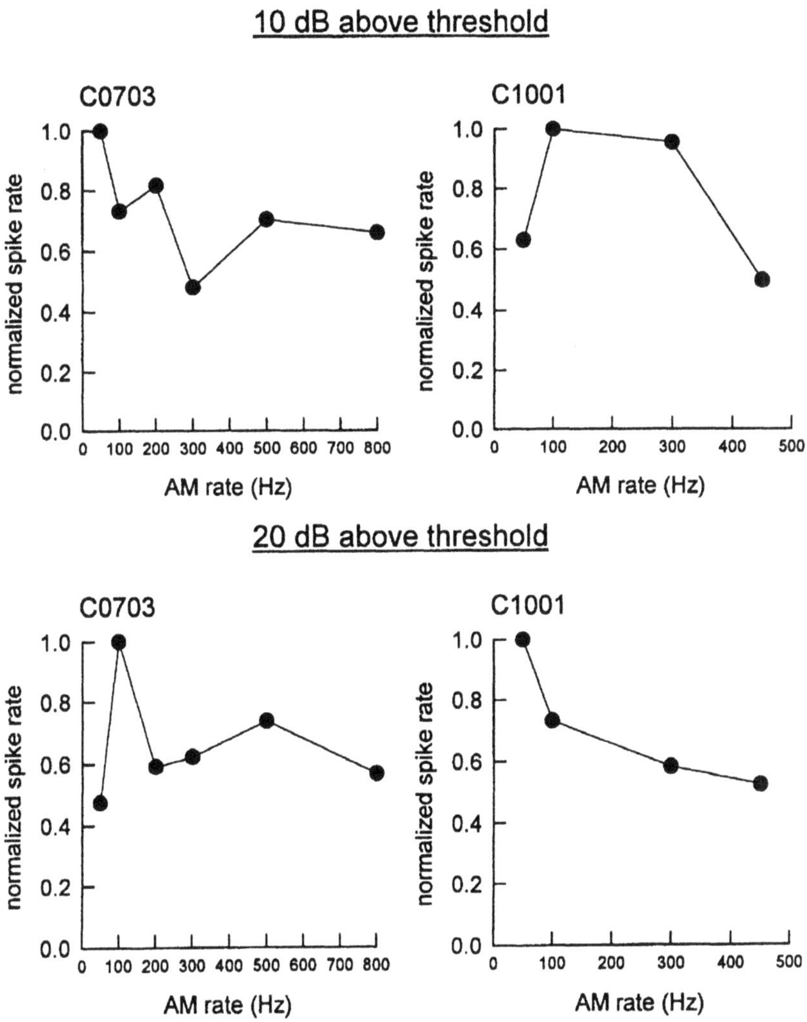

FIGURE 9.17 Tuning curves of 2 cells in the TS in response to AM noise presented at different modulation rates and at different intensities (10 and 20 dB above threshold). For both of these cells, the tuning curve changes its shape with stimulus intensity.

intensity-dependent changes in AM tuning may be the result of averaging across activity from several individual cells; but it implies that AM-tuned cells with different best modulation frequencies are not spatially segregated within the TS. Whether cells tuned to particular AM rates are clustered in the principal or magnocellular nuclei to form a kind of periodotopic or pitch map has not been explicitly tested. Such a spatial arrangement has been seen in the cat inferior colliculus (Langner and Schreiner, 1988).

Consistent with previous suggestions (Gooler and Feng, 1992; Rose and Capranica, 1985), these data suggest that one way in which the time structure of neural firing in the auditory periphery might be preserved in the auditory midbrain is by conversion from a temporal code based on phase-locked discharges into a rate code, represented by activation of different cells tuned to different AM rates. This implies that there is a transformation in the mechanism of coding of temporal cues from the auditory nerve to the auditory midbrain, but the means by which this transformation is carried out are not yet known. It is not known, for example, if cells in the TS show intrinsic oscillations as have been identified in cats and mynah birds (Langner, 1981, 1992). An examination of neural response properties in the frog central auditory system consistent with mechanisms of periodicity extraction proposed by Licklider (1951) and by Langner (1981) have yet to be conducted.

Summary and Conclusions

The robust synchronization of bullfrog eighth nerve fibers to the amplitude envelope of complex sounds, whether these sounds be synthetic advertisement calls, AM noise or AM tones, suggests that the anuran auditory periphery extracts the periodicity of complex sounds by time-domain processing. The fine-structure of complex stimuli is not typically represented in the discharge patterns of these fibers; rather, fibers are extracting information about periodicity from the waveform envelope. Amphibian papilla fibers encode information about the envelope and the fine-structure of inharmonic stimuli and of stimuli whose phase spectra have been varied to produce stimulus envelopes with less well-defined peaks of maxima and minima. The coding of periodicity information in the bullfrog's auditory periphery appears to be less complex and more homogeneous across fibers than is seen in the mammalian auditory periphery, where both envelope and fine-structure information are robustly encoded. The differences between the anuran and the mammalian system may reflect differences between a periodicity extraction algorithm based primarily on timing information, and one based on both place and timing information. The striking similarity in processing between the amphibian papilla, which supports tonotopy, and the basilar papilla, which does not operate on a place principle, suggests that a place code does not play a major role in signalling periodicity in the anuran auditory periphery. Rather, the periphery is transmitting periodicity information through arrays of phase-locked activity. The existence of AM tuned cells in the torus semicircularis suggests that the auditory midbrain is computing from these arrays an estimate of the envelope periodicity based on which type of cell is activated. In this way, place information might be used to extend the limited range of phase-locking activity in the midbrain. The exact nature of the neural computations by which this is carried out is unknown.

Acknowledgments

The research reported in this paper was supported by NIH grant NS-28565 to AMS. We thank Andrew Moiseff and Cynthia F. Moss for their comments on the manuscript. We thank Susan E. Allan for providing the spectrograms of green treefrog vocalizations, and Mark Sanderson for preparing the figures. Fred Bouffard, Gita Reese, Michael Ferragamo, and Joshua Schwartz assisted in collection of much of the data presented in this paper.

References

Allan, S.E. and Simmons, A.M. (1994) Temporal features mediating call recognition in the green treefrog, *Hyla cinerea*: Amplitude modulation. *Anim. Behav. 47:* 1073-1086.

Attneave, F. and Olson, R. K. (1971) Pitch as a medium: A new approach to psychophysical scaling. *Am. J. Psych. 84:* 147-166.

Bekesy, G. von (1960) *Experiments in Hearing*. New York: McGraw-Hill.

Bilsen, F. A. (1966) Repetition pitch: Monaural interaction of sound with the repetition of the same but phase shifted sound. *Acustica 77:* 295-300.

de Boer, E. (1956) Pitch of inharmonic signals. *Nature 178:* 535-536.

de Boer, E. (1976) On the 'residue' and auditory pitch perception. In W. D. Keidel and W. D. Neff (Eds.) *Handbook of Sensory Physiology (V/3)*. New York: Springer-Verlag, pp. 479-583.

Burns, E. M. and Viemeister, N. F. (1976) Nonspectral pitch. *J. Acoust. Soc. Am. 60:* 863-868.

Capranica, R.R. (1965) The Evoked Vocal Response of the Bullfrog: *A Study of Communication by Sound*. Cambridge, MA.: MIT Press.

Capranica, R.R. (1976) Morphology and physiology of the auditory system. In R. Llinas and W. Precht (Eds.) *Frog Neurobiology*. New York: Springer-Verlag, pp. 551-575.

Capranica, R.R. and Moffat, A.J.M. (1983) Neuroethological principles of acoustic communication in anurans. In J-P Ewert, R.R. Capranica and D.J. Ingle (Eds.) *Advances in Vertebrate Neuroethology*. New York: Plenum Press, pp. 701-730.

Delgutte, B. and Kiang, N.Y.S. (1984) Speech coding in the auditory nerve: I. Vowel-like sounds. *J. Acoust. Soc. Am. 75:* 908-918.

Deng, L., Geisler, C.D. and Greenberg, S. (1987) Responses of auditory-nerve fibers to multiple-tone complexes. *J. Acoust. Soc. Am. 82:* 1989-2000.

Diekamp, B. and Schneider, H. (1988) Neuronal processing of conspecific and related calls in the torus semicircularis of *Rana r. ridibunda Pall.* (Anura): single-unit recordings. *J. Comp. Physiol. A 13:* 301-315.

Dunia, R. and Narins, P.M. (1989) Temporal resolution in frog auditory-nerve fibers. *J. Acoust. Soc. Am. 85:* 1630-1638.

Eggermont, J.J. (1990) Temporal modulation transfer functions for single neurons in the auditory midbrain of the leopard frog. Intensity and carrier-frequency dependence. *Hear. Res. 43:* 181-198.

Epping, W.J.M. and Eggermont, J.J. (1986) Sensitivity of neurons in the auditory midbrain of the grassfrog to temporal characteristics of sound. I. Stimulation with acoustic clicks. *Hear. Res. 24:* 37-54. (a)

Epping, W.J.M. and Eggermont, J.J. (1986) Sensitivity of neurons in the auditory midbrain of the grassfrog to temporal characteristics of sound. II. Stimulation with amplitude modulated sound. *Hear. Res. 24:* 55-72. (b)

Evans, E. F. (1978) Place and time coding of frequency in the peripheral auditory system: Some physiological pros and cons. *Audiology 17:* 369-420.

Feng, A.S. and Lin, W. (1991) Differential innervation patterns of three divisions of the frog auditory midbrain (torus semicircularis). *J. Comp. Neurol. 307:* 1-18.

Feng, A.S. and Lin, W. (1994) Phase-locked response characteristics of single neurons in the frog "cochlear nucleus" to steady-state and sinusoidal-amplitude-modulated tones. *J. Neurophys. 72:* 2209-2221.

Feng, A.S., Narins, P.M. and Capranica, R.R. (1975) Three populations of primary auditory fibers in the bullfrog (*Rana catesbeiana*): Their peripheral origins and frequency selectivities. *J. Comp. Physiol. A 100:* 221-229.

Feng, A.S., Hall, J.C. and Gooler, D.M. (1990) Neural basis of sound pattern recognition in anurans. *Prog. Neurobiol. 34:* 313-329.

Feng, A.S., Hall, J.C. and Siddique, S.S. (1991) Coding of temporal parameters of complex sounds by frog auditory nerve fibers. *J. Neurophys. 65:* 424-445.

Flanagan, J. L. and Guttman, N. (1960) On the pitch of periodic pulses. *J. Acoust. Soc. Am. 32:* 1308-1319.

Fletcher, H. (1934) Loudness, pitch, and the timbre of musical tones and their relation to the intensity, the frequency, and the overtone structure. *J. Acoust. Soc. Am. 6:* 59-69.

Freedman, E.G. Ferragamo, M. and Simmons, A.M. (1988) Masking patterns in the bullfrog (*Rana catesbeiana*). II. Physiological effects. *J. Acoust. Soc. Am. 84:* 2081-2091.

Gerhardt, H.C. (1978) Mating call recognition in the green treefrog (*Hyla cinerea*): The significance of some fine-temporal properties. *J. Exp. Biol. 61:* 229-241.

Gerhardt, H.C. (1981) Mating call recognition in the barking treefrog (*Hyla gratiosa*): responses to synthetic mating calls and comparisons with the green treefrog (*Hyla cinerea*). *J. Comp. Physiol. 144:* 17-25.

Gerhardt, H.C. (1982) Sound pattern recognition in some North American treefrogs (*Anura: Hylidae*): implications for mate choice. *Am. Zool. 22:* 581-595.

Gerhardt, H.C. (1988) Acoustic properties used in call recognition by frogs and toads. In B. Fritzsch, W. Wilczynski, M.J. Ryan, T. Hetherington and W. Walkowiak (Eds.) *The Evolution of the Amphibian Auditory System.* New York: Wiley, pp. 455-483.

Gerhardt, H.C. and Klump, G.M. (1988) Phonotactic responses and selectivity of barking treefrogs (*Hyla gratiosa*) to chorus sounds. *J. Comp. Physiol. A 163:* 795-802.

Gerhardt, H.C., Allan, S. and Schwartz, J.J. (1990) Female green treefrog (*Hyla cinerea*) do not selectively respond to signals with a harmonic structure in noise. *J. Comp. Physiol. A 166:* 791-794.

Goldstein, J. L. (1973) An optimum processor theory for the central formation of the pitch of complex tones. *J. Acoust. Soc. Am. 54:* 1496-1516.

Goldstein, J. L., and Srulovicz, P. (1977) Auditory-nerve spike interval as an adequate basis for aural frequency discrimination. In E. F. Evans and J. P. Wilson (Eds.) *Psychophysics and Physiology of Hearing.* New York: Academic, pp. 337-348.

Gooler, D.M. and Feng, A.S. (1992) Temporal coding in the frog auditory midbrain: The influence of duration and rise-fall time on the processing of complex amplitude-modulated stimuli. *J. Neurophys. 67:* 1-22.

Hainfeld, C.M., Boatright-Horowitz, S.L., Boatright-Horowitz, S.S. and Simmons, A.M. (1995) Discrimination of phase spectra in complex sounds by the bullfrog (Rana catesbeiana). *J. Comp. Physiol. A.*

Hall, J.C. and Feng, A.S. (1986) Neural analysis of temporally patterned sounds in the frog's thalamus: processing of pulse duration and pulse repetition rate. *Neurosci. Lett. 63:* 215-220.

Hall, J.C. and Feng, A.S. (1990) Classification of the temporal discharge patterns of single auditory neurons in the dorsal medullary nucleus of the northern leopard frog. *J. Neurophys. 64:* 1460-1473.

Hartmann, W.M. (1988) Pitch perception and the segregation and integration of auditory entities. In G.M. Edelman, W.E. Gall and W.M. Cowan (Eds.) *Auditory Function: Neurobiological Bases of Hearing.* New York: Wiley, pp. 623-645.

Hillery, C.M. and Narins, P.M. (1984) Neurophysiological evidence for a travelling wave in the amphibian inner ear. *Science 225:* 1037-1039.

Hillman, D.E. (1976) Morphology of peripheral and central vestibular systems. In R. Llinas and W. Precht (Eds.) *Frog Neurobiology.* New York: Springer-Verlag, pp. 452-480.

Horst, J.W., Javel, E. and Farley, G.R. (1986) Coding of spectral fine-structure in the auditory nerve. I. Fourier analysis of period and interspike interval histograms. *J. Acoust. Soc. Am. 79:* 398-416.

Javel, E. (1980) Coding of AM tones in the chinchilla auditory nerve: Implications for the pitch of complex tones. *J. Acoust. Soc. Am. 68:* 133-146.

Javel, E. and Mott, J. B. (1988) Physiological and psychophysical correlates of temporal processes in hearing. *Hear. Res. 34:* 275-294.

Kiang, N. Y. S. (1965) *Discharge Patterns of Single Fibers in the Cat's Auditory Nerve.* Cambridge, MA: M.I.T. Press.

Langner, G. (1981) Neuronal mechanisms for pitch analysis in the time domain. *Exp. Brain Res. 44:* 450-454.

Langner, G. (1992) Periodicity coding in the auditory system. *Hear. Res. 60:* 115-142.

Langner, G. and Schreiner, C. E. (1988) Periodicity coding in the inferior colliculus of the cat. I. Neuronal mechanisms. *J. Neurophys. 60:* 1799-1822.

Lazzaro, J. and Mead, C. (1989) Silicon modeling of pitch perception. *Proc. Nat. Acad. Sci. USA 86:* 9597-9601.

Lewis, E. R. (1992) Convergence of design in vertebrate acoustic sensors. In D. B. Webster, R. R. Fay and A. N. Popper (Eds.) *The Evolutionary Biology of Hearing.* New York: Springer Verlag, pp. 163-183.

Lewis, E. R., Leverenz, E. L. and Koyama, H. (1982) The tonotopic organization of the bullfrog amphibian papilla, an auditory organ lacking a basilar membrane. *J. Comp. Physiol. 145:* 437-445.

Licklider, J. C. R. (1951) A duplex theory of pitch perception. *Experientia 7:* 128-134.

Licklider, J. C. R. (1956) Auditory frequency analysis. In C. Cherry (Ed.) *Information Theory.* New York: Academic, pp. 253-268.

Loeb, G. E., White, M. W., and Merzenich, M. M. (1983) Spatial cross-correlation: A proposed mechanism for acoustic pitch perception. *Biol. Cybernet. 47:* 149-183.

Loftus-Hills, J.J. and Littlejohn, M.J. (1971) Pulse repetition rate as a basis for mating call discrimination by sympatric species of Hyla. *Copeia 1971:* 154-156.

Lundeen, C. and Small, A. M. (1984) The influence of temporal cues on the strength of periodicity pitches. *J. Acoust. Soc. Am. 75:* 1578-1587.

Meddis, R. and Hewitt, M. J. (1991) Virtual pitch and phase sensitivity of a computer model of the auditory periphery. I: Pitch identification. *J. Acoust. Soc. Am. 89:* 2866-2882.

Megela, A.L. and Capranica, R.R. (1983) A neural and behavioral study of auditory habituation in the bullfrog, *Rana catesbeiana. J. Comp. Physiol. 151:* 423-434.

Megela-Simmons, A., Moss, C.F. and Daniel, K.M. (1985) Behavioral audiograms of the bullfrog (*Rana catesbeiana*) and the green treefrog (*Hyla cinerea*). *J. Acoust. Soc. Am. 78:* 1236-1244.

Moore, B.C.J. (1989) *An Introduction to the Psychology of Hearing.* New York: Academic.

Narins, P.M. and Hillery, C.M. (1983) Frequency coding in the inner ear of anuran amphibians. In R. Klinke and R. Hartmann (Eds.) *Hearing - Physiological Bases and Psychophysics.* New York: Springer, pp. 70-76.

Narins, P.M. and Wagner, I. (1989) Noise susceptibility and immunity of phase locking in amphibian auditory-nerve fibers. *J. Acoust. Soc. Am. 85:* 1255-1265.

Palmer, A.R. (1990) The representation of the spectra and fundamental frequencies of steady-state single- and double-vowel sounds in the temporal discharge patterns of guinea pig cochlear-nerve fibers. *J. Acoust. Soc. Am. 88:* 1412-1426.
Patterson, R.D. (1987) A pulse ribbon model of peripheral auditory processing. In W. A. Yost and C. S. Watson (Eds.) *Auditory Processing of Complex Sounds.* Hillsdale, N.J.: Erlbaum, pp. 167-179.
Pitchford, S. and Ashmore, J.F. (1987) An electrical resonance in hair cells of the amphibian papilla of the frog *Rana temporaria. Hear. Res. 27:* 75-83.
Plomp, R. (1976) *Aspects of Tone Sensation.* New York: Academic.
Richards, D.G. and Wiley, R.H. (1980) Reverberations and amplitude fluctuations in the propagation of sound in a forest: Implications for animal communication. *Amer. Nat. 115:* 381-399.
Ritsma, R. J. (1962) Existence region of the tonal residue. I. *J. Acoust. Soc. Am. 34:* 1224-1229.
Ritsma, R. J. (1963) Existence region of the tonal residue. II. *J. Acoust. Soc. Am. 35:* 1241-1245.
Ritsma, R. J. (1967) Frequencies dominant in the perception of the pitch of complex sounds. *J. Acoust. Soc. Am. 42:* 191-198.
Rose, G.J. and Capranica, R.R. (1984) Processing amplitude-modulated sounds by the auditory midbrain of two species of toads: matched temporal filters. *J. Comp. Physiol. 154:* 211-219.
Rose, G.J. and Capranica, R.R. (1985) Sensitivity to amplitude modulated sounds in the anuran auditory nervous system. *J. Neurophys. 53:* 446-465.
Rose, G.J., Brenowitz, E.A. and Capranica, R.R. (1985) Species specificity and temperature dependency of temporal processing by the auditory midbrain of two species of treefrogs. *J. Comp. Physiol. 157:* 763-769.
Rose, J. E., Brugge, J. F., Anderson, D. J. and Hind, J. E. (1967) Phase-locked response to low-frequency tones in single auditory nerve fibers of the squirrel monkey. *J. Neurophys. 30:* 769-793.
Semal, C. and Demany, L. (1980) The upper limit of musical pitch. *Music Percep. 8:* 165-175.
Schouten, J. F. (1940) The residue and the mechanism of hearing. *Proc. K. Ned. Akad. Weten. 43:* 991-999.
Schouten, J. F., Ritsma, R. J. and Cardozo, B. L. (1962) Pitch of the residue. *J. Acoust. Soc. Am. 34:* 1418-1424.
Schwartz, J.J. (1987) The importance of spectral and temporal properties in species and call recognition in a neotropical treefrog with a complex vocal repertoire. *Anim. Behav. 35:* 340-347.
Schwartz, J.J. and Simmons, A.M. (1990) Encoding of a spectrally-complex communication sound in the bullfrog's auditory nerve. *J. Comp. Physiol. A 166:* 489-499.
Schwartz, J.J. and Wells, K.D. (1984) Interspecific acoustic interactions of the neotropical treefrog *Hyla ebraccata. Behav. Ecol. Sociobiol. 14:* 211-224.
Schwartz, J.J. and Wells, K.D. (1985) Intra- and interspecific vocal behavior of the neotropical treefrog *Hyla microcephala. Copeia 1985:* 27-38.
Siebert, W. M. (1970) Frequency discrimination in the auditory system: Place or periodicity mechanisms. *Proc. IEEE 58:* 723-730.

Shofner, W.S. and Feng, A. S. (1983) A quantitative light microscopic study of the bullfrog amphibian papilla tectorium: correlations with the tonotopic organization. *Hear. Res. 11:* 103-116.

Simmons, A.M. (1988) Selectivity for harmonic structure in complex sounds by the green treefrog (*Hyla cinerea*). *J. Comp. Physiol. A 162:* 397-403. (a)

Simmons, A.M. (1988) Masking patterns in the bullfrog (*Rana catesbeiana*), I: Behavioral effects. *J. Acoust. Soc. Am. 83:* 1087-1092. (b)

Simmons, A.M. and Ferragamo, M. (1993) Periodicity extraction in the anuran auditory nerve, I: 'Pitch-shift' effects. *J. Comp. Physiol. A 172:* 57-69.

Simmons, A.M., Schwartz, J.J. and Ferragamo, M. (1992) Auditory nerve representation of a complex communication sound in background noise. *J. Acoust. Soc. Am. 91:* 2831-2844.

Simmons, A.M., Reese, G. and Ferragamo, M. (1993) Periodicity extraction in the anuran auditory nerve, II: Phase and temporal fine-structure. *J. Acoust. Soc. Am. 93:* 3374-3389. (a)

Simmons, A.M., Buxbaum, R.C. and Mirin, M. (1993) Perception of complex sounds by the green treefrog, *Hyla cinerea*: Envelope and fine-structure cues. *J. Comp. Physiol. A 173:* 321-327. (b)

Stevens, S. S. (1935) The relation of pitch to intensity. *J. Acoust. Soc. Am. 6:* 150-154.

Stevens, S. S., Davis, H., and Lurie, M. H. (1935) The localization of pitch perception on the basilar membrane. J. *Gen. Psychol. 13:* 297-315.

Straughan, I.R. (1975) An analysis of the mechanisms of mating call discrimination in the frogs *Hyla regilla* and *H. cadaverina*. *Copeia 1975:* 415-424.

Thurlow, W. R., and Small, A. M. (1953) Pitch perception for certain periodic acoustic stimuli. *J. Acoust. Soc. Am. 27:* 132-137.

Walkowiak, W. (1984) Neuronal correlates of the recognition of pulsed sound signals in the grass frog. *J. Comp. Physiol. 155:* 57-66.

Wells, K.D. (1988) The effect of social interactions on anuran vocal behavior. In B. Fritzsch, W. Wilczynski, M.J. Ryan, T. Hetherington and W. Walkowiak (Eds.) *The Evolution of the Amphibian Auditory System*. New York: Wiley, pp. 433-454.

Wever, E. G. (1949) *Theory of Hearing*. New York: Wiley.

Wever, E. G. (1985) *The Amphibian Ear.* Princeton: Princeton Univ. Press.

Wier, C., Jesteadt, W. and Green, D. (1977) Frequency discrimination as a function of frequency and sensation level. *J. Acoust. Soc. Am. 61:* 178-184.

Wightman, F. L. (1973) The pattern-transformation model of pitch. *J. Acoust. Soc. Am. 54:* 407-416.

Young, E.D. and Sachs, M.B. (1979) Representation of steady-state vowels in the temporal aspects of the discharge patterns of populations of auditory-nerve fibers. *J. Acoust. Soc. Am. 66:* 1381-1403.

Zakon, H.H. and Wilczynski, W. (1988) The physiology of the anuran eighth nerve. In B. Fritzsch, W. Wilczynski, M.J. Ryan, T. Hetherington, and W. Walkowiak (Eds.) *The Evolution of the Amphibian Auditory System*. New York: Wiley, pp. 125-155.

Zwislocki, J. (1965) Analysis of some auditory characteristics. In R. D. Luce, R. R. Bush, and E. H. Galanter (Eds.) *Handbook of Mathematical Psychology, Vol. III*. New York: Wiley, pp. 1-97

10

Recognition of Visual Signals: Eyes Specialize

Russell D. Fernald

Abstract

Since the beginning of biological evolution, light has been a dominant selective force in evolution. As a consequence, there exists a remarkable array of biological adaptations to light ranging from circadian rhythms to vision. This chapter addresses how vision has evolved in the service of behavior. Visual systems are responsible for collecting and reducing visual information, sifting the important from the unnecessary. By examining how environmental constraints have shaped the evolution of visual signals, important selective pressures have been identified. Reproduction and predation appear to be the most important determinants of visual processes. A manifestation of these selective forces can be seen in extant eyes. Eyes must capture light and resolve it into images. Virtually all the possible optical designs can be found in existing eyes from the simplest photoreceptive pit to complex eyes with several fovea. Despite their vast differences at the level of the entire organ, all these eyes share a common photoreceptive molecule, opsin. In contrast, other features of eyes are not homologous but arise through common descent. The use of eyes to recognize and respond to visual signals is common throughout the animal kingdom. As an example, the complex array of visual signals in one species is presented and changes in the brain that result from social interactions outlined.

Introduction

Since biological evolution began over 5 billion years ago, the earth has been alternately bathed in sunlight and darkness, fueling life and defining biological time. Light and the light/dark cycle are arguably the most potent selective forces ever to have acted on biological organisms. Light is used both as the source of energy in the plant kingdom and as a source of information for most animals. The consequences of this selective pressure can be seen in the phenomenal array of adaptations to light, ranging from biological rhythms to vision itself. Approximately 10^{15} sunrises and sunsets have occurred since biological time began (Fernald, 1989) and have shaped the visual systems and biological rhythms of all living organisms. Eyes are thought to have evolved independently in different organisms at least 40 times and probably as many as 65 times (Salvini-Plawen and Mayr, 1977), demonstrating their importance to animals. Image forming eyes arose in six of the 33 extant metazoan phyla (Cnidaria, Mollusca, Annelida, Onychophora, Arthropoda, and Chordata) which contribute about 96% of the known species (Land and Fernald, 1992). Even though light has clearly been a central force in visual evolution, understanding how selective pressures have led to particular adaptations in the visual system has proven difficult. However, standard comparative analyses along with recently developed molecular techniques have provided many important insights.

Von Uexkill (1909) first realized that every animal species lives its own Umwelt", meaning particular world and referring to the unique perceptual world of each species. A bat using sonic echoes to delineate the world in darkness surely perceives its surroundings differently than does a giraffe which relies on its eyes, looking down on the scene from above. Animals possess different sensory capabilities which limit and define the stimuli they can sense. This restriction on their perceived world correspondingly restricts their responses. At the turn of the century, von Uexkill could not have anticipated the discovery of magnetic, electric or pressure senses, nor could he imagine vision that extends into the infrared and ultraviolet. These discoveries make his prescient writing all the more remarkable.

This chapter will address the question of how vision has evolved in the service of animal behavior. What do animals need to see and how do they see it? Where possible, it will also address the selective pressures that appear to have shaped the evolution of animal vision. This analysis of animal vision will first describe what needs to be seen and how vision is an integral part of understanding an organism. Then, there is a short introduction to the evolution of visual systems and visual system pathway organization and finally some model systems are presented in which knowledge about how vision works has been extracted from experiments.

I. Visual Stimuli: Collecting and Reducing Information

Any scene contains an enormous amount of visual information of which only a fraction may be biologically important. Early ethologists recognized this fact and conceptualized the role of the visual (and other sensory systems) as relatively simple filters', evolved to sort important from non-important information (e.g. Lorenz, 1978). Although filtering may play an important role in processing of visual information, seeing' is a much more active process. Interpreting a scene requires selecting relevant elements, classifying them and thereby reducing uncertainty about events in that scene. For example, as humans, we can recognize objects, motion, depth, color and brightness, all without apparent conscious thought. Such recognition or classification radically reduces the amount of information to be processed. Instead of the many thousand bits of information which characterize all the features of say, a face, the act of recognition reduces this to one bit of information: is or isn't it Auntie Em - yes or no?

In addition to recognition, a second important requirement of seeing is localizing important objects in space and estimating their movement or trajectory. Is that predator moving towards or away from me and how fast is it moving? Reduced to its essence, processing visual information about objects in space requires quick, reliable reporting of their significance, location and trajectory. This may require more time than recognition of objects *per se* and may be processed differently.

Successful animals process visual information in different ways, exploiting many different types of eyes to collect information and visual systems to process it. Natural selection ensures that each such system serves its species effectively. The challenge is to discover how such variation in visual systems arose and how it now functions.

One way to explore the functioning of vision is to study how the visual signals that have evolved in the service of communication among animals are processed. Most visual signals appear to have evolved to allow the maximum signal to be detected relative to background noise. Endler (1992) has argued eloquently that visual signals, receptors and behavior are functionally related and, thus, influence the evolution of one another. Visual signals that are easily detected and clearly perceived will be selected, as will behavior that favors using these signals in situations in which they will be effective.

There are several selective pressures that are known to have influenced the evolution of visual signals. First are the premier, competing selective forces of reproduction and predation. Visual signals evolve as a balance between the relative importance to the species of predation and mate selection (Endler, 1980). In conditions in which predation dominates, coloration will be more cryptic, whereas when mate selection dominates, coloration will be more conspicuous. Second, the evolution of visual signals is greatly influenced by the visual environment. Illuminating light can vary in the relative intensity and distribution of wavelengths it contains. The intensity and chromatic content of incident light may also vary as a function of location, time of day, and season. All these factors bias the properties of visual signals and hence their visibility. Most signalling is done by physical structures on the animal that reflect light in some fashion. The environment itself modifies the transmission of signals by modulating, decreasing or even disrupting the information sent. Communicatory signals generated by living animals are the result of a combination of all these selective forces.

Every habitat can have a characteristic spectrum of illumination with distinct temporal and spatial variation. Animals in that environment may use visual signals at specific times and places in which their effectiveness is greater than at other times and places. This suggests that finding a match between environmental characteristics and visual signals might be quite difficult (e.g. Endler, 1992). One exception is found among animals living underwater that experience a stable, well defined optical habitat due to the strong and effective filtering properties of water. Moreover, most underwater habitats including large lakes and oceans, are not subject to transient spectral filtering produced by forest canopies or other features of the terrestrial landscape. As a consequence, most aquatic animals have evolved and now live in relatively stable optical environments, making them ideal candidates for analysis of the evolution of visual signalling.

The general notion that there might be a match between environmental lighting conditions and the visual systems of fish was first fully explored by Munz and McFarland (1977). They found that the spectral sensitivities of many species of fish occupying a common habitat had similar characteristics. Tropical marine and freshwater fish living in clear water tend to have more blue sensitive retinal photopigments than those living in colored waters (Munz and McFarland, 1977; Levine and MacNichol Jr., 1979; Lythgoe, 1988). Fish living in deep water are less sensitive to reds than shallow water species, as would be expected based on the filtering properties of water (e.g. Fernald, 1988). In several instances those matches have been traced to the molecular properties of visual photopigments responsible for detecting light (see below).

As specific lighting conditions shape the visual system through selection, many features of the animal may reflect these selective forces. For example, detection of food may depend on the lighting conditions which will, in turn, determine where and how the animal spends its time. Endler (1992) has written in some detail about the idea that sensory system characteristics will both influence and be influenced by the environmental conditions. An opportunity for understanding this interplay exists in the case of pollinating insects.

Many insects, both social and solitary, are dependent on the pollen and nectar diet offered by flowers. Since flowers are scattered and unreliable food sources, insects should have sensory systems that favor efficient foraging strategy. Ultimately, this requires that insects be able to learn colors associated with a reward, an ability that has been shown in a wide variety of insects (e.g. von Frisch, 1914, 1967; see chapter by Gould). An essential prerequisite for this capacity is a color vision system that allows effective discrimination among flowers of different species. Many angiosperm plant species compete with one another for animals as pollen vectors (e.g. Darwin, 1876), and the optimal version of this process occurs when individual pollinators visit conspecific flowers exclusively. In this case, the visual signals provided by flowers must be easily detectable but also readily distinguishable from those of competing species (e.g. von Frisch, 1914, 1967; Menzel, 1967). Although many investigators have proposed such an evolutionary relationship between flower signals and insect vision, Chittka and Menzel (1992) have provided a quantitative evolutionary explanation for this relationship, by collecting evidence at three levels. First, they showed that the flower color spectra contain sharp steps precisely where the insect photoreceptor sets can best distinguish such steps. Second, they showed that the insect photoreceptors are distributed across the color spectrum such that they generate the largest possible range of different excitation for flower signals. And, third, they showed that the photoreceptor signals in insects are combined to maximize the perceived differences among flower colors. Taken together, these data strongly support the idea that the selective pressure from feeding have shaped the spectral sensitivity of vision among pollinating insects.

Clearly, selective pressures such as these do not exist in isolation, and a constellation of structures and processes are shaped during the course of evolution. However, the potency of light among the variety of selective pressures has resulted in making its consequences more evident.

II. Evolution of Eyes: No Shortage of Surprises

Eyes provide the preeminent source of sensory information for the brain in most species, and the many features of the eye reflect particular selective pressures, both from the environment and from other animals. As a result, the evolution of eyes, among all the sense organs, has attracted the most attention from scientists. In writing about the eye, very few authors can resist invoking a famous passage from Darwin (1859) in which he anticipated that the evolution of the eye would become a target for attacks on his theory because eyes are "such organs of extreme perfection and complication." He wrote: "...that the eye... could have been formed by natural selection seems, I freely confess, absurd in the highest possible degree." More than a century later, the complex and diverse structures and functions of eyes still astonish scientists. However, there has been progress in understanding how eyes are shaped by natural selection both at the macroscopic level, from the recent discovery of wholly new kinds of eyes, and at the microscopic level, from the conservation of key molecules and developmental processes during evolution.

One common feature of visual systems is their sensitivity to only a narrow range of wavelengths relative to the broad spectrum of energy produced by sunlight. An explanation for this limit is that eyes first evolved in animals living in water which strongly filters all but two narrow ranges of electromagnetic (EM) radiation (Fernald, 1989). As shown in Figure 10.1, the range of EM radiation visible" for most organisms is a narrow, sharply defined band with an attenuation 6 orders of magnitude lower than its flanking wavelengths. Only EM in this narrow band penetrates water and hence must have shaped the eyes of early organisms that evolved biochemical mechanisms sensitive within this range. Interestingly, though many animal species have long since moved onto land where they are exposed to the full spectrum of EM radiation from the sun, animal eyes remain limited to this narrow band. A large number of insects can see well into the ultraviolet side of the visible spectrum, and recently it has been shown that certain fish and birds do as well (e.g. Viltala et al. 1995). The other range of EM radiation which might allow transmission through water is at the very low frequency end of the spectrum ($< 10^3$ Hz). There is no radiation from the sun at this low frequency but several groups of animals have evolved mechanisms to exploit this long wavelength energy. For example, weakly electric fish evolved independently in both Africa and South America, and species from both groups exploit the low frequency end of the EM spectrum for signalling conspecifics. This signalling window is available in murky water and is used effectively by the small group of electric fish species. EM radiation of such low frequencies, however, means that the wavelengths are very long (Figure 10.1), making image formation impossible since physical laws dictate that any image forming receptor must be large relative to the wavelengths detected. Consequently, electric fish have an array of small receptors distributed along their bodies and images' of the world must be computed from the collective responses of these organs.

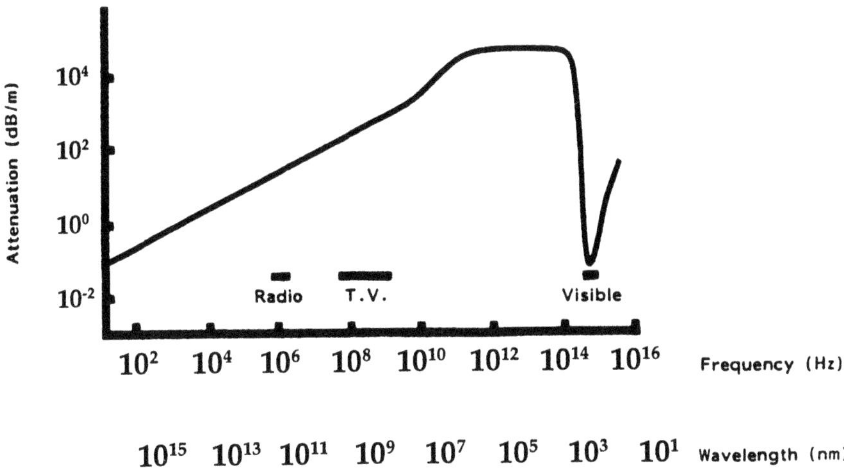

FIGURE 10.1 Attenuation (dB/m) of electromagnetic radiation (EM) in seawater plotted as a function of frequency (Hz) and wavelength (nm) of that radiation. The narrow band of electromagnetic energy which corresponds to visible light is shown as are the bands used for radio and television transmission. The band of EM radiation we now consider visible light is transmitted through water with an attenuation 6 orders of magnitude less than adjacent wavelengths. (Redrawn from Fernald, 1988).

To be useful to their owners, eyes must capture light and, in image forming eyes, resolve it into images. Eye design is fundamentally limited by the physical properties of light: it travels in straight lines, can be reflected, and varies in both wavelength (color) and intensity. Many of the principles of design and even apparent errors we find in existing eyes reflect constraints arising directly from these physical properties of light. Visual information from the environment is detected by specialized cells called photoreceptors contained in a thin retinal layer at the back of the eye. The light energy collected by the lens from the environment is contained in photons that are captured by two molecules acting together within the photoreceptor. Opsin molecules, held in association with a visual pigment or chromophore, change shape in response to excitation by photons and then convert photon energy into electrical energy that is interpretable by the nervous system. Photoreceptors with a peak sensitivity to a particular wavelength differ in certain characteristics of the opsin molecule they contain (Nathans et al. 1987). The molecular structure of opsins consists of seven transmembrane helices with short loops on both sides of the membrane; the helices are arranged like barrels roughly in a circle within the membrane. The chromophore, which is one of a family of only four close relatives of vitamin A, is covalently attached in the center of the helices of the opsin molecule at key amino acids. When the chromophore accepts a photon of light, the molecule flips from the 11-*cis* to the all-*trans* form, which triggers a biochemical cascade leading to excitation of the receptor cell (Yau and Baylor, 1989). Recent work has related specific amino acids in the sequence of the opsin molecule to the peak wavelength absorbance (Neitz et al. 1991; Chang et al. 1995). The discovery and clarification of a direct causal link between a molecular structure and its consequences for a perceptual process is remarkable in its own right, but also because features described here are common to all metazoan opsins. Based on the degree of similarity in their DNA, they must share a common origin which predates anything we might call an eye (Land and Fernald, 1992). Two regions of the molecule, the cytoplasmic loop between helices 1 and 2 and the site of attachment of the chromophore molecule in helix 7, show very close similarity in opsins from vertebrates, insects and Octopus, whose ancestries diverged in the Cambrian. Recently Goldsmith (1990) has reviewed the molecular phylogeny of the opsin molecules.

The evidence that closely related opsins are used throughout phylogeny to convert photons to electrons might suggest that other features of eyes would also be conserved. Surprisingly, the other major protein in the eye that has been carefully studied, the lens protein, is not conserved across species. Instead, a variety of proteins serve as lenses, including enzymes (Land and Fernald, 1992). This suggests that the evolution of eyes proceeds at several levels, and even though the gross eye structures are not homologous, some of their constituents clearly are.

Tracing the evolution of eyes requires deciding whether similarity in structure is a result of evolutionary convergence or of common descent (Land and Fernald, 1992). Because there are only a few ways to produce an eye that gives a usable image, similar structures occur in unrelated animals. The best example is the phylogenetically unrelated eyes of fish and squid (Packard, 1972). The eyes of these two aquatic animals are similar in many details, presumably because production of large, camera-type eyes to be used underwater requires a spherical lens, eye-muscles, iris, etc. (Fernald, 1990a). However, the fact that these eyes are unrelated is clearly demonstrated by comparing the structure of the simpler, noninverted retina of the cephalopod with the inverted and multilayered retina in the fish. In contrast, human and fish eyes are related by common descent though in some respects, they are quite different from one another (Fernald, 1993).

Evolution of ocular structures has proceeded in two stages (Land and Fernald, 1992). The first step, production of simple eye-spots, is found in nearly all the major animal groups. Such eye-spots are comprised of

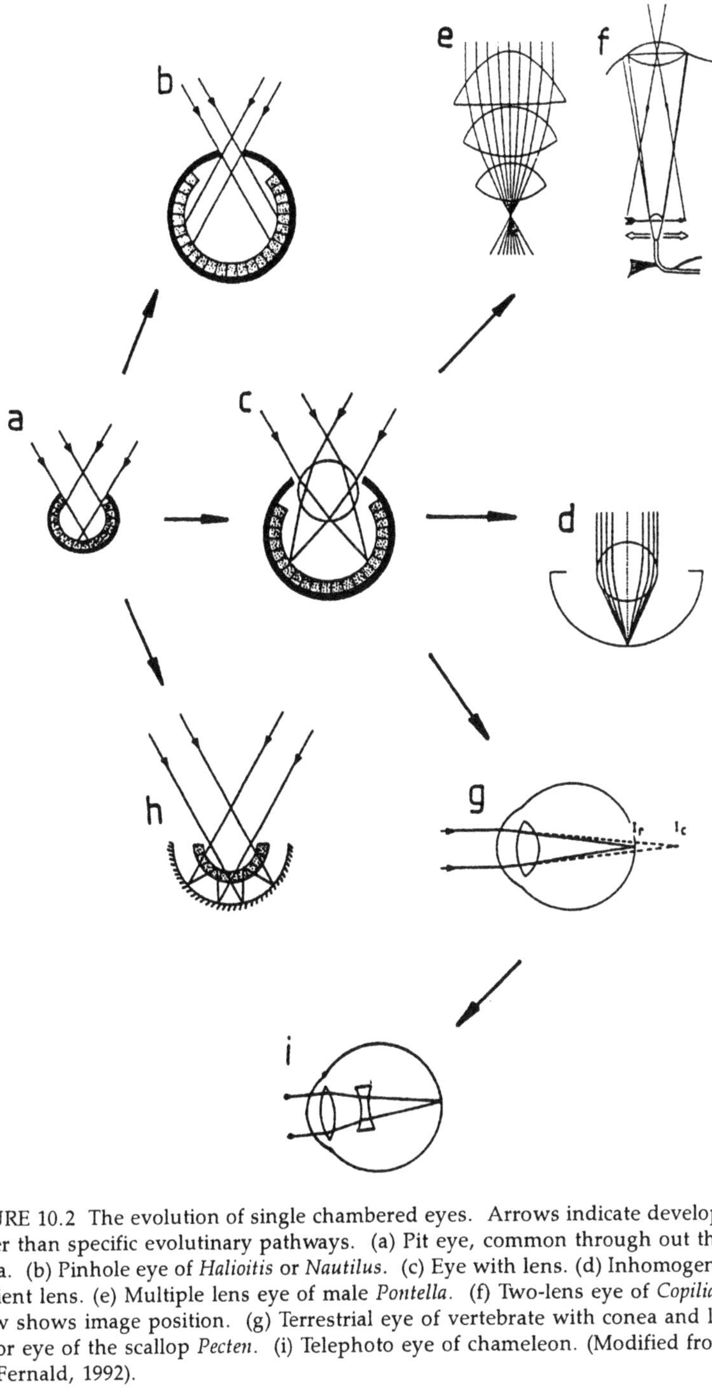

FIGURE 10.2 The evolution of single chambered eyes. Arrows indicate developments, rather than specific evolutinary pathways. (a) Pit eye, common through out the lower phyla. (b) Pinhole eye of *Halioitis* or *Nautilus*. (c) Eye with lens. (d) Inhomogeneous or gradient lens. (e) Multiple lens eye of male *Pontella*. (f) Two-lens eye of *Copilia*. Solid arrow shows image position. (g) Terrestrial eye of vertebrate with conea and lens. (h) Mirror eye of the scallop *Pecten*. (i) Telephoto eye of chameleon. (Modified from Land and Fernald, 1992).

a small number of receptors in an open cup of screening pigment (Figure 10.2). Though useful for informing an animal about the distribution of light and dark, such eye-spots can play no role in pattern recognition or control of locomotion. The second stage in eye evolution is the addition of an optical system which provides an image. As noted above, such eyes occur in 96% of known species distributed among 6 phyla (Land and Fernald, 1992). At least eleven distinct optical methods of producing images have evolved, the most recently described is a telephoto lens, which was identified in 1995 (Ott and Schaefel, 1995). Indeed, it is remarkable that five or six of the optical mechanisms have been discovered only in the past 25 years. The puzzle is why all animals do not have camera-type eyes, which are demonstrably superior in several respects (Nilsson, 1989). Perhaps conversion of one eye type to another requires intermediates that are either useless or much worse than the existing design, making such a switch essentially lethal to animals that depend on sight. Though this argument makes a certain amount of sense, some existing cases of novel optical combinations suggest this is probably not the whole story.

Textbooks tend to group animal eyes into two groups, the camera-style or simple' eyes and the compound eyes. Although this dichotomy reflects a real and fundamental difference in optical mechanisms, it conceals a remarkable diversity of optical systems subsumed under each heading (e.g. Land and Fernald, 1992; Nilsson, 1989). There is not space here to enumerate the differences and possible evolutionary sources of each, but two recent surprise discoveries suggest that the evolutionary paths to each kind of eye may not be as divergent as once thought.

Nilsson and Modlin (1994) have recently described a mysid shrimp (*Dioptromysis paucispinous*) that has a combined simple and compound eye. This eye is partly compound, with multiple facets exactly like the eye of an insect and partly simple, with a single lens focussing an image on a sheet of receptors like that of a human. The shrimp are about 5 mm long and the eyes, which sit at the ends of eyestalks, are roughly spherical. The surface of the eye is largely faceted, with the characteristic honeycomb of hexagonal facets (ca. 800-900). But at the back there is a single giant facet peering towards the shrimp's tail. Although these giant facets normally point backward, the shrimp frequently rotates them forward, probably to get a better look at something since the giant facet has ca. five times the acuity (but much lower sensitivity) than the remainder of the eye. It is as if the shrimp is carrying a pair of binoculars for the occasional detailed look at something ahead of it. One serious problem with this combination eye is that the optics of compound eyes produce an upright image while the optics of simple eyes produce an inverted image. This discrepancy has been solved optically by this animal so that the visual images produced by the different parts of the eye are continuous. The discovery that simple and compound eye types can be found in a single animal raise a question of how a developmental program could produce this outcome. Another recent discovery about ocular development may bear on this question.

Recently, Halder et al. (1995) discovered a gene named eyeless that is expressed in the eye during development of the fruit fly (*Drosophila*) and that is homologous to a developmental gene expressed in mouse and humans. When this gene is missing in a fruit fly, no eyes develop, and, conversely, when the gene is expressed at different locations on the fruit fly body it initiates the construction of an eye there, complete with all the requisite structures. Perhaps more remarkable, a fruit fly that is missing a copy of this gene can be given the homologous gene from a mouse and develop a normal, fruit fly eye. This is a particularly stunning result because the compound eyes of fruit flies and chambered eyes of mice are radically different. The discovery of homologous genes guiding the development of decidedly non-homologous eyes suggests that eye development is a homologous process, a new kind of evolutionary distinction. In the shrimp with binoculars, this gene may be controlling two different developmental trajectories within a single eye. Perhaps this explains why even the most pessimistic estimate of the time required for an eye to evolve is remarkably short. Even with generous assumptions, Nilsson and Pelger (1994) estimate that it would take less than 364,000 years for a camera-eye to evolve from a light-sensitive patch, meaning that there would have been time enough for eyes to evolve more than 1500 times since they first appeared in the Cambrian (Land and Fernald, 1992). This implies that eyes can be expected to respond very rapidly to evolutionary changes in the lifestyle of a species. Indeed, eye design of a species may say less about its phylogenetic relationships than about its need for vision. Eye structure thus reflects primarily the animal's present requirement. Given what we know now, it appears the evolution of the eye was never a real threat to Darwin's theory of evolution.

III. Model Systems: Visual Information Controlling Behavior

A triumph of early ethologists was to understand the order underlying animals behaving. They did this by concentrating on species with somewhat less complex behavioral interactions during particularly important life events such as territorial behavior, courtship and feeding. Using this tactic, early ethologists Konrad Lorenz, Niko Tinbergen and others, identified what are now two central dogmas of Ethology: the fixed action pattern and the sign stimulus. Fixed action patterns are complex behavioral acts, elicited by specific stimuli and performed by all or nearly all members of a species of a given age and sex. Though qualifications have been raised about how exactly to characterize fixed action patterns (e.g. Camhi, 1984), there is broad agreement that this categorization scheme applies to most behaving animals. Many of the fixed action patterns studied are stereotypical gestures that serve as communication signals among members of the same species. Since such motor patterns often involve complex body movements as well as exposure of brightly colored patches on the body, ethologists began to explore exactly which of their features served as the signal to other animals. Experiments intended to clarify which elements of the stimulus were important led to the notion of a sign or key stimulus, a collection of special features of the general stimulus pattern responsible for eliciting particular behavioral actions. Sign stimuli have been characterized and studied in many species under a variety of conditions. Ethologists learned that many such key stimuli could be studied effectively in the laboratory by constructing models on which elements of the stimulus could be presented. The most famous of these models were constructed by Tinbergen to analyze the features causing territorial fights in the three spined stickleback. Through systematic variation of parameters of the signal, Tinbergen showed that the color red, only when located on the belly of a dummy, caused neighbor animals to initiate a fight. This classic example showed that both the stimulus pattern and its context define its role in producing a response. Although the great majority of signals that have been studied are visual, there are many important sign stimuli in other modalities.

Discoveries about fixed action patterns and sign stimuli have provided considerable insight about how and why animals behave as they do. However, these elemental features of behavioral activity need to be understood in the context of the natural setting of the animal. Animals behave and interact continuously so that there is a seamless interrelationship among the fixed action patterns and sign stimuli. Many of these interactions rely on visual information as can be seen in experiments performed using a teleost fish as a model system.

In the African cichlid fish, *Haplochromis burtoni*, there are two kinds of males: those with territories and those without (Fernald, 1977). Territorial males are brightly colored, having a blue or yellow body color, dramatic black stripe through the eye, vertical black bars on the body, a black spot on the tip of the gill cover and a large red patch just behind it. In contrast, nonterritorial males are cryptically colored, making them difficult to distinguish from the background and from females that are similarly camouflaged (Figure 10.3). Social communication among these

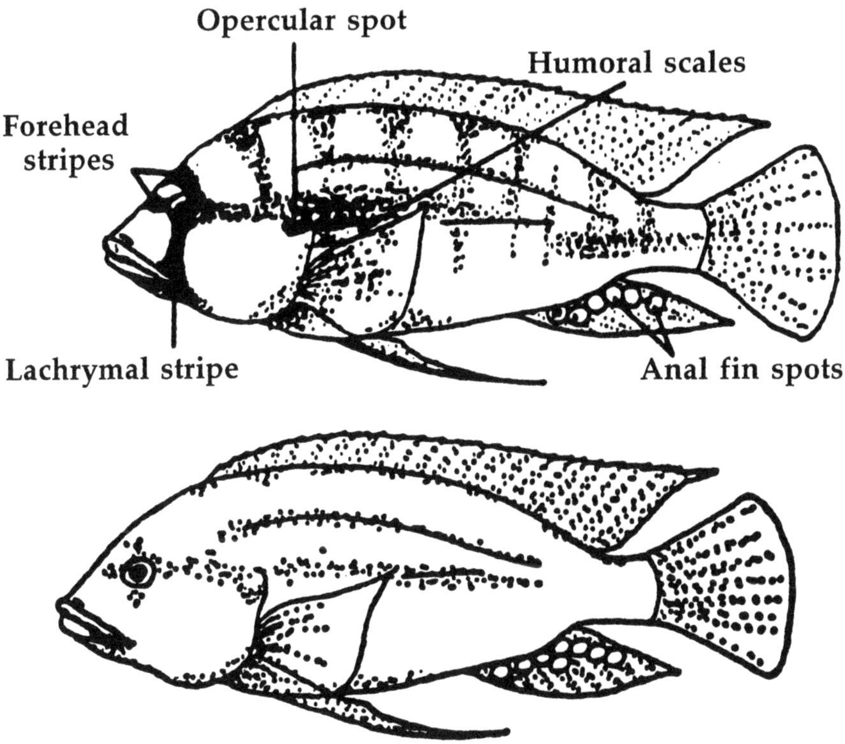

FIGURE 10.3 Schematic illustration of the body patterns for typical territorial and nonterritorial males. The territorial male (above) has distinctive anal fin-spots and dark forehead and lacrymal stripes (eye-bar) and is brightly colored including orange humeral scales. The overall body color may be either yellow or blue. Nonterritorial males (below) lack the robust markings of their territorial counterparts and are sand-colored overall. (Adapted from White et al. 1995)

fish depends primarily on visual signals (Fernald, 1984). In their natural habitat, Lake Tanganyika, *H. burtoni* live in a lek-like social system in which brightly colored territorial males vigorously defend contiguous territories arrayed over a food supply (Fernald and Hirata, 1977 a,b). They are very active, performing at least 19 distinct fixed action patterns during fast paced social encounters (Fernald, 1977). They spend time digging a pit in the center of their territory, fighting with neighbors at their common territorial boundaries, chasing nonterritorial animals away and soliciting and courting females. Solicitation and courtship behaviors are easily identified since the male displays bright coloration patterns towards the courted female. Courtship includes "leading" the female toward the territory and during "courting" the male quivers his opened, brightly colored anal fin in front of the female. Females led into the territory will feed by nipping at, and sifting through the bottom cover. Nonterritorial males mimic this behavior accurately enough so the territorial males will allow them to eat before the deception is discovered and the female impersonator is chased off. If the female responds to these entreaties by the male, he will lead her into his pit and continue the elaborate courtship movements, swimming to the front of the female and rapidly quivering his entire body with his spread anal fin in her view. As the pair disappears into the spawning pit out of direct view of the territory, other animals exploit this opportunity to feed energetically. The spawning male repeatedly interrupts his courtship behavior to chase off the intruders to preserve his food supply. If sufficiently stimulated, the female lays her eggs at the bottom of the pit, collecting them in her mouth almost immediately. After she lays several eggs, the male swims in front of her, again displaying the anal fins spots, his body quivering. It appears that the male displays this fin because the spots may appear to the female like eggs not yet collected. So, while attempting to "collect" the spots, the female ingests the milt ejected near them by the male and ensures fertilization. After several bouts of this alternating behavior, the female may go to the territory of another male to lay more eggs or depart from the territorial arena with the fertilized eggs to brood them (Fernald, 1984).

This brief description of the natural behavior of *H. burtoni* reveals the extensive role of visual signals in social interactions and how much the social scene governs the behavior of individual animals. Each behavioral act influences the next, both in the observed individual and in the animals involved in the interaction. During the behavior, a great deal of information is being exchanged between individuals. What does the animal attend to and what are the consequences of that observation? Several specific fixed action patterns and key stimuli have been studied in this species. In particular, the role of the black eyebar has been studied extensively. Leong (1969) tested the response of territorial males to dummies with various configurations of the distinctive body patterns. She showed that the black eyebar (lachrymal stripe) alone increased the attack readiness toward blinded target fish and the orange-red patch of humeral scales alone decreased attack readiness. Subsequent experiments using similar dummy presentations showed that the response was dependent on the orientation of the eyebar relative to the body and on the angular orientation of the animal in space (Heiligenberg and Kramer, 1972; Heiligenberg et al. 1972). Taken together, these experiments support the idea that the black eyebar and the red humeral patch constitute key stimuli that cause behavioral responses of opposite types. When males were reared from hatching in complete isolation, they showed the same response to the presentation of dummies with an appropriately placed black eyebar and humeral patch (Fernald, 1980). The selective response to the appropriate key stimuli in inexperienced animals suggests that the pattern recognition capability for these key stimuli is genetically controlled.

In *H. burtoni*, the visual system has also been examined for evidence that these signals are important in the natural habitat (Fernald and Hirata, 1977a). In the primary habitat, the shore pools along Lake Tanganyika, the filtering properties of the water match the color patterns on the body and appear to maximize the detection of the visual signal (Fernald and Hirata, 1977a). Moreover, the visual system itself is well matched to both the environment and to the signals. *H. burtoni* has a retina with three types of cone photoreceptors sensitive to distinct wavelengths (Fernald and Liebman, 1980), a prerequisite for trichromatic vision. Further analysis revealed that the photoreceptors are arranged in an array which is optimal for color vision (Fernald, 1981). The spectral sensitivity measured behaviorally (Allen and Fernald, 1985) showed that the animal could distinguish three wavelengths as predicted from the morphological measurements. Thus the visual system is matched to the environment and to the social signals produced by conspecifics including the red humeral patch and black eyebar.

Both the detection of visual signals and their production appear to have been subject to natural selection in this species. The black eyebar can appear and disappear within seconds, independently of other coloration patterns. This eyebar is controlled by a specialized branch of a cranial nerve, which regulates localized changes in the pigment concentration (Muske and Fernald, 1987a). Interestingly, there is a striking difference in the anatomical and physiological processes controlling the eyebar between non-territorial and territorial animals (Muske and Fernald, 1987b). Eyebars located on territorial males show much greater sensitivity to the neurotransmitter that controls changes in expression of that social signal. Moreover, there is a long term morphological change in the eyebar of territorial males. In such males, the eyebar tissue adds iridiphores that enhance the contrast and hence effectiveness of the signal (Muske and Fernald, 1987b). Thus the production of the visual displays is distinctly different at the physiological level in territorial males when compared with nonterritorial males. These discoveries about the regulation of social interactions in the adults raise questions about how social signals and responses to interact with the development of the organism.

As young fish grow, social behavior of conspecifics regulates their growth (Fraley and Fernald, 1982). For the first 7 to 8 weeks of life, living in a group facilitates growth of males as compared to broodmates reared in total isolation. After this time, group-reared males that do not acquire and defend territories grow more slowly than those with territories. Males that form territories develop their color patterns faster, weigh more and have large and more highly developed gonads than do animals reared under any other conditions. The growth of the animals under optimal conditions is dramatic and has resulted in novel developmental solutions. These include adding new cells to the lens and retina (Fernald and Wright, 1983; Fernald, 1983, 1989; Johns, 1981). Such social control of maturation and growth is widespread among animals and takes a variety of forms in different species. In *H. burtoni*, there are some unusual effects of this social regulation of growth.

Males reared with adults show delayed maturation relative to those reared without adults, and moreover, these animals have smaller gona-

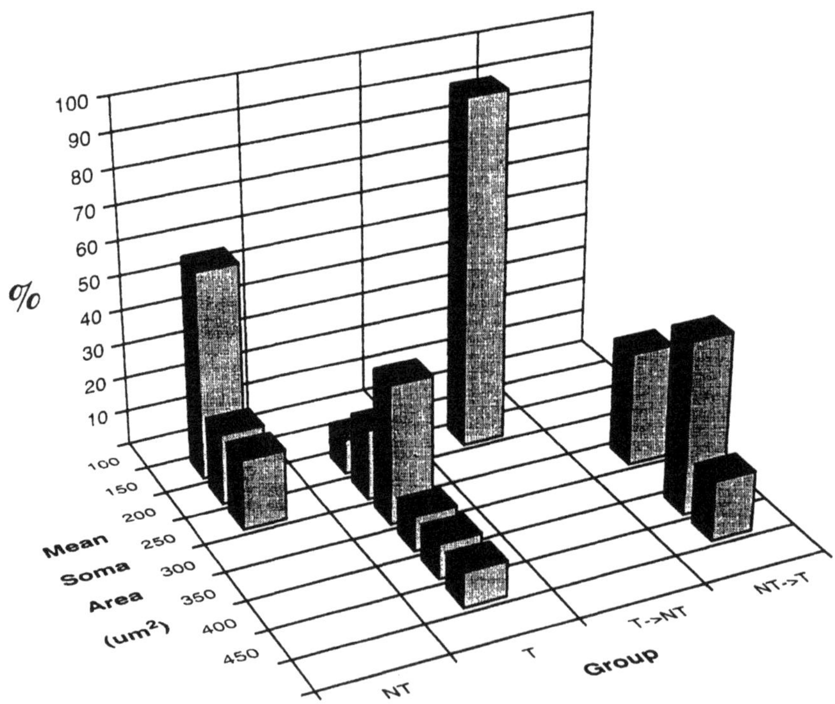

FIGURE 10.4 Three dimensional plot of the mean soma sizes of POA irGnRH neurons, showing significant differences between T and T->NT males and between NT and NT->T males. The percentage of cells in each size class (%) is plotted as a function of mean soma size (m2) for each condition. There are no significant differences (Mann-Whitney U test) between T males and their experimental counterparts (NT->T), or between NT males and their experimental counterparts (T->NT). (Adapted from Francis et al. 1993)

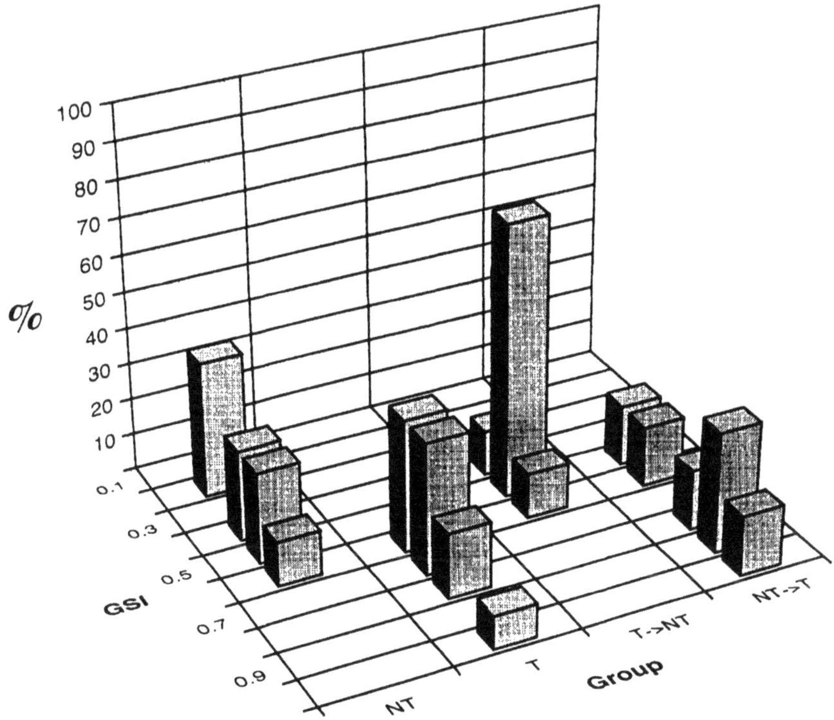

FIGURE 10.5 Three dimensional representation of the gonadosomatic index data, showing significant differences between T and T->NT males and between NT and NT->T males. The percentage of animals of each size class (%) is plotted as a function of GSI. There are no significant differences (Mann-Whitney U test) between T males and their experimental counterparts (NT->T) or between NT males and their experimental counterparts (T->NT). (Adapted from Francis et al. 1993)

dotropin releasing hormone-containing (GnRH) neurons in the preoptic area (POA) of the ventral hypothalamus (Davis and Fernald, 1990). This finding suggests that the social control of maturation is reflected in changes in structures in the brain. To understand the role of social status in determining reproductive state, males were changed from territorial (T) to nonterritorial (NT) or vice versa. To do this, T males were moved into communities with larger T males, as a result of which they became NT (T->NT). Correspondingly, NT males were moved to new communities consisting of females and smaller males which they could dominate, as a result of which they became T (NT->T). In each case the subjects remained in the altered social setting for four weeks, after which the size of GnRH containing cells was measured (Francis et al. 1993).

Mean soma size of both the POA immunoreactive gonadotropin-releasing hormone (irGnRH)-containing neurons (Figure 10.4) and gonadosomatic index (GSI) (Figure 10.5) were significantly larger in the control T males than in the control NT males. In another irGnRH cell group located in the terminal nerve in the telencephalon there was no difference in mean soma sizes between T and NT males, indicating that the status-linked variation in soma size is not a general property of GnRH containing neurons, but is, rather, confined to the POA population.

Social status clearly determines both soma size of POA irGNRH neurons and GSI, and these effects are reversible. The relatively large testes and irGnRH neurons characteristic of T males is a consequence of their social dominance, and when this dominance advantage is lost, both neurons and testes shrink. Much of this social control of neuron size appears to be mediated by visual signals. Additional studies have shown that hormones which play a role in mediating these changes are also under social control (Francis et al. 1993). Interestingly, female *H. burtoni* do not exhibit similar kind of response to social interactions (White and Fernald, 1993).

In *H. burtoni*, the regulation of growth and development may be adaptive in their habitat, where territorial space is limited. In the shore pools where these animals live, only a fraction of the males can breed at any time. These breeding males appear to be particularly vulnerable to avian predators (Fernald and Hirata, 1977b), and hence territorial ownership may be relatively brief. Thus there may be a selective advantage for males to have a retarded growth rate until they have an opportunity to become territorial, whereupon they grow rapidly.

Understanding how visual information exchanged in a social context is transduced into structural changes in the brain is a major challenge. It seems likely that these processes will be common to many species, particularly those in which social interactions regulate reproduction. Just how visual recognition of social opportunity is translated into physiological change remains to be discovered. New insights about this will require combining social interactions with physiological measurements (White et al. 1995).

Conclusions

Visual systems appear to be able to evolve rather rapidly in the service of animal behavior. The need for animals to see and interpret social signals reliably has led to the evolution of highly specialized eyes, remarkably well matched to the behavioral requirements of the species. Visual systems process information so that the information is collected and reduced. Special cases of fish and pollinating insects reveal the match between the visual system characteristics and visual tasks. The changes in eyes during evolution seem to be driven primarily by visual requirements of the species. Recent discoveries suggest that eyes may develop under the control of homologous regulatory genes and contain homologous molecules even though the structures themselves are non-homologous. This novel insight may require a new understanding of the evolutionary history of ocular structures.

In the model teleost fish system studied, the visual system has been tested for the presence of key stimuli and fixed action patterns, both of which exist and play central roles in the behavior of the animals. Social interactions mediated by visual signalling can result in structural changes in the brain. How does an animal's recognition of social opportunity become translated into physiological change? This too may ultimately understood in its evolutionary context.

Acknowledgments

I would like to thank the students and colleagues who collaborated on the experiments from my laboratory and L. Fernald, K. Hoke and I. Yun for valuable comments on the manuscript. Supported in part by NIH EY 05051.

References

Allen, E. A., and Fernald, R. D. (1985). Scotopic visual threshold in the African cichlid fish, *Haplochromis burtoni*. *J. Comp. Physiol. 157*: 247-253.

Camhi, J. M. (1984). *Neuroethology.* Sunderland, MA: Sinauer Associates, Inc.

Chang, B. S. W., Crandall, K. A., Carulli, J. P., and Hartl, D. L. (1995). Opsin phylogeny and evolution: A model for blue shifts in wavelength regulation. *Molecular Phylogenetics and Evolution 4(1)'*: 31-43.

Chittka, L., and Menzel, R. (1992). The evolutionary adaptation of flower colors and the insect pollinators' color vision. *J. Comp. Physiol. A: Sensory, Neural and Behavioral Physiology 171*: 171-181.

Darwin, C. (1859). *On the origin of species.* London: Murray.

Davis, M. R., and Fernald, R. D. (1990). Social control of neuronal soma size. *J. Neurobiol. 21(8)*: 1180-1188.

Endler, J. A. (1980). Natural selection on color patterns in *Poecilia reticulata*. *Evolution 34*: 76-91.

Endler, J. A. (1992). Signals, signal conditions, and the direction of evolution. *The American Naturalist - Supplement 139*: S125-S153.

Fernald, R. D. (1977). Quantitative behavioral observations of *Haplochromis burtoni* under semi-natural conditions. *Anim. Beh. 25*: 643-653.

Fernald, R. D., and Hirata, N. (1977a). Field study of Haplochromis burtoni: Habitats and co-habitants. *Eniv. Biol. Fishes 2*: 299-308.
Fernald, R. D., and Hirata, N. (1977b). Field study of *Haplochromis burtoni*: Quantitative behavioral observations. *Anim.Beh. 25*: 964-975.
Fernald, R. D. (1980). Response of male cichlid fish, *Haplochromis burtoni*, reared in isolation to models of conspecifics. *Z. Tierpsych. 54:*: 85-93.
Fernald, R. D., and Liebman, P. (1980). Visual receptor pigments in the African cichlid fish, *Haplochromis burtoni. Vision Res. 20*: 857-864).
Fernald, R. D. (1981). Chromatic organization of a cichlid fish retina. *Vision Res. 21*: 1749-1753.
Fernald, R. D., and Wright, S. (1983). Maintenance of optical quality during crystalline lens growth. *Nature 301*: 618-620.
Fernald, R. D. (1984). Vision and behavior in an African cichlid fish. *Am. Sci. 72(1)*: 58-65.
Fernald, R. D. (1988). Aquatic adaptations in fish eyes. In: J. Atema, R. R. Fay, A. N. Popper, and W. Tavolga (Eds.) *Sensory Biology of Aquatic Animals* (pp. 185-208). New York: Springer-Verlag.
Fernald, R. D. (1989). Seeing through a growing eye. In T. J. Carew and D. B. Kelley (Ed.) *Perspectives in Neural Systems and Behavior* (pp. 151-174). New York: Liss.
Fernald, R. D. (1990a). Teleost Optics. In R. H. Douglas and M. B. Djamgoz (Eds.) *The Visual System of Fish* (pp. 45-62). London: Croon Helm, Ltd.
Fernald, R. D. (1990b). *Haplochromis burtoni*: A case study. In R. H. Douglas and M. B. Djamgoz (Eds.) *The Visual System of Fish* (pp. 443-464). London: Croon Helm, Ltd.
Fernald, R. D. (1991). Teleost vision: Seeing while growing. *J. Exp. Zool. 5*: 167-180.
Fernald, R. D. (1993). Vision. In D. H. Evans (Ed.) *The Physiology of Fishes* (pp. 161-189). Boca Raton: CRC Press.
Fraley, N. B., and Fernald, R. D. (1982). Social control of developmental rate in the African cichlid fish, *Haplochromis burtoni. Z. Tierpsychol. 60*: 66-82.
Francis, R. C., Soma, K., and Fernald, R. D. (1993). Social regulation of the brain-pituitary-gonadal axis. *Proc. Nat. Acad. Sci. (USA) 90*: 7794-7798.
Frisch, K. v. (1914). *Demonstration von Versuchen zum Nachweis des Farbensinnes bie angeglich total farbblinden Tieren.* Verh. Dtsch. Zool. Ges. in Freiburg, Berlin.
Frisch, K. v. (1967). *The dance language and orientation of bees.* Cambridge: Harvard University Press.
Goldsmith, T. H. (1990). Optimization, constraint, and history in the evolution of the eyes. *Quarterly Review of Biology 65*: 281-322.
Halder, G., Callaerts, P., and Gehring, W. J. (1995). Induction of ectopic eyes by targeted expression of the eyeless gene in *Drosophila. Science 267*: 1788-1792.
Heiligenberg, W., and Kramer, U. (1972). Aggressiveness as a function of external stimulation. *J. Comp. Physiol. 77*: 332-340.
Heiligenberg, W., Kramer, U., and Schulz, V. (1972). The angular orientation of the black eye bar in *Haplochromis burtoni* (Cichlidae, Pisces). *Z. vergl. Physiol. 76*: 168-176.

Johns, P. R., and Fernald, R. D. (1981). Genesis of rods in teleost fish retina. *Nature 293*: 141-142.

Land, M. F., and Fernald, R. D. (1992). The Evolution of Eyes. *Ann. Rev. Neuro. 15*: 1-29.

Leong, C.-Y. (1969). The quantitative effect of releasers of the fish *Haplochromis burtoni* (Cichlidae, Pisces). *Z. vergl. Physiol. 65*: 29-50.

Levine, J. S., and MacNichol, E. F. (1979). Visual pigments in teleost fishes: effects of habitat, microhabitat, and behavior on visual system evolution. *Sensory Processes 3*: 95-131.

Lorenz, K. Z. (1978). *Vergleichende Verhaltensforschung: Grundlagen der Ethologie.* Springer-Verlag, Wien, 380pp.

Menzel, R. (1967). Untersuchungen zum Erlernen von Spektralfarben durch die Honigbiene, *Apis mellifica. Z. Vergl. Physiol. 56*: (22-62)

Munz, F. W., and MacFarland, W. N. (1977). Evolutionary adaptations of fishes to the photic environment. In F. Crescitelli (Ed.) *Handbook of Sensory Physiology* (pp. 193-274). Berlin: Springer-Verlag.

Muske, L., and Fernald, R. D. (1986a). Control of a teleost social signal: Neural basis for differential expression of a color pattern. *J. Comp. Phys. 160*: 89-97.

Muske, L., and Fernald, R. D. (1986b). Control of a teleost social signal: Anatomical and physiological specializations of chromataphores. *J. Comp. Physiol. 160*: 99-107.

Nathans, J. (1987). Molecular biology of visual pigments. *Ann. Rev. Neurosci. 10*: 163-194.

Neitz, M., Neitz, J., and Jacobs, G. H. (1991). Spectral tuning of pigments underlying red-green color vision. *Science 252*: 971-174.

Nilsson, D. E. (1989). Vision, optics and evolution. *BioScience 39(5)*: 298-307.

Nilsson, D. E., and Modlin, R. F. (1994). A mysid shrimp carrying a pair of binoculars. *J. Exp. Biol. 189*: 213-236.

Nilsson, D. E., and Pelger, S. (1994). A pessimistic estimate of the time required for an eye to evolve. *Proc. Royal Soc. London 256*: 53-58.

Ott, M., and Schaeffel, F. (1995). A negatively powered lens in the chameleon. *Nature 373*: 692-694.

Packard, A. (1972). Cephalopods and fish: The limits of convergence. *Biol. Rev. 47*: 241-307.

Salvini-Plawen, L. V., and Mayr, E. (1977). On the evolution of photoreceptors and eyes. *Evol. Biol. 10*: 207-63.

Uexkull, J. v. (1909). *Umwelt und Innenleben der Tiere.* Berlin.

Viltala, J., Korpimaki, E., Palokangas, P., and Koivula, M. (1995). Attraction of kestrels to vole scent marks visible in ultraviolet detection. *Nature 373*: 425-427.

White, S. A., Kasten, T. L., Bond, C. T., Adelman, J. P., and Fernald, R. D. (1995). Three gonadotropin-releasing hormone genes in one organsim suggest novel roles for an ancient peptide. *Proc. Natl. Acad. Sci. (USA)* 92: 8363-8367.

Yau, K.-Y., and Baylor, D. A. (1989). Cyclic GMP-activated conductance of retinal photoreceptor cells. *Ann. Rev. Neurosci. 12*: 289-327.

PART THREE

Specializations for Spatial Perception and Orientation

11

Perception Along the Axis of Target Range in the Echolocating Bat

Cynthia F. Moss and James A. Simmons

Abstract

Echolocating bats have evolved a remarkable biological sonar system which they use to orient in the environment and capture insect prey. The bat produces ultrasonic vocalizations and listens to echoes that reflect off targets in the path of the sound beam. From the features of echoes, the bat gathers information about a target's size, range, and velocity of movement. This chapter reviews research over the last thirty years on target ranging in bats, and details a series of recent behavioral experiments with the FM-bat, Eptesicus fuscus that examine the bat's perception along the range axis. Psychophysical data demonstrate that bats estimate target distance from the time delay between sonar emissions and returning echoes, and that FM-bats can discriminate changes in echo delay of less than 0.5 μs, corresponding to changes in distance of less than a millimeter. When the bat encounters multiple echoes from closely-spaced reflecting surfaces, the bat uses the echo delay from the nearest surface to estimate the absolute target distance and uses the composite echo spectrum to determine the range profile of the complex target. The range profile of a target is initially encoded in the spectral interference pattern of overlapping echoes; however, it is ultimately represented along the same perceptual dimension the bat uses for estimating the absolute distance to a target.

Introduction

Sensory information processing provides the foundation for our perceptions of the world, and it is widely accepted that information about particular kinds of stimuli can be obtained only through particular sensory modalities. For example, the taste and smell of a stimulus arise through activation of the chemical senses, namely gustation and olfaction. By contrast, information about an object's shape, surface structure, size and relation to other objects can be obtained through vision or touch. One might therefore conclude that each sensory modality can be characterized by the kind of information it conveys.

In most animals, spatial information about objects is derived from the direct correspondence between the physical world and the topography of the brain. For example, the size, shape and location of a visual stimulus is conveyed by the relative activation of neighboring retinal areas that send projections to the retinotopically organized thalamic nuclei and primary visual cortical areas. Indeed, changes across a visual scene produce concomitant changes in activity across the surface of the brain (e.g. Tootell et al. 1982). However, for some animals, spatial information about the physical world cannot be obtained reliably through visual channels. One example is the echolocating bat, an animal that perceives the world through sensory experiences far removed from our own. The echolocating bat has evolved a biological acoustic imaging system which permits it to orient in the environment and capture flying insect prey in the dark. The bat emits ultrasonic sounds and listens to the echoes of these sounds reflecting off of objects in the path of the sound beam (Griffin, 1958). Echoes provide the bat with information about the size, location, and perhaps even the shape of a target, in effect allowing the bat to "see" with its ears.

The focus of this chapter is on the bat's target-ranging behavior and what psychophysical data on range discrimination can tell us about spatial information processing in bats. Specifically, it describes research findings that have allowed us to map the perceptual image of a sonar target perceived by bats, albeit the image of the simplest possible target, a point target that reflects an exact replica of the bat's sonar emission. This chapter begins by providing a general overview of sonar signals used by bats and echolocation behavior in these animals. This introduction also briefly covers some theory on echo information processing, emphasizing research findings that pertain to the bat's perception of a point target along the range axis.

Echolocation Behavior

Sonar sounds emitted by bats can be described broadly in terms of their frequency modulated or constant frequency components. Figure 11.1 shows some spectrograms of echolocation sounds made by different species of bats. Frequency is plotted on the Y-axis and time on the X-axis. Note that the bat's echolocation signals are ultrasonic, i.e. above the upper limit of human hearing at around 20 kHz. In the spectrograms of *Eptesicus fuscus's* and *Phyllostomus hastus's* echolocation sounds, there is a rapid decrease in sound frequency over time. These are frequency modulated (FM) sounds. By contrast, the spectrograms of *Pteronotus's* and *Rhinolophus's* echolocation sounds show a constant frequency (CF) signal followed by an FM sweep. Overall, CF sounds are best suited to carry information in the frequency domain, referring to a representation based on which neural elements are activated in a tonotopically organized auditory system. By contrast, FM sounds are best suited to carry information in the time domain, referring to a representation based on the time-of-occurrence of neural discharges and the temporal pattern of neural firing in response to an auditory stimulus. The FM sweep of *Eptesicus*, for example, is particularly well suited for time domain processing, as each frequency represented in the sound provides a marker for its time-of-occurrence.

Many echolocating bats are insectivorous, and an important task faced by these animals is the pursuit and capture of insect prey. Insect hunting by bats can be divided into three different stages: search, approach and terminal. This is illustrated in Figure 11.2, which plots sound spec-

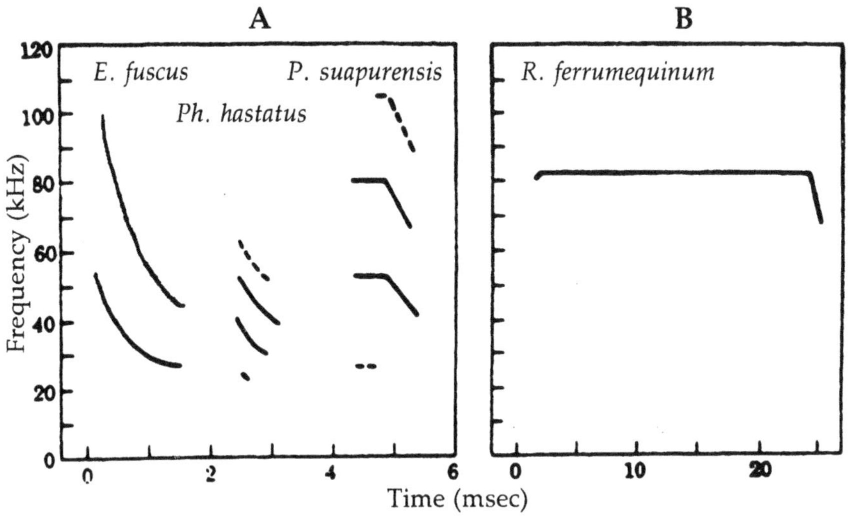

FIGURE 11.1 Spectrograms of echolocation sounds produced by four different species of bat. A. FM-bats, *Eptesicus fuscus* and *Phyllostomus hastatus*, and a short CF-FM bat, *Pteronotus suapurensis*. B. A long CF-FM bat, *Rhinolophus ferrumequinum*. Note that the time axes are different in A and B. Adapted from Simmons, 1973.

trograms recorded from the big brown bat, *Eptesicus fuscus*, during the different stages of insect pursuit. Represented along the abscissa is the time before insect capture, and time wraps around from one panel to the next. Note how the animal's FM sweep becomes steeper as the bat moves in on its target. Also, the sounds become progressively shorter as the bat moves from the search to approach to terminal stages. At the final stage of insect capture, the sounds are less than 0.5 ms in duration.

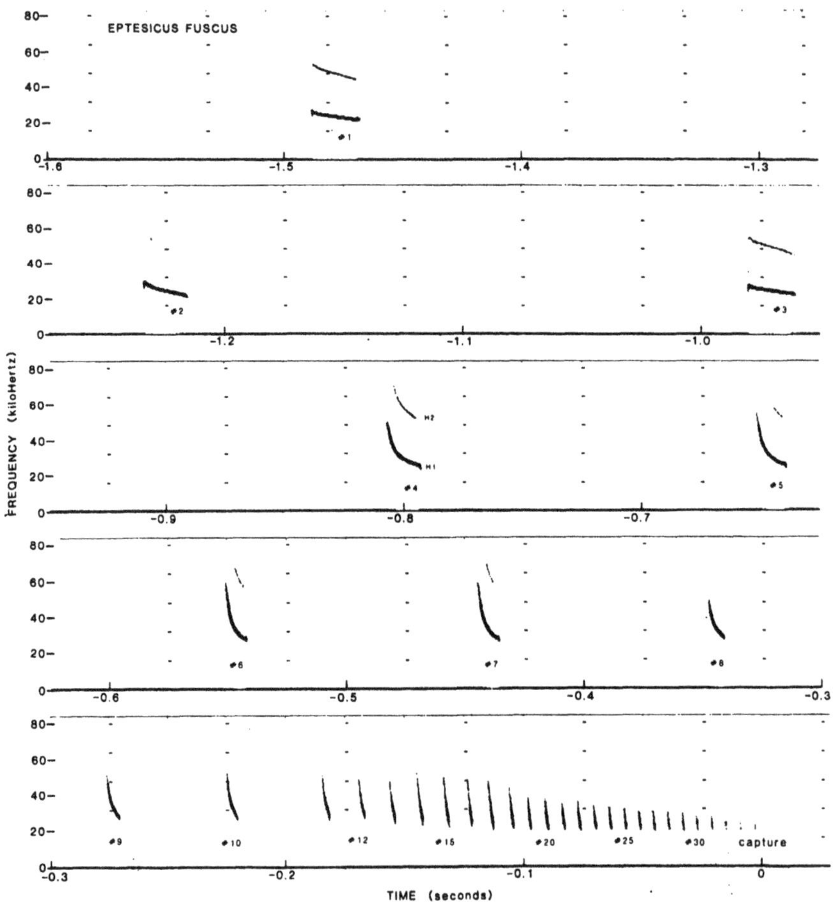

FIGURE 11.2 Spectrogram representation of a sound sequence recorded from the big brown bat, *Eptesicus fuscus*, as it approaches a target. Sounds #1- #3 are characteristic of those produced by the bat during the search phase, sounds #4 - #18 of those during the approach phase and sounds #19- #34 of those during the terminal phase. From Simmons, 1989.

As the sonar sounds produced by a bat approaching a target decrease in duration, they increase in sound repetition rate. During the search stage, the bat emits sounds at a repetition rate of about 5-10/sec. Once the bat detects an insect and begins to approach it, the sound repetition rate increases to about 40/sec. And finally, when the bat enters the terminal stage of insect capture, it emits sounds at a rate of over 100/sec. This sequence happens rapidly, and the time between detection and capture of prey occurs in less than a second (Griffin, 1958).

In order to capture its prey, a bat must track an insect's location in space: It must continuously measure the insect's azimuth, elevation and range. The big brown bat (*Eptesicus fuscus*) can discriminate the angular separation of adjacent targets as small as 1.5° in the horizontal plane (Simmons et al. 1983) and 3.0° in the vertical plane (Lawrence and Simmons, 1982). Haresign et al. (this volume) describe the cues bats use in horizontal and vertical localization. This chapter focuses on the third perceptual dimension, that of target range.

Target Ranging in Bats

Bats generally capture insects by scooping them up in the wing or the tail membrane and bringing the insect to the mouth. In order to successfully capture prey, bats require an accuracy of about 1-3 cm (Webster and Griffin, 1962; Trappe, 1982). However, as described below in more detail, they are capable of making distance discriminations of a fraction of a millimeter, far greater than that needed to catch flying insects.

The earliest hypothesis on the bat's perceptual cue for target range was put forth by Hartridge (1945). He proposed that bats may rely on echo delay to estimate the distance to a target. Since the time delay between an echolocation sound and the returning echo varies systematically with the distance to the reflecting target (a close target reflects an echo at a shorter delay than a more distant target), the bat could measure echo arrival-time to determine target range. Around this time, Donald Griffin (1944), compared man-made sonar and radar systems with bats, suggesting that both may rely on the same principle (using echo delay) to determine the distance to a target.

The bat's perceptual cue for target range remained an open question for many years. Pye (1960) entertained the hypothesis that outgoing sounds overlapped with returning echoes, producing beat-notes that changed with the distance of the reflecting target. He proposed that a specified beat-note signaled a particular target distance, and bats interpreted changes in these beat-notes as changes in distance. Cahlander et al. (1964) showed, however, that bats using FM signals (commonly referred to as FM-bats) shorten sounds as they approach a target, apparently avoiding overlap between outgoing sounds and returning echoes. They trained the echolocating FM-bat, *Myotis lucifugus*, to catch mealworms tossed into the air, and the flying animals learned to intercept them with the tail membrane or wing and transfer the insect to the mouth. Using synchronized high-speed photography and audio recordings, they measured the duration of the bat's echolocation sounds as a function of its distance from the insect. They found that the duration of the bat's sounds decreased systematically as the distance between the bat and the mealworm grew shorter. In fact, the length of the sound used by the bat approximated the travel time (distance) to the worm. If a beat-note were to exist, the length of the sound would have to exceed twice the travel time (and corresponding distance) to the target. Therefore, the results of this study provide compelling evidence against the beat-note hypothesis. More recent studies of echolocation behavior in the field are consistent with Cahlander's early report that FM-bats actively avoid overlap of outgoing sounds and returning echoes by shortening the duration of the sonar cries as they approach a target (see Schnitzler et al. 1987).

Data showing that FM-bats avoid overlap of sonar emissions and echoes, hence arguing against the beat-note hypothesis, led researchers to favor the alternative hypothesis that bats rely on direct estimates of echo delay to determine the distance to a target. The bat's use of echo delay for target ranging was confirmed over 20 years ago in a psychophysical experiment (see Simmons, 1973). Bats were first trained to discriminate between two targets, one at a fixed distance (e.g. 30 cm) and another target at a closer, variable distance. This is illustrated in Figure 11.3A. Bats learned to approach the closer target (S+) and received a food reward for each correct response. The distance of the closer target was initially 10 cm from the bat, and the difference in distance between the two targets was systematically decreased. In this way, the bat's distance discrimination threshold (range difference producing 75% correct performance) was estimated to be 12-14 mm.

Psychophysical data replotted from Simmons (1973) are shown in Figure 11.3B. Performance for range difference discrimination in bats tested with targets located at absolute ranges of 30 and 60 cm is shown, and the two curves are essentially overlapping. To determine if indeed the bat's cue for target distance is echo delay, the apparatus was modified. The real targets were replaced with phantom targets, simulated by picking up the bat's sounds with two ultrasound microphones (one to the bat's left, the other to the bat's right), delaying these sounds and playing them back to the bat through two speakers (left/right). This is illustrated in Figure 11.4A. The delays were selected to simulate echo

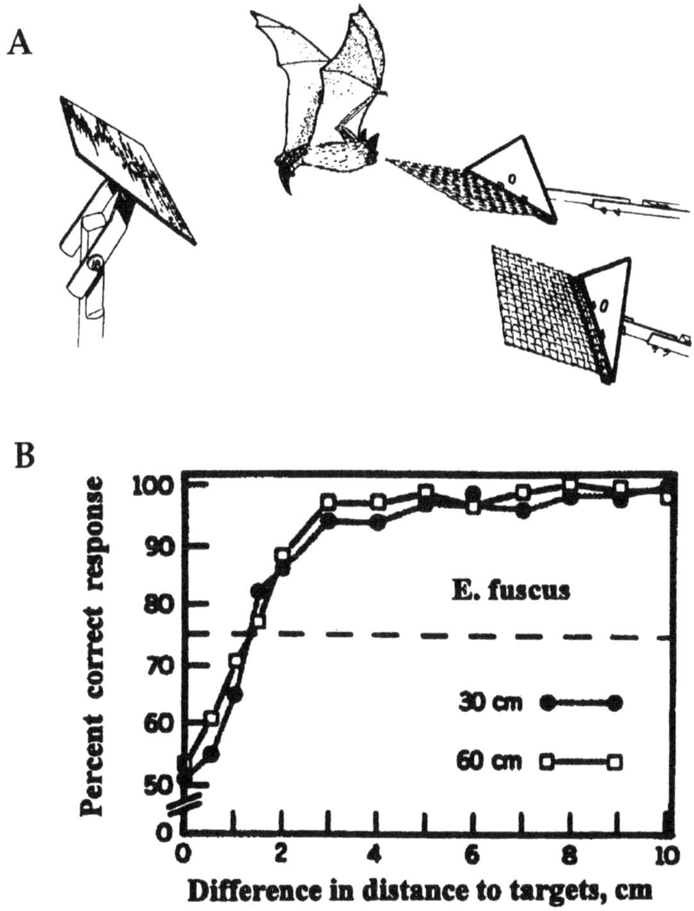

FIGURE 11.3 A. Schematic of the real-target range discrimination apparatus used by Simmons (1973). The targets differ only in their distance from the bat's position on the starting platform. B. Behavioral range discrimination data using the real-target apparatus. From Simmons, 1973.

reflections from the fixed 30-cm target and the variable-distance target. As in the first experiment, the bat was required to discriminate between the two targets, but this time the only cue available to the animal was echo delay. The bat's performance for the phantom target distance discrimination and that for the real target discrimination are shown in Figure 11.4B. The two behavioral performance curves are essentially overlapping, strongly suggesting that the FM-bat uses echo delay to discriminate target range.

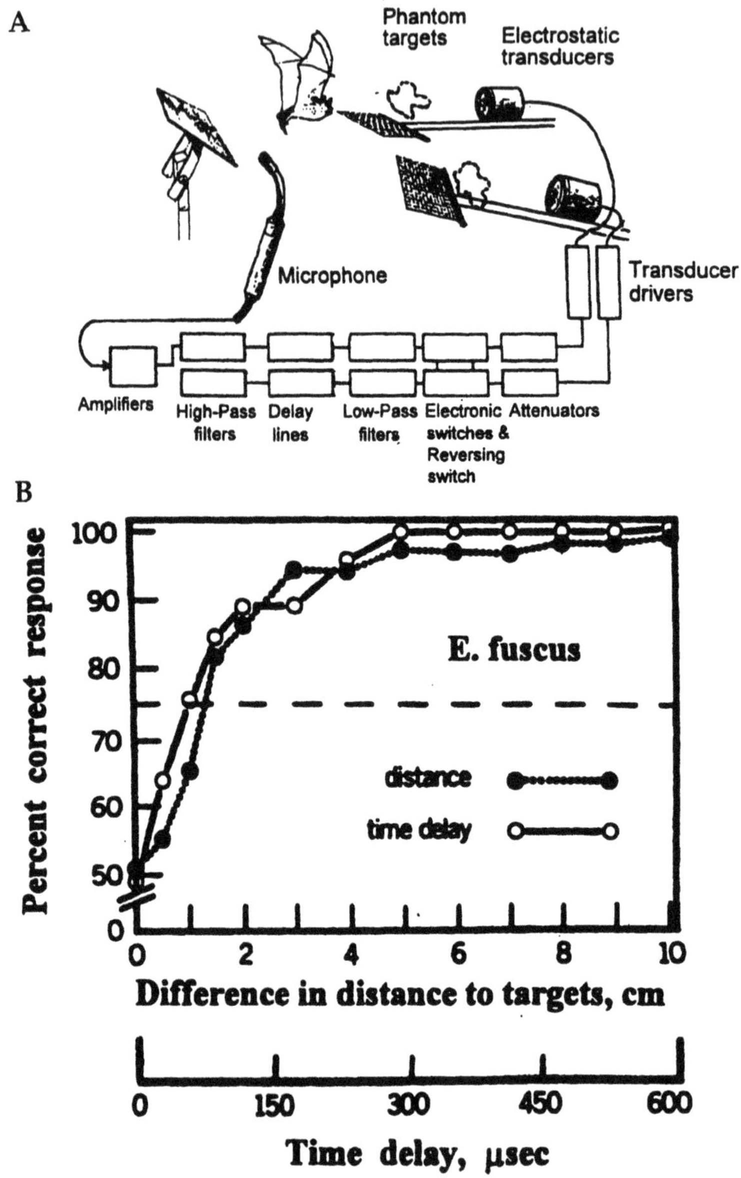

FIGURE 11.4 A. Schematic of the phantom-target apparatus for the study of echo delay discrimination by the bat. The bat's sounds are picked up by two microphones and broadcast back to the bat through electrostatic speakers after a pre-set electronic delay. B. Psychophysical data on range difference discrimination using the real-target apparatus (solid symbols) and echo delay discrimination using the phantom-target apparatus (open symbols). Discrimination data for both tasks were collected at a target distance of 30 cm. Note that the functions overlap. From Simmons, 1973.

Ideal Receiver Model for Target Ranging by Sonar

The data on target distance discrimination in bats have led to a lively debate over receiver models of echolocation. Specifically, the question has been raised whether the bat performs as an ideal coherent receiver (Schnitzler et al. 1985; Menne and Hackbarth, 1986). An ideal coherent receiver achieves high ranging accuracy by cross-correlating the sonar emission and returning echo, then reading a display of the echo that precisely registers its time-of-occurrence.

Figure 11.5 illustrates how the correlation function allows for a much more precise registration of a sound's time-of-occurrence than a spectrogram. A spectrogram representation of the sound is shown above (A) and the corresponding autocorrelation function below (B and C). (For a point target that reflects a replica of the outgoing sound, the autocorrelation function is equivalent to the cross-correlation function.) Since the bat's cue for target range is echo delay, a single distinct peak along the time axis allows for a more accurate estimate of echo delay or target range.

One might imagine that the bat uses the time delay at which it perceives the peak of the correlation function as its best estimate of target range. The central peak of the correlation function is less than 10 μs wide, and cross-correlation of the sonar sound and echo would therefore allow the bat to make a much more accurate estimate of target range than a spectrogram representation whose temporal smear is many hundreds of microseconds (horizontal width of dark area).

In 1979 Simmons collected behavioral data which led him to propose that *Eptesicus* may perceive the fine-structure of the cross-correlation function of sonar cries and echoes. That is, the bat may perceive the phase of sonar cries relative to echoes. If the bat perceives the fine structure of the correlation function, it may be used to describe the bat's perceptual image of a point target. One can think of this image as closely spaced peaks and valleys occurring over a very short time interval. The correlation function shows a narrow central peak and side lobes at 30-μs intervals (the average period of the signal). If the bat perceives the fine structure of the cross-correlation function, it perceives the details of the peaks and valleys.

What led Simmons (1979) to propose that *Eptesicus* may perceive the fine structure of the correlation function? He carried out an experiment in which he trained the bat to discriminate between two phantom targets, a stationary target that returned echoes at a fixed delay and a jittering target that returned echoes that alternated between two delays (schematic diagram in Figure 11.6A). This experiment showed the bat successfully discriminated changes in echo delay of less than 1 μs (change

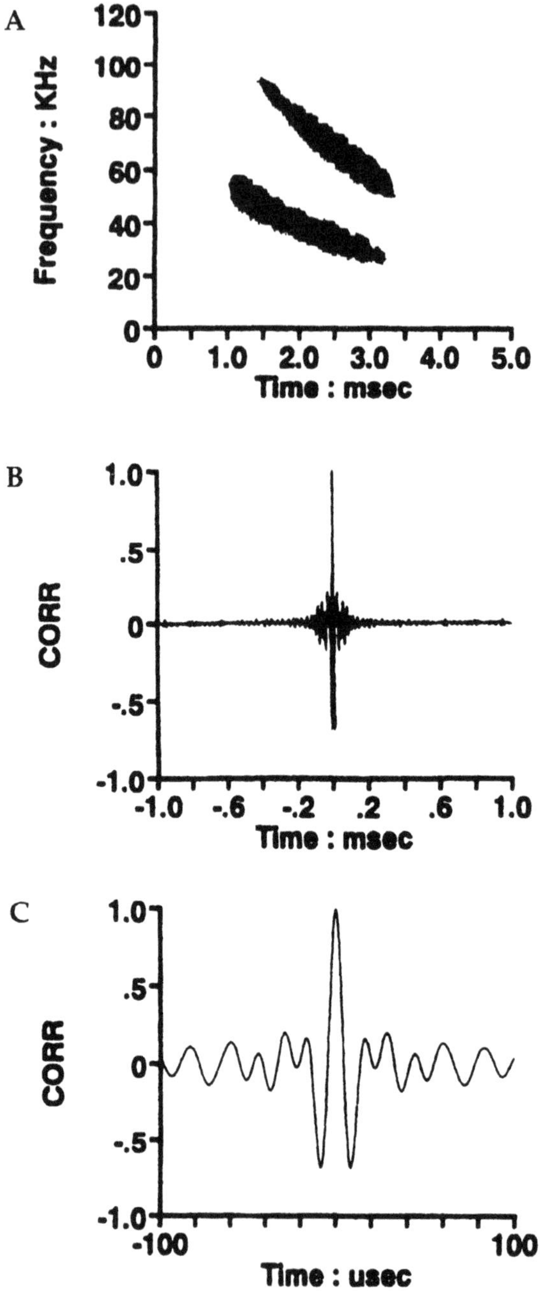

FIGURE 11.5 A. Spectrogram function of an echolocation sound emitted by the big brown bat, *Eptesicus fuscus*. B. Autocorrelation of the same signal. The autocorrelation function shown in B is presented with an expanded time scale in C.

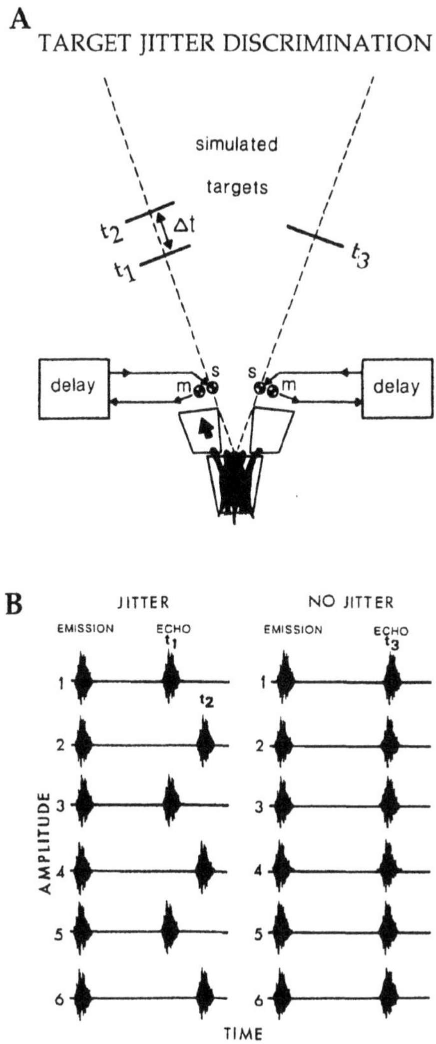

FIGURE 11.6 A. Schematic of the apparatus used for the jitter discrimination experiments. Jittering (left) and non-jittering (right) phantom targets were presented to the bat. A jittering phantom target returned echoes to the bat that alternated in delay, and a non-jittering phantom target returned echoes to the bat at a fixed delay (see Figure 7). Phantom target echoes were simulated by picking up the bat's sounds with a microphone (m), digitizing the signal, and playing it back after a pre-set delay through a speaker (s). In the *discrimination* experiment the bats were trained to approach the side of the platform on which the jittering target was presented (also indicated by arrow on the platform). From Simmons et al. 1990a. B. Schematic illustration of jittering and non-jittering phantom target echoes. Each sonar emission triggered a playback echo. For the jittering target, the time delay between the emission and echo alternated between two set values (t_1 and t_2). For the non-jittering target the time delay between sonar emission and echo playback was fixed (t_3). From Moss and Schnitzler, 1989.

in target range of less than 0.1 mm). He also noted that the bat showed a rise in errors for discriminating 30-μs jitter compared to smaller and larger values. Recall that the correlation function shows a narrow central peak and side lobes at 30 μs. If the bat perceives the fine structure of the correlation function, it would experience ambiguity for a jitter of 30 μs, confusing the central peak and the side peak, thus accounting for the rise in errors at 30-μs jitter. From this, Simmons (1979) argued that this peak in the error curve suggests that *Eptesicus* perceives the equivalent of the fine structure of its correlation function.

Recent Target Discrimination Experiments

The jitter discrimination data presented by Simmons in 1979 stimulated a great deal of debate among researchers in the field. In particular, the echo delay discrimination threshold of less than 0.5 μs was called into question, because Simmons's experiment was conducted with an analog delay line which produced a delay-dependent spectral cue that the bats may have used to perform the task. To address this criticism, the target jitter discrimination experiment was repeated with a digital system, ensuring that changes in echo delay did not also produce spectral changes in the echoes (see Moss and Simmons, 1993). Bats were trained to sit on a Y-shaped platform and emit sounds into two microphones, positioned at the end of the left/right arms of the platform. The bat's echolocation sounds were digitized, delayed, amplified, and played back to the animal through one of two speakers located adjacent to the microphones at a distance of 20 cm. Bats were trained to discriminate between two sonar targets, one that returned echoes at a fixed delay, simulating a stationary target at a distance of 57 cm and the other that returned echoes that alternated between two delays, simulating a target that jittered in space about a mean distance of 57 cm. Echoes from jittering and stationary targets are illustrated in Figure 11.6B. The bat received a food reward for crawling down the arm of the platform toward the jittering target.

When the bat scanned for a target, the strength of the sonar sounds picked up by the left and right microphones changed, depending on the aim of the bat's head. An electronic controller built into the target simulator determined which of the two microphones received the stronger signal (the side towards which the bat's head was aimed) and returned a replica of the bat's emission through the speaker on only that side. If the bat's head was aimed to the right (so that the right microphone received the greater signal), the sound was played back through the speaker on the right side, and conversely, if the bat's head was aimed to the left (so that the left microphone received the stronger signal), the sound was played back through the speaker on the left. The sound level of the playback echoes was set at 15 dB above detection threshold for a stationary target, and the magnitude of the echo jitter ranged from 0 to 50 μs in 5-μs steps.

The results of the jitter discrimination experiment are presented in Figure 11.7 (solid lines). For 5-μs temporal jitter, the smallest value tested in this study, bats showed reliable discrimination performance. Other

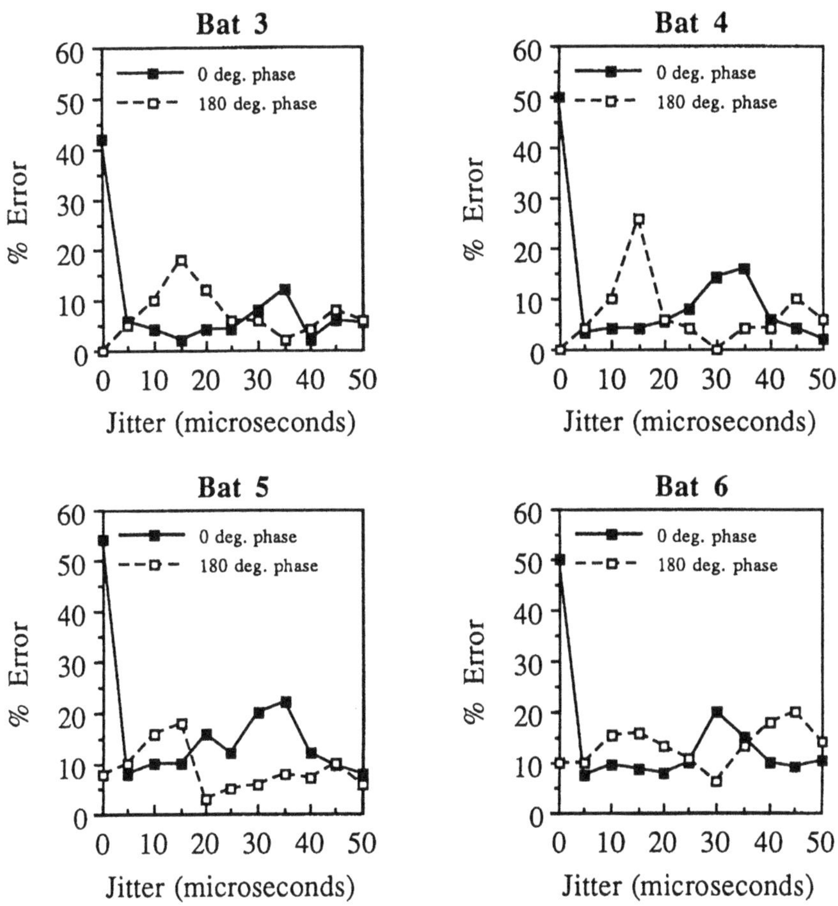

FIGURE 11.7 Percentage errors in the *jitter discrimination* task as a function of echo jitter for four individual bats. Data for the 0° phase shift condition are plotted (solid line) along with data for the 180° phase shift condition (dashed line). In the 180° phase shift condition, echo t_1 (shorter delay) of the jittered echo pair was phase-reversed. From Moss and Simmons, 1993.

experiments in which bats were tested with temporal jitter smaller than 1 μs report that *Eptesicus* can discriminate echo-jitter of 0.4 μs (Menne et al. 1989; Moss and Schnitzler, 1989), replicating Simmons's (1979) original finding. Figure 11.7 also shows that the bat's performance declined at 30 μs relative to smaller and larger jitter values. Recall, if the bat's perceptual image of a point target is represented by the fine structure of the signal autocorrelation function, the bat may confuse the central and side peaks in the jitter discrimination task, producing ambiguity of target jitter at about 30 μs.

While these data support the hypothesis that the bat performs the neural equivalent of cross-correlation processing of emissions and echoes, they are not conclusive, and further studies have been conducted to test this idea. If the bat perceives the equivalent of the fine structure of the correlation function of its sonar emissions and echoes, a 180° phase shift of one of the echoes in the jittered-target echo-pair should produce a predictable change in the performance curve (Altes, 1981). That is, a 180° phase shift of the echo with the shorter delay (t_1; see Figure 11.6), should result in an error curve with peaks at about 15 and 45 μs rather than at 30 μs. Using the same attenuation and jitter values, the jitter discrimination experiment was repeated with one modification: The jittering target reflected echoes that alternated between two delays, and the phase of the echoes delivered at the shorter delay (t_1, Figure 11.6) was inverted. The phase of the jittering target echo at the longer delay (t_2, Figure 11.6) and the phase of the non-jittering target echo were both left unchanged.

The data from this echo condition in the jitter discrimination experiment are also presented in Figure 11.7 (dotted lines). As predicted by the cross-correlation sonar receiver model, the performance curve for the discrimination of target jitter is shifted along the time axis by 180° (see Altes, 1981). It is also interesting to note that a 180° phase shift in echo t_1, without additional temporal jitter, was discriminated by the bat.

Humans cannot hear the ultrasonic vocal signals of bats and can only monitor these signals electronically. As a consequence, one may question what cues available to the bat could potentially influence an animal's performance. Some data from a control experiment address this issue and suggest that the bat is using changes in echo delay to discriminate target jitter. The logic behind this control experiment stems from the following observations. The bat's auditory system initially encodes the FM sweeps of individual sonar emissions and echoes in terms of the time-of-occurrence of each frequency. This is illustrated in Figure 11.8, which comes from Simmons and Kick (1984) on the physiological mechanisms for spatial filtering and enhancement in the sonar of bats and is based on data collected in the lab of George Pollak at the University of

Texas (Pollak et al. 1977). In Figure 11.8, each dot represents the on-response of a primary auditory nerve fiber. As the FM sonar sound stimulates the ear, it activates fibers tuned to different frequencies, sweeping from high-to-low. Thus, the sound first elicits neural discharges in

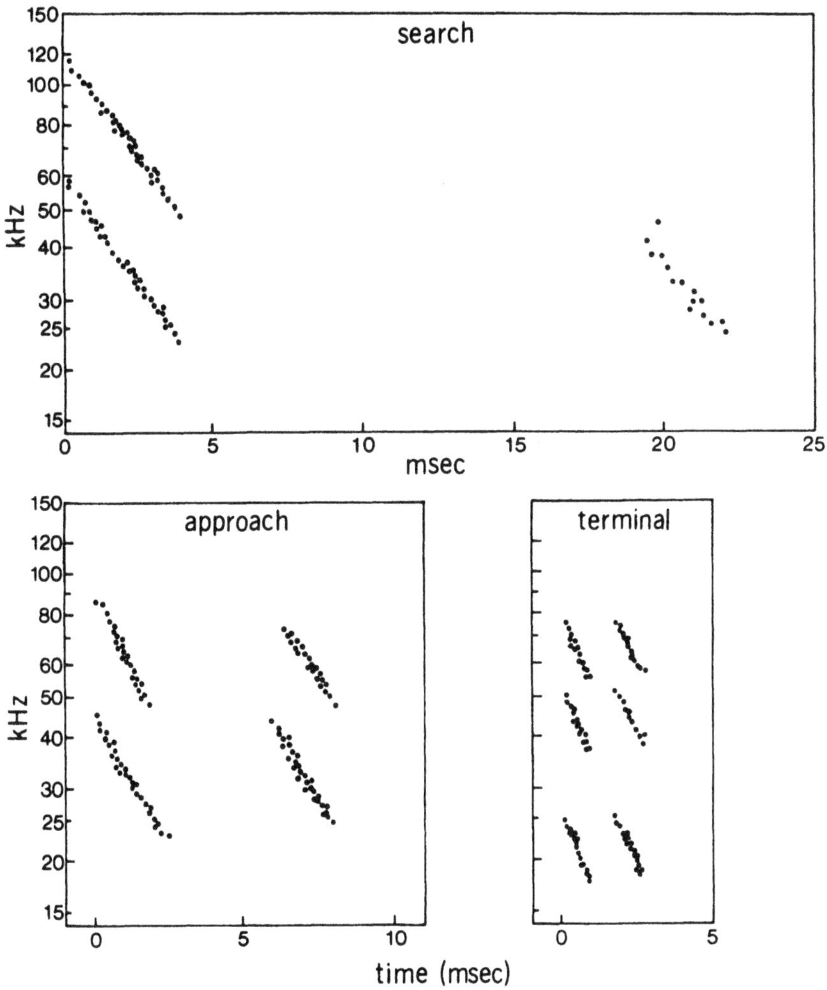

FIGURE 11.8 Neural spectrogram representation of sonar emissions and echoes. Each dot represents the on-response of a primary auditory nerve fiber tuned to a particular frequency. The bat's FM sweep evokes neural activity from nerve-fibers, first tuned to high frequencies and then to low. Examples are shown for the echolocation sounds produced by *Eptesicus* during the three phases of insect pursuit: search, approach, terminal. From Simmons and Kick, 1984. Reproduced with permission from the *Annual Review of Physiology*.

fibers tuned to high frequencies and later elicits discharges from those tuned to lower frequencies. The result is a neural spectrogram representation of the response to the sound (frequency vs. time), and examples are shown in Figure 11.8 for sonar cries produced by *Eptesicus* during the search, approach and terminal phases of insect pursuit.

The representation of the echo delay is affected by the amplitude of the stimulus such that the latency of the on-response of the primary auditory nerve fibers changes with the amplitude of the stimulus. A relatively high amplitude sound elicits responses at a shorter latency than a relatively low amplitude sound. Thus, these neural spectrograms can shift in their time-of-occurrence with changes in stimulus amplitude as well as changes in echo delays.

We examined the magnitude of this effect by recording evoked potentials from an electrode placed in the inferior colliculus of the bat. We used FM stimuli that varied in amplitude from 10 to 90 dB SPL and found that the latency of the N_1 response (reflecting the on-response of the primary auditory nerve fibers) and the N_4 response (a longer latency peak, which may reflect the response at the level of the nuclei of the lateral lemniscus) shortened with an increase in sound level. This is shown in Figure 11.9A as evoked potential records, where latency increased as the amplitude of the sound decreased. At 15 dB about threshold the N_1 response shifted by approximately 13 μs for each dB change in sound pressure level and the N_4 response shifted by approximate 18 μs/dB. Since we know that time of echo arrival is the bat's cue for target distance, a latency change of neural firing therefore translates into a perceived change in target distance (see Simmons et al. 1990a,b).

Returning to the control study designed to evaluate the cues that the bat might use to discriminate a change in target distance, the jitter discrimination experiment was repeated with minor modifications. The amplitude of the first echo of the jittering target (t_1, shorter delay) was reduced by 1 dB. If the bat's perceptual cue for target range is echo delay, this 1-dB change in echo amplitude should increase the perceived distance of the target by 2-3 mm . (Recall that each decibel change in sound level produces a 13-18 μs change in neural response latency and presumably a 13-18 μs change in perceived echo delay; 5.8 μs echo delay/mm of target range).

This experiment was also conducted with the amplitude of the second echo (t_2) increased by 1 dB, presumably decreasing the latency of neural discharge by approximately 13-18 μs (Simmons et al. 1990a). Again, if the bat's perceptual cue for target range is echo-arrival-time, this 1-dB change in the more distant echo's amplitude would reduce its perceived range by about 2-3 mm. The results of this behavioral experiment are summarized in Figure 11.9B, which shows a rise in errors in

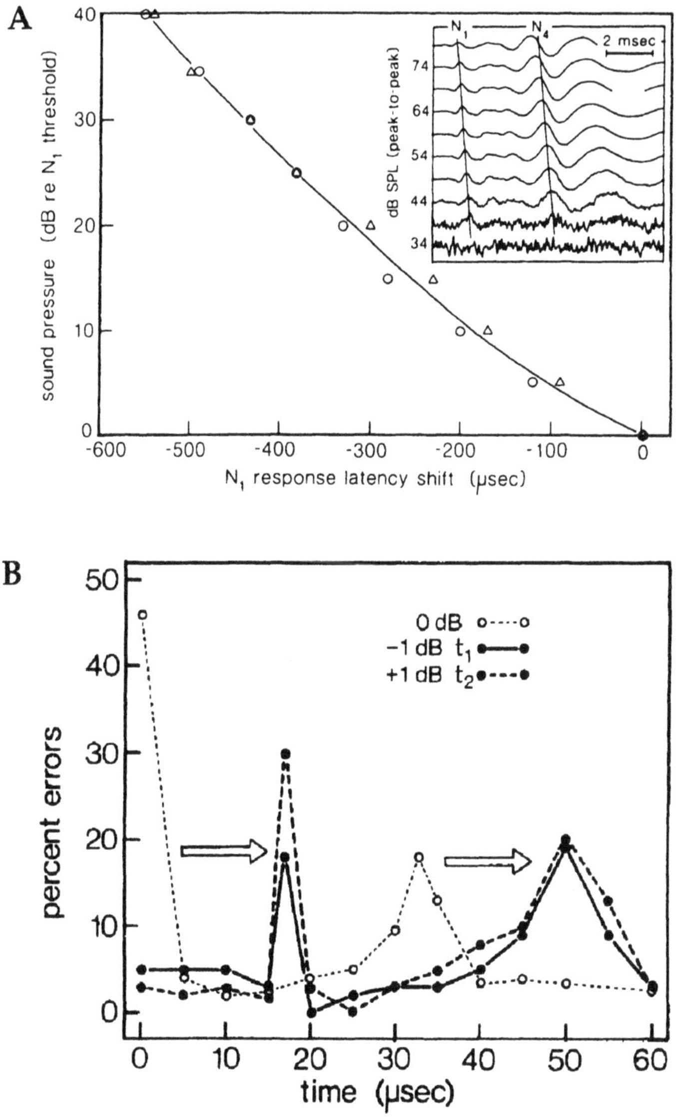

FIGURE 11.9 A. Latency of the N_1 response relative to threshold plotted for two different animals (triangles and circles). Note that the slope of the N_1 latency-intensity function is approximately -13μs/dB at sound levels 15 dB above threshold. The slope of the N_4 latency-intensity function is approximately -18 μs/dB. Evoked potential records for an individual animal are shown in the insert. From Simmons et al. 1990b. B. Target jitter discrimination performance when t_1 was changed by +1 dB (dashed line) or t_2 is changed by -1 dB (solid line). Under these conditions, errors rise for a jitter of approximately 18 μs and again for jitter of 50 μs. Baseline data from Figure 8 (no change in echo playback level in the jitter discrimination experiment) were replotted for a single bat (dotted line). From Simmons et al. 1990a.

the bat's discrimination performance for a jitter value of about 18 μs. Time-intensity trading interferes with the bat's perception of target jitter for this time value. That is, when the amplitude of the first echo (t_1; see Figure 11.6) of the jitter-echo pair is reduced by 1 dB or the second is echo (t_2; Figure 11.6) is increased by 1 dB, the bat's discrimination performance for 18-μs jitter in echo delay is close to chance levels, because to the bat, the two echoes appear to arrive at the same time. Recall that the slope of the latency-intensity function for the N_4 evoked potential response is -18 μs/dB at about 15 dB above threshold.

Also note that a rise in errors still occurs for larger jitter values, but the peak in the error curve is shifted along the abscissa by approximately 18 μs (here at about 50 μs). These data lend further support to the notion that the bat relies on echo delay alone to discriminate target range and that the rise in errors for approximately 30-μs jitter is tied to the bat's perception of a point target along the range axis.

Why would bats be able to discriminate distance on the order of a mere fraction of a mm when capturing an insect requires only 1-3 cm? It may be that bats use fine range information to assess the three dimensional structure or shape of a target. We will now present an experiment designed to assess the bat's perception of a range-extended target, and the results of this experiment suggest that the neural representation of a depth-structured target involves a hybrid of time-domain and frequency-domain processing (see Simmons et al. 1990b).

In this experiment, bats were trained in a two alternative forced choice procedure to discriminate between a simple target (with one echo reflection) and a complex target (with two echo reflections, or glints, separated in time by 100 μs, corresponding to a distance of slightly less than 2 cm). The complex target represents a stimulus with depth structure, and the bat's task was to select the complex target. The simple target was presented at different distances with respect to the complex target. For example in some trials, the time-of-arrival of the simple target was the same as the first echo (or glint) of the complex target, and in other trials, the time-of-arrival of the simple target was the same as the second glint of the complex target. Also, the amplitude of the simple target was changed; the amplitude of the simple target varied from 6 dB below to 9 dB above that of each of the glints of the complex target. Performance was measured as the distance of the simple target was moved relative to the complex target under two different conditions, a simultaneous discrimination in which both the complex and simple targets were presented together and a sequential discrimination in which the complex and simple targets were presented separately. A schematic of this experiment is shown in Figure 11.10. These two conditions, sequential and simultaneous discriminations, and changing the amplitude and delay (or distance) of the simple target allowed us to access the perceptual image of the range extended target and hypothesize about the neural representation of a complex acoustic image.

Figure 11.11A shows the results for the simultaneous discrimination when the simple (single-glint) target was the same amplitude as each of the two glints of the complex target (above) and when it was 6 dB below that of the two glints (below). Plotted are percentage errors (Y-axis) against the distance (or delay) of the simple target relative to the two-glint complex target. Zero represents the condition in which the simple target and the first glint of the complex target arrived at the same time. When the simple target was of the same amplitude as each of the glints of the complex target, errors occurred when the simple target was at the same distance as either of the two glints. The echo of the single-glint target and either of the glints of the complex target arrived at the bat's ears at the same time, interfering with discrimination performance. Thus, the location of the errors along this time axis provided an index of the perceived location of the simple target with respect to the first and sec-

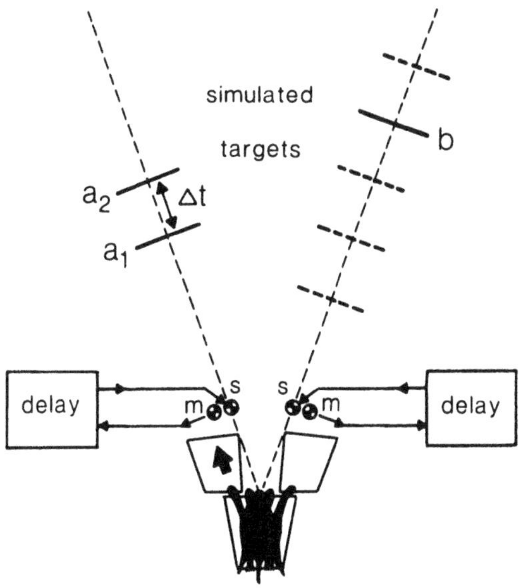

FIGURE 11.10 Schematic of the apparatus for the complex target discrimination task. Phantom target echoes were simulated by picking up the bat's sounds with a microphone (m), digitizing the signal, and playing it back after a pre-set delay through a speaker (s). Bats were trained in this experiment to approach the side of the Y-shaped platform on which the complex target (a_1 and a_2) was presented (indicated by arrow on the platform). The simple target (b) was presented at different amplitudes and delays relative to a_1 and a_2. From Simmons et al. 1990b.

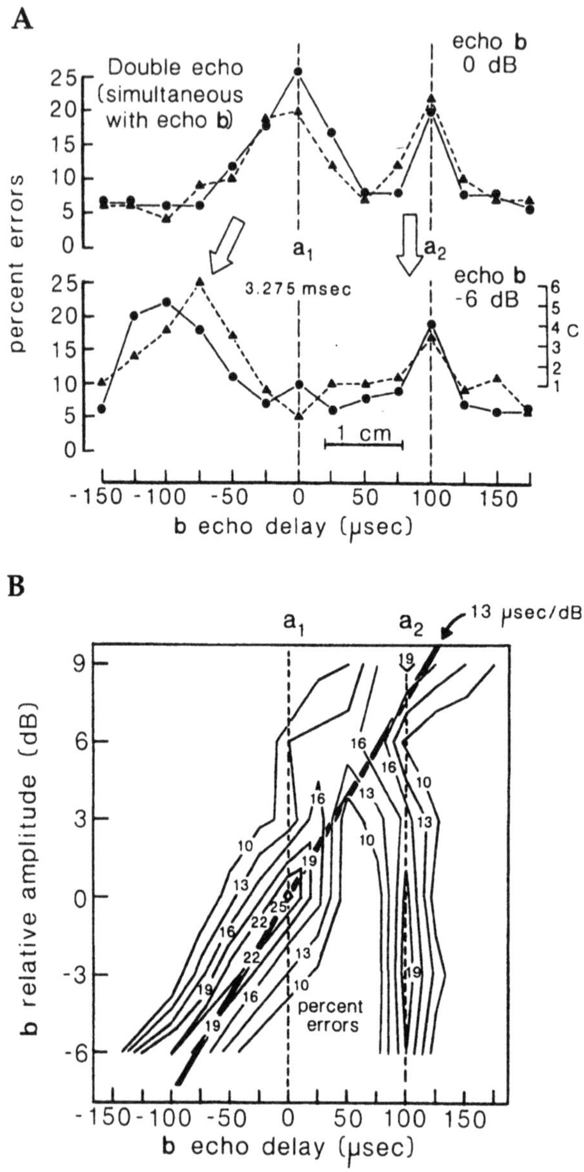

FIGURE 11.11 Simultaneous discrimination data. A. Error curves for 2 bats trained to discriminate between the complex (a_1 and a_2) and simple targets (b) when the simple target was presented at the same echo level as a_1 and a_2 (0 dB, above) and 6 dB weaker (below) than the individual components of the complex target. B. Contour plot of errors for discriminating between the complex target (a_1 and a_2) and the simple target (b) over a range of levels and delays for the simple target. The solid line plots the 13 μs/dB change in the N_1 evoked potential latency at approximately 15 dB above threshold. From Simmons et al. 1990b.

ond glints of the complex target. When the amplitude of the simple target was reduced by 6 dB, an interesting change in the data appears. The simple target and the first glint of the complex target were perceived by the bat to be at the same distance when the simple target was actually presented earlier in time. The rise in errors for the first glint of the complex target occurred approximately 78 μs earlier in time than in the zero-dB (equal amplitude) condition. This can be explained by the fact that higher amplitude sounds elicit discharges of primary auditory nerve fibers at shorter latencies than lower amplitude sounds (see Figure 11.9A). Thus, the lower amplitude single-glint target would appear to be at the same distance as the first glint of the complex target when it occurred earlier in time. This again illustrates that time and intensity can be traded for the perception of distance to a target under experimentally controlled conditions.

It is important to note that the perceived distance of the simple target did not change with respect to the distance of the *second glint* of the complex target when the amplitude of the simple target was reduced by 6 dB. Discrimination errors occurred when the simple target was at the same delay (or range) as the second glint of the complex target even though the simple target was 6 dB weaker than that of the second glint of the complex target. This may be explained in the following way: The sounds emitted by the bats were about 1.5-3 ms in duration. Given the 100-μs separation of the two glints of the complex echo, the two sounds were largely overlapping. This created an interference pattern of cancellation and reinforcement at particular frequencies of the composite echo spectrum. For a particular time separation between the two sounds, there is a predictable change in the composite spectrum. For example, a 100-μs separation produces spectral notches at 10-kHz intervals, starting at 25 kHz (see Figure 11.12), and the bat appears capable of converting this spectral information into distance information. In other words, this spectral information may be used by the bat to measure the depth-structure of the complex target, and coding of range by spectral processing is largely resistant to changes in echo amplitude.

Figure 11.11B is a contour plot, showing percent errors in the simultaneous discrimination across a range of amplitudes of the single-glint. Errors related to the first glint of the complex target changed with the amplitude of the single-glint simple target (as marked by the sloping ridge) with a slope of about 13 μs/dB. Thus, for every decibel change in amplitude of the simple target, the first glint of the complex target had to be moved by 13 μs (approximately 2 mm) for the two targets to appear at the same distance. The solid line plots the N_1 evoked potential latency change with changing stimulus amplitude, and this line falls right on top of the behavioral data. This suggests that the range of the

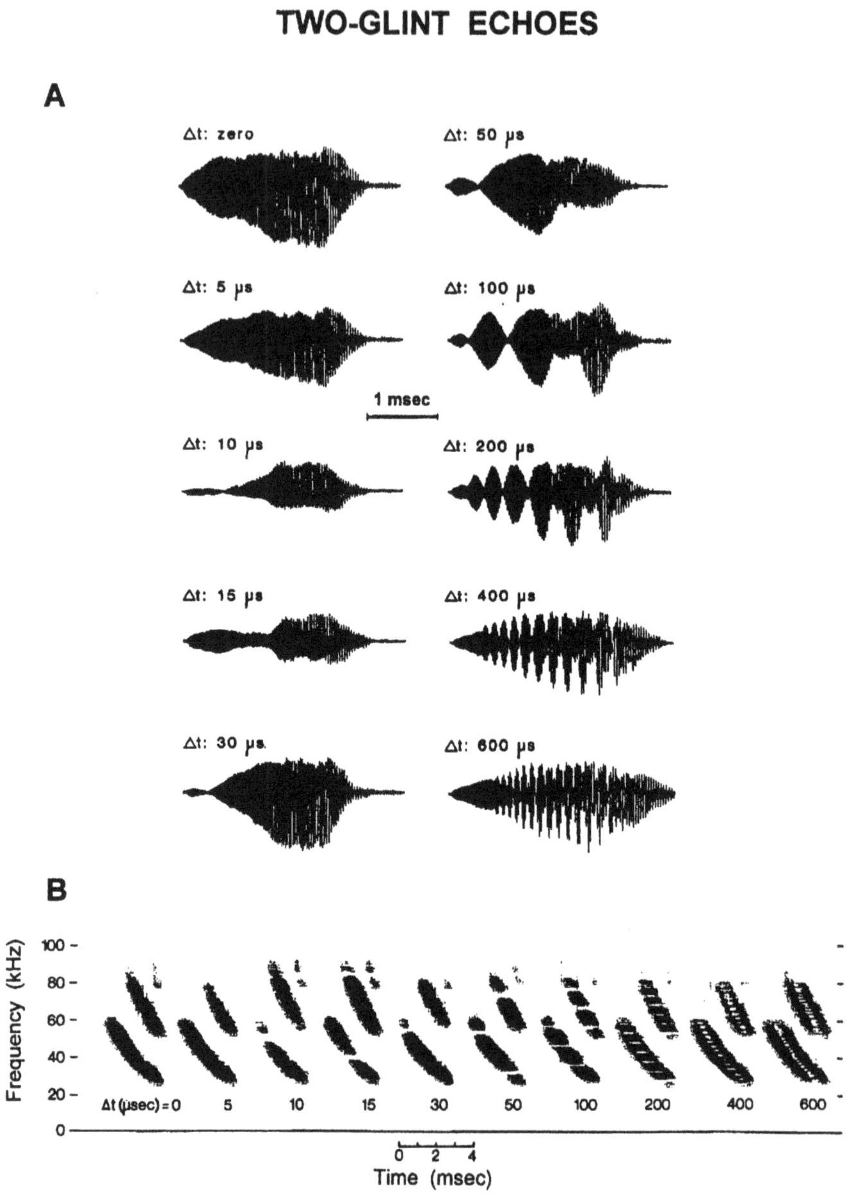

FIGURE 11.12 A. Time waveforms of two overlapping FM echolocation sounds offset in time by 0-600 μs. From Simmons, 1989. B. Spectrograms of the same set of overlapping FM echolocation sounds. From Simmons, 1992.

first glint was coded by the timing of neural discharges in the primary auditory nerve fibers. The second ridge remains constant as the amplitude of the simple target changed from -6 to +9 dB. Thus, the perceived distance of the simple target with respect to the second glint of the complex target was not affected by the amplitude of the simple target. This indicates that the distance of the second glint of the complex target was coded by frequency information derived from the spectral structure of the overlapping echoes, and in a simultaneous discrimination, this information was resistant to changes in echo amplitude.

When hunting insects, a gain control system operates to minimize the changes in amplitude of echoes that are due to changes in the distance to a target. As the bat gets closer to a target, the echoes are stronger. However, the bat's middle ear muscles contract with each sonar emission, and this regulates the perceived strength of the returning echoes. Over a range of about a meter and a half, the echoes from the first reflecting surface of a target stimulate the bat's ear at a roughly constant sensation level with changes in target distance (Kick and Simmons, 1984; Hartley, 1992; Simmons et al. 1992). This gain control mechanism may serve to isolate amplitude changes due to features of a target, such as relative size and separation of body parts of an insect.

Now consider the sequential discrimination between the single-glint and two-glint targets in the psychophysical experiment described above. In the sequential discrimination, the microphone receiving the higher amplitude sound activated the speaker on that side. The bat scanned back and forth between the left and right speakers, and sounds were played back through the speaker on the side towards which the bat's head was aimed. In this case, the bat relied on a stored representation of the complex and simple stimuli to make its discrimination. Under these conditions, the perceived range of the single-glint simple target shifted with respect to the perceived range of both the first and second glints of the complex target when the amplitude of the simple target was reduced by 6 dB (Figure 11.13A). The contour plot for this condition shows this phenomenon across a range of amplitudes for the single-glint simple target. The ridges for *both* glints in the complex target slope upward to the right (Figure 11.13B). Thus when the bat must rely on memory to make the correct response, the second glint of the complex target appeared to shift with the first glint. This suggests that the bat converts the frequency information present in the spectrum of the complex echo into the same perceptual dimension it uses for determining the distance to a target (Simmons et al. 1990b).

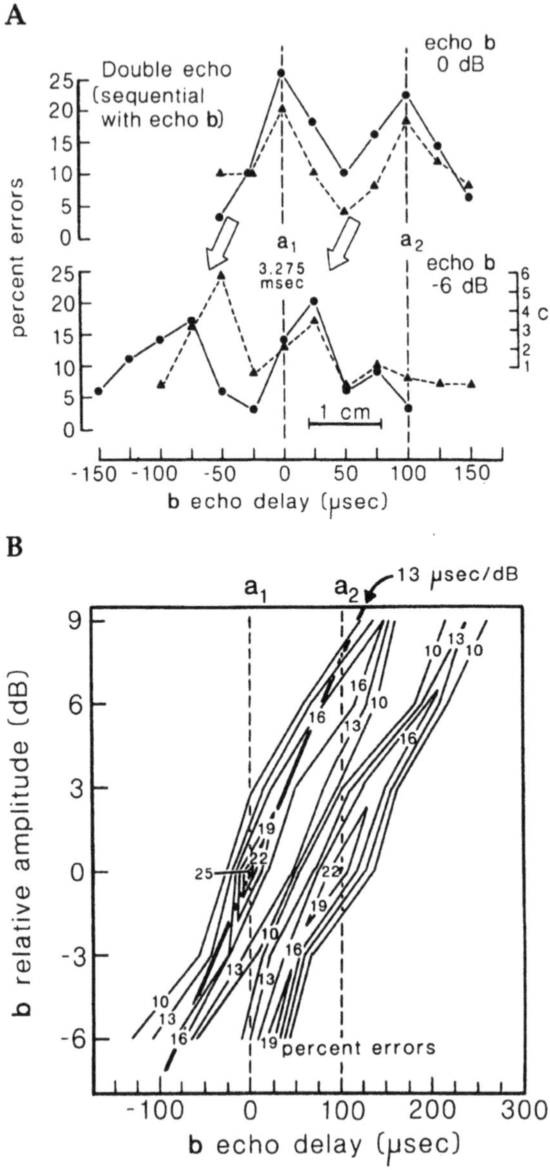

FIGURE 11.13 Sequential discrimination data. A. Error curves for 2 bats trained to discriminate between the complex (a_1 and a_2) and simple targets (b) when the simple target was presented at the same echo level as a_1 and a_2 (0 dB, above) and 6 dB weaker (below) than the individual components of the complex target. B. Contour plot of errors for discriminating between the complex target (a_1 and a_2) and the simple target (b) over a range of levels and delays for the simple target. The solid line plots the 13 μs/dB change in the N_1 evoked potential latency at approximately 15 dB above threshold. From Simmons et al. 1990b.

Conclusions

Spatial information processing is important for many behaviors in most animal systems. The echolocating bat, for example, relies on spatial information to orient in the environment and intercept insect prey. By emitting high frequency sounds and listening to echoes of these sounds, the bat extracts spatial information about the environment through acoustic channels. The acoustic information arriving at the bat's ears allows it to determine the azimuth, elevation and range of a target with high accuracy. While differences in the level and arrival time of a sound at the two ears allow most hearing animals to identify the vertical and horizontal location of a sound source, auditory signals are not generally used to extract spatial information about the distance and depth structure of objects. The auditory system of the echolocating bat, however, can extract information about a target's range from acoustic signals, and this chapter reviews the findings from several behavioral experiments in which we attempted to map out the perceptual image of an acoustic target. The bat's perception of a target along the range axis can inform us about the neural computations taking place in the auditory system that give rise to spatial information through acoustic channels, and the underlying neural mechanisms may share common features with those in the visual and somatosensory systems.

Acknowledgments

This work was supported by an NSF Young Investigator Award to C.F. Moss and the Office of Naval Research Contract Number N00014-86-K-0401 awarded to J.A. Simmons. We thank J. Park and M. Ferragamo for assistance with the behavioral experiments, C. Gounden and A. Grossetête for assistance preparing the figures and D. Valentine and S. Shettleworth for comments on the manuscript.

References

Altes, R.A. (1981) Echo phase perception in bat sonar? *J. Acoust. Soc. Am. 69*: 505-509.

Cahlander, D.A., McCue, J.J.G. and Webster, F.A. (1964) The determination of distance by echolocation. *Nature 201*: 544-546.

Griffin, D. (1944) Echolocation by blind men, bats and radar. *Science 100*: 589-590.

Griffin, D. (1958) *Listening in the dark*. Yale University Press: New Haven.

Hartridge, H. (1945) Acoustic control in the flight of bats. *Nature 156*: 490-494.

Hartley D.J. (1992) Stabilization of perceived echo amplitudes in echolocating bats. I. Echo detection and automatic gain control in the big brown bat, *Eptesicus fuscus*, and the fishing bat, *Noctilio leporinus*. *J. Acoust. Soc. Am. 91*: 1120-1132.

Kick S.A. and Simmons J.A. (1984) Automatic gain control in the bat's sonar receiver and the neuroethology of echolocation. *J. Neurosci. 4*: 2725-2737.

Lawrence, B.D. and Simmons, J.A. (1982) Echolocation in bats: The external ear and perception of the vertical positions of targets. *Science 218*: 481-483.

Menne, D. and Hackbarth, H. (1986) Accuracy of distance measurement in the bat *Eptesicus fuscus*: Theoretical aspects and computer simulations. *J. Acoust. Soc. Am. 79*: 386-397.

Menne, D., Kaipf, I., Wagner, I., Ostwald, J. And Schnitzler, H.-U. (1989) Range estimation by echolocation in the bat *Eptesicus fuscus*: trading of phase versus time cues. *J. Acoust. Soc. Am. 85*: 2642-2650.

Moss, C.F. and Schnitzler, H.-U. (1989) Accuracy of target ranging in echolocating bats: Acoustic information processing. *J. Comp. Physiol. A. 165*: 383-393.

Moss, C.F. and Simmons, J.A. (1993) Acoustic image representation of a point target: Evidence for sensitivity to echo phase in bat sonar. *J. Acoust. Soc. Am. 93*: 1553-1562.

Pollak, G.D., Marsh, D., Bodenhamer, R., Souther, A. (1977) Characteristics of phasic on neurons in the inferior colliculus of unanesthetized bats with observations relating to mechanisms for echo ranging. *J. Neurophys. 40*: 926-941.

Pye, J.D. (1960) A theory of echolocation by bats. *J. Laryngol. Otol. 74*: 718-729.

Schnitzler, H.-U., Menne, D. and Hackbarth, H. (1985) *Range determination by measuring time delays in echolocating bats*. In A. Michelsen (Ed) Time resolution in auditory systems, Springer, Berlin, pp. 180-204.

Schnitzler, H.-U., Kalko, E., Miller, L. and Surlykke, A. (1987) The echolocation and hunting behavior of the bat, *Pipistrellus kuhli*. *J. Comp. Physiol. 161*: 267-274.

Simmons, J.A. (1973) The resolution of target range by echolocating bats. *J. Acoust. Soc. Am. 54*: 157-173.

Simmons, J.A. (1979) Perception of echo phase information in bat sonar. *Science 204*: 1336-1338.

Simmons, J.A. (1987) Acoustic images of target range in the sonar of bats. *Naval Res. Rev. 39*: 11-26.

Simmons, J.A. (1989) A view of the world through the bat's ear: the formation of acoustic images in echolocation. *Cognition 33*: 155-199.

Simmons J.A. (1992) *Time-frequency transforms and images of targets in the sonar of bats*. In Bialek, W. (Ed) Princeton Lectures on Biophysics-1991, NEC Research Institute, Princeton, NJ.

Simmons J.A., Moffat, A.J.M., and Masters W.M. (1992) Sonar gain control and echo detection thresholds in the echolocating bat, *Eptesicus fuscus*. *J. Acoust. Soc. Am. 91*: 1150-1163.

Simmons, J.A. Ferragamo, M., Moss, C.F., Stevenson, S.B. and Altes, R.A. (1990a) Discrimination of jittered sonar echoes by the echolocating bat, *Eptesicus fuscus*: The shape of target images in echolocation. *J. Comp. Physiol. A. 167*: 589-616.

Simmons, J.A. and Kick, S.A. (1984) Physiological mechanisms for spatial filtering and image enhancement in the sonar of bats. *Ann. Rev. Physiol. 46*: 599-614.

Simmons, J.A., Kick, S.A., Lawrence, B.D., Hale, C., Bard, C. and Escudié, B. (1983) Acuity of horizontal angle discrimination by the echolocating bat, *Eptesicus fuscus. J. Comp. Physiol. A. 153*: 321-330.

Simmons, J.A., Moss, C.F., and Ferragamo, M. (1990b) Convergence of temporal and spectral information into acoustic images of complex sonar targets perceived by the echolocating bat, *Eptesicus fuscus. J. Comp. Physiol. A. 166*: 449-470.

Simmons, J.A. and Vernon, J.A. (1971) Echolocation: Discrimination of targets by the bat, *Eptesicus fuscus. J. Exp. Zool., 176*: 315-328.

J.A. Simmons R.A. and Stein, R.A. (1980) Acoustic imaging in bat sonar: Echolocation signals and the evolution of echolocation. *J. Comp. Physiol. A. 135*: 61-84.

Trappe, M. (1982) Verhalten und Echoortung der Grossen Hufeisennase (*Rhinolophus ferrumequinum*) beim Insektenfang. Ph.D. Thesis, University of Marburg, Germany.

Tootell, R., Silverman, M., Switkes, E. and DeValois, R. (1982) Deoxyglucose analysis of retinotopic organization in primate striate cortex. *Science 218*: 902-904.

Webster, F.A. and Griffin, D.R. (1962) The role of the flight membrane in insect capture by bats. *Anim.Behav. 10*: 332-340.

12

Converging Approaches to Determining Sound Localization Mechanisms in the Big Brown Bat, *Eptesicus fuscus*

Tim Haresign, Janine M. Wotton, Michael Ferragamo, and James A. Simmons

Abstract

*The ability to determine the location of a sound source is one of the major functions performed by the auditory system. Echolocating bats use sound localization as their primary means of determining the position of objects in the environment. Using the big brown bat (*Eptesicus fuscus*) as a model animal, several experimental techniques were used to investigate both vertical and horizontal localization. Physical measurements of the sounds arriving at the bat's head were made to determine the potential acoustic cues that might be used for localization of sound. Behavioral experiments were conducted to begin to determine how localization of sound might be affected by manipulation of acoustic parameters. Neurophysiological recordings were conducted to look at the neural processing of the acoustic information used in determining sound source location. The combination of these three approaches have provided results that lead to a better understanding of how these bats are able to localize sounds.*

Introduction

Sensory systems perform two major, interrelated roles. They serve to permit the identification and classification of stimuli and they also provide a means by which stimulus sources can be located. It is this second aspect, stimulus localization, that will be discussed in this chapter, specifically the localization of auditory stimuli. The mechanisms of localization used by the auditory system are quite different from the mechanisms used by the two other systems in which localization is a major function: the visual and somatosensory systems. In the visual system the eye is constructed so that an organized topographical representation of light sources is projected onto the sensory surface (the retina). Thus, the receptor array at the sensory surface forms a "map" of the location of stimulus sources by registering which areas of the sensory surface are activated by light sources. The topographical organization of the receptor array in the retina is projected, intact, to higher levels of the nervous system. Computational enhancements of this map occur at higher levels of the nervous system, but the basic features of the map are the result of maintaining the organized projection of the receptor surface. The same is true in the somatosensory system, where an orderly projection of the sensory surface(s) onto higher levels of the nervous system results in maps of somatosensory space. The sensory interface in audition is the cochlea, which serves to transmit information about the spectral content of sounds to the nervous system. The sensory surface of the cochlea, the basilar membrane, can be thought of as an array, with individual elements of the array being differentially sensitive to frequency (for a review of cochlear function see Dallos, 1984). Thus, a projection of the sensory surface does not contain *a priori* information about the location of the sound source. Auditory localization depends instead on neural processing of cues to create a neural representation of auditory space. These cues include binaural cues: interaural time differences (ITDs), interaural intensity differences (IIDs), interaural spectral differences (ISDs) and monaural cues: changes in the spectrum of sounds as a function of sound source location (for a review of auditory localization see Blauert, 1983).

There are three major questions that need to be addressed when examining how animals determine the location of sound sources: (1) What are the potential auditory cues available at the ear(s) for determining sound source location? (2) What is the behavioral relevance of these cues? and (3) What are the neural mechanisms by which the incoming sounds are processed to produce the observed localization behavior? We have attempted to begin to address each of these questions using the big brown bat, *Eptesicus fuscus*, as a model animal. In this chapter we will describe work that has been done towards addressing these questions in bats and relate the work done in bats to sound localization in general.

Echolocating bats such as *Eptesicus* are ideal for studies of auditory localization. *Eptesicus* emits frequency-modulated (FM) sounds that sweep from high to low frequencies (approximate range 100 kHz - 20 kHz). The bat uses the echoes that return from this emission not only for short range navigation during flight, but also to detect, track, identify, and capture prey (small flying insects). The image the bat perceives of its immediate environment is based on auditory cues in the returning echoes. One specific question we are interested in is: What parameters in the returning echo allow the bat to determine the location of the echo- producing object in terms of azimuth, elevation, and range? It is well known from both behavioral and electrophysiological studies that the bat uses the time-delay between its emission and returning echoes to determine the distance to objects that produce the echoes (Griffin, 1958; Simmons, 1973; Suga, 1984; Moss and Simmons, this volume), and that range can be determined with extremely high acuity (Simmons, 1979; Simmons et al. 1990; Moss and Schnitzler, 1989). Less is known about how the bat determines azimuth and elevation of objects.

To determine the nature of the cues available to *Eptesicus* we measured how the head and external ears affect incoming sounds as a function of sound source location. This was done by placing microphones (Brüel and Kæjer model 4138) in both ear canals of the head of a dead bat at the level of the tympanic membrane (effectively replacing each tympanum with the diaphragm of a microphone). The head was positioned at the center of a hoop-system. Sounds were delivered from a speaker mounted on the hoop and the position of the speaker could be varied in both azimuth and elevation relative to the bat's head, without changing the radial distance from the head. The resulting sounds picked up at each microphone were digitized and stored for analysis. A short duration broad-band sound containing energy from 100 kHz to 10 kHz was used (linear FM sweep from 100 kHz to 10 kHz, 0.5 ms duration, trapezoidal envelope). Sounds were recorded from a large number of positions, and the resulting records were analyzed to quantify spatial variation of interaural intensity differences (IIDs), interaural time differences (ITDs) and both monaural and binaural spectral cues.

An interesting result of these measurements was obtained by examination of the transfer functions. The transfer function is a spectral representation of the filter characteristics of the external ear for a given sound source location. We corrected for the filter characteristics of the microphones and all electronic equipment when calculating transfer functions (see Wotton et al. 1995, for details of methods). In Figure 12.1a a series of transfer functions is shown which were obtained by varying the vertical position of the sound source while keeping the horizontal position constant. This *vertical slice* shows how sound spectrum is shaped by the ear as a function of vertical position. It can be seen that there is

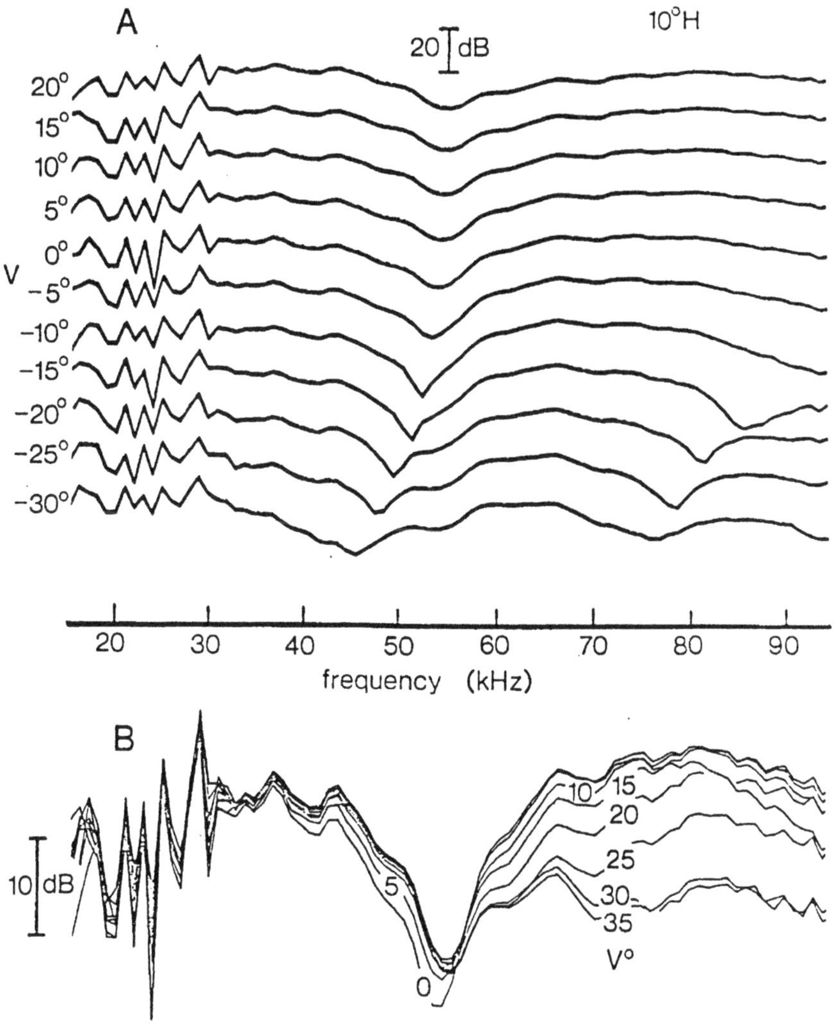

FIGURE 12.1 Vertical slice showing transfer functions of the external ear of *Eptesicus fuscus* for a series of vertical positions at a fixed horizontal position (data shown in this Figure was taken 10° ipsilateral to the vertical midline). (A) Transfer functions shown sequentially for vertical positions from 20° above the horizontal midline to 30° below the midline. Note the prominent notch in the spectrum. At vertical positions above the midline the center frequency of spectral notch is approximately 53 kHz. For vertical positions below the midline the center frequency of the notch systematically decreases. (B) Overlay of transfer functions for vertical positions above the midline. There is not much shift in the position of the spectral notch as the elevation of the sound source increases but there is a large and systematic change in the amplitude of the region from 65 to 90 kHz.

a prominent "notch" in the spectrum, centered at 55 kHz for a sound source located at +10° vertical. The center frequency of this notch moves systematically as the elevation of the sound source is decreased, until it is centered at 45 kHz for a sound source located at -30° vertical. For vertical positions above the horizon (10° to 35°, Figure 12.1b) the position and the depth of the spectral notch do not vary much, while the amplitude of the region from 60 kHz to 90 kHz decreases systematically. Vertical slices for horizontal positions from +30° to -30° show approximately the same pattern, with spectral notches decreasing in center frequency for positions below the horizon and systematic increases in the amplitude of the high frequency region for positions above the horizon. The spectral notches can be examined more quantitatively by using local minima to determine the center frequency of the notches and then plotting notch center frequency as a function of sound source location. The

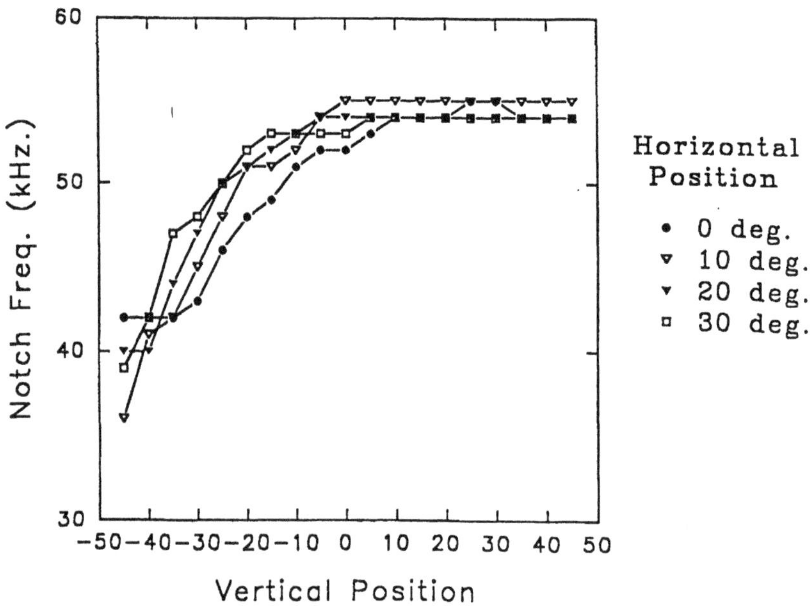

FIGURE 12.2 Spectral notch center frequency plotted as a function of vertical position for 4 vertical slices. Center frequency of the notch was determined by finding the local minimum of the transfer function (between the values of 25 and 90 kHz). Lines through the different symbols represent vertical slices taken at different horizontal positions (date is shown for vertical slices taken at the midline and 10°, 20°, and 30° ipsilateral to the midline). For positions above the midline (positive values on the x-axis) there is little or no change in the center frequency of the spectral notch. Below the midline, notch center frequency decreases as a function of vertical position. Note that there is also a change in notch center frequency as a function of horizontal position, although the magnitude of this change is not as great as the change with vertical position.

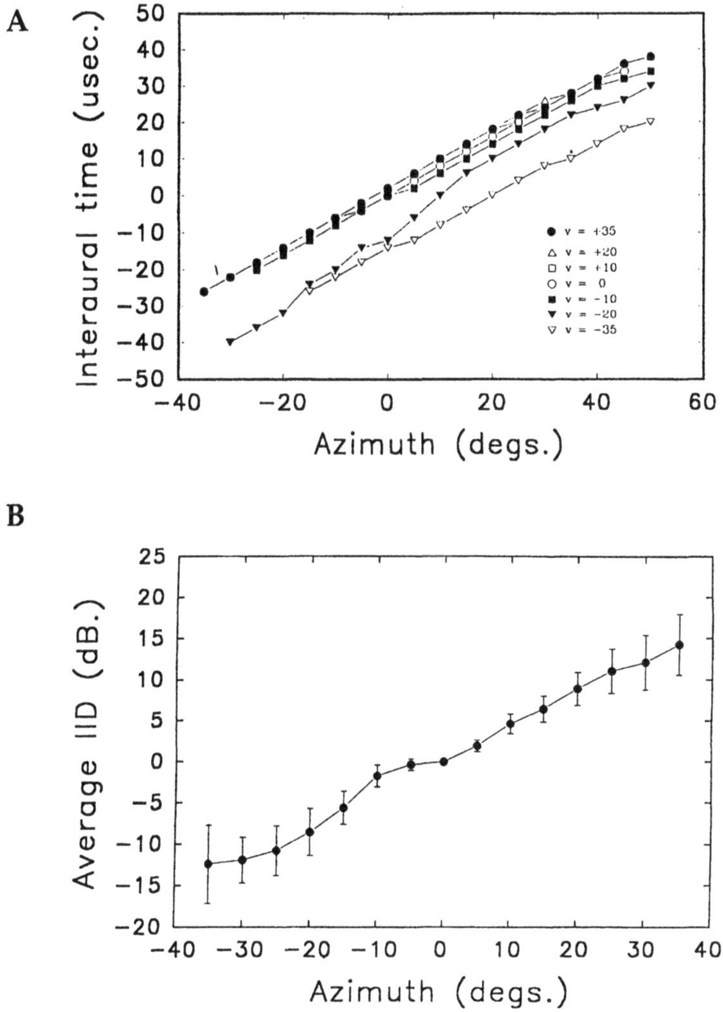

FIGURE 12.3 (A) Interaural time difference (ITD) as a function of the azimuth of the sound source for a number of horizontal slices (vertical position of the sound source constant, horizontal position varied). ITD was determined by recording sounds from both ears as the sound source was moved around the head and then performing a cross-correlation between the waveforms recorded at each ear. The largest peak in the cross-correlation function was used as the ITD for a given sound source location. The results are fairly linear and consistent for each of the horizontal slices, and the maximum the bat is exposed to is about 40 μsec. (B) Average interaural intensity difference (IID) as a function of azimuth for a horizontal slice (vertical constant at +5°). It each position the IID was determined by averaging the IID across all frequencies in the range of 50-25 kHz. The change in IID was fairly constant as a function of azimuth, although there was more variation of IID with frequency for more extreme positions.

series of plots in Figure 12.2 show that the center frequency of the notch varies primarily with vertical position and very little with horizontal position. There is a small change in the center frequency of the notch as a function of horizontal position, but these changes are small compared to the changes seen as a function vertical position. These results were consistent for all ears tested. From this data it is clear that the external ears create spectral cues that vary as a function of the vertical location of the sound source. Thus, potentially strong monaural cues for vertical localization are contained in spectral changes introduced into returning echoes by the location dependent transforms of the external ear.

As shown in Figure 12.2 there is also variation in spectral content of sound reaching the ears as a function of horizontal position (at least as far as notches are concerned). Since notch position is slightly dependent on horizontal location, the vertical location of the sound source will be slightly ambiguous unless the horizontal location can be determined. There are a number of binaural cues that could provide the bat with information about horizontal location. There are spectral differences between sounds from the ipsilateral and contralateral sides of the head (Figure 12.2), thus unless a sound is coming directly from the midline there will be spectral differences between the two ears. It is possible that the bat could use these interaural spectral differences as a horizontal localization cue. Two other potential cues for horizontal localization are interaural intensity differences (IIDs) and interaural time differences (ITDs). IID cues are created by the acoustic shadowing effects of the head — Sounds coming from the left of the head are louder in the left ear than the right ear and *vice versa*. ITD cues are created by differences in pathlength for sound at the two ears — Sounds from the left of the head will arrive at the left ear before arriving at the right ear, and sounds from the right of the head will arrive at the right ear before the left. The IIDs and ITDs obtained from the ear measurements are shown in Figure 12.3. Both vary in a fairly linear manner as a function of horizontal location of the sound source. Thus, we have quantified a number of potential cues the bat could use for horizontal and vertical localization and now we can begin to discuss how the bat might utilize the information in these cues.

Horizontal Localization

Models of horizontal localization in mammals consider two major cues: ITDs for frequencies less than 2-3 kHz (up to 10 kHz in owls, Moiseff and Konishi, 1983, see Moiseff et al. this volume) and IID cues for higher frequencies. At low frequencies (< 3 kHz) the mammalian cochlea can encode cycle-by-cycle phase information at each ear. Higher centers of the nervous system can then perform a phase comparison of the signals at the two ears across frequency and extract an estimate of ITD (Yin and Kuwada, 1983). At higher frequencies the cochlea is presumed to no longer encode phase information so IID cues are used for localization of azimuth. Thus, models of horizontal localization for echolocating bats have been built on the assumption that bats use only IID cues, because it is presumed bats cannot determine ITD at ultrasonic frequencies. These models have been developed both for bats that use predominant constant-frequency (CF) components of their biosonar emission (*Pteronotus parnelli*, Fuzessery and Pollak, 1985), and for bats that use frequency-modulated (FM) calls (*Eptesicus*, Poon et al. 1990; *Tadarida*, Pollak, 1988). In both cases, measurements of the directional properties of the external ears using pure tones show that intensity across frequencies varies as a function of the azimuth of the sound

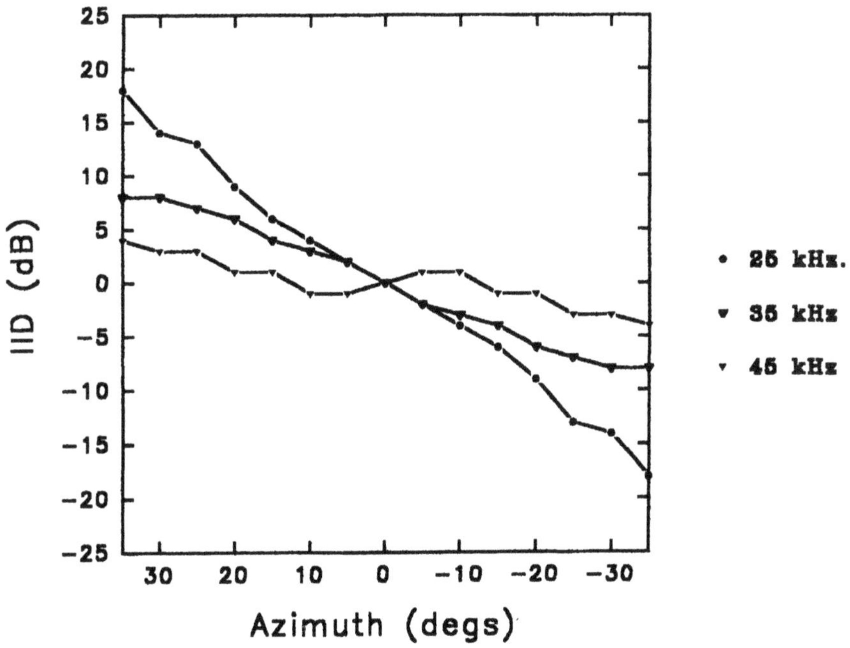

FIGURE 12.4 IID plotted as a function of 3 specific first harmonic frequencies. The IID plots shown were obtained from a from the same data set use to calculate the average IID shown in Figure 3A. The IID for each of the frequencies shown is less than 10 dB. for horizontal positions between +/- 20° horizontal. At more extreme positions there is greater variation in the IID between frequencies. Note that there is no "peak" IID value for nay of the frequencies shown, rather they all exhibit a continuous increase.

source relative to the ear. In general, the maximum frequency sensitivity of the ear varies in a fairly linear manner with direction — from maximum sensitivity at low frequencies when the sound was near the midline, to maximum sensitivity at high frequencies when the sound was coming from more lateral positions (Fuzessery and Pollak, 1984; Jen and Chen, 1988). This suggests the possibility that horizontal location could be resolved simply by determining which frequency band in the returning echo is the most intense (or where the IID was greatest). According to this frequency-specific postulate, the horizontal location of sound sources is determined mainly by the directional filter properties of the external ears.

There are several drawbacks to this postulate that make it unlikely the bat would place a large reliance on intensity at specific frequencies for determination of horizontal location in space. The emitted biosonar sound of bats is directional, and best described as a cone of sound directly in front of the bat. Measurements of the directional properties of the biosonar emission of *Eptesicus* have been made (Hartley and Suthers, 1989). Their results show that directly in front of the bat the high frequency components of the emission were more intense than lower frequency components, while lower frequencies were relatively more intense at more lateral positions. Thus, the frequency-specific directional properties of the emission have a roughly reciprocal relationship to the frequency-specific directional properties of the external ears. This will tend to flatten out frequency-specific IID cues created by the directional properties of the ears. Using the transfer function measurements it is possible to determine the IID values for specific frequencies at specific horizontal locations. Figure 12.4 shows that the IID values vary in a fairly monotonic manner as a function of horizontal position. There are no clear peaks in the IID values at specific horizontal locations for specific frequencies as would be expected for a coding of horizontal space based on individual frequencies. As mentioned above, these curves would be even flatter if the directional properties of the emission are factored in.

Even if specific frequencies do not encode specific spatial locations, it is still possible that information about IIDs could be combined across frequencies to get an estimate of the horizontal position of the sound source. In Figure 12.4, for horizontal positions between +20° and -20° there are relatively small differences between the IID values at different frequencies, indicating that at these positions IID differences between frequencies would be a poor cue for horizontal localization (for the first harmonic frequencies shown). Another problem with this model arises in conditions of clutter, when multiple echo-producing objects are present, creating multiple and in some cases overlapping echoes. In a cluttered environment, echoes from objects to the left and right of the bat will sometimes produce echoes that overlap in arrival time at the two ears. In this situation, reliance on solely IID cues will create erroneous estimates of azimuth, in which several real targets would be combined into spurious "phantom" targets located between the real ones. For example, an echo from an object 25° to the left of a bat's head might be 15 dB louder at the left ear than the right ear, while the overlapping arrival of an echo from an object 25° to the right of the head could be 15 dB louder at the right ear compared with the left ear. The sum of the IIDs from the two objects would be 0 dB, thus, from IID cues alone it would appear to the bat that there is a single target straight ahead. While IID is clearly an important cue for localization in azimuth, there are cases, such as the above example, where IID is not sufficient, and may actually create false representations of object location. Few behavioral experiments have been done to attempt to directly determine the nature of the cues the bat is using for horizontal localization. However, both obstacle-avoidance and aerial discrimination experiments show clearly that bats do not suffer from phantom target misdirections that an IID model would produce (Simmons et al. 1993).

If IID cues alone are not sufficient for horizontal localization then what other cues might the bat be using for horizontal localization? An interesting possibility is that the bat might be using ITD cues. Behavioral studies have shown that *Eptesicus* can perceive echo-delay differences of less than 1 μsec (Simmons, 1979; Menne et al. 1989; Moss and Schnitzler, 1989; Simmons, 1990). If fine range determination could be performed separately at each ear then an interaural comparison of echo-delay at the two ears would yield an estimate of ITD. The possibility that the bat uses ITD cues has not been considered, mainly because no neurophysiological evidence has been found to show that the nervous system of FM bats can encode the phase information or the small time differences necessary to make ITD estimates at ultrasonic frequencies (Pollak et al. 1977; Bodenhamer and Pollak, 1981); although a small number of cells have been reported to be sensitive to ITDs (Harnischfeger et al. 1985). Later in the chapter we will present neurophysiological evidence for sensitivity to ITDs.

Vertical Localization

Most models of mammalian vertical localization postulate spectral cues as a major information bearing parameter used for determining the elevation of a sound source (Blauert, 1969; Blauert, 1983; Asano et al. 1990). In bats spectral cues are also thought to be used (Grinnell and

Grinnell,1965). Studies using pure tones to measure ear directionality have shown that the ear of the bat has differential vertical sensitivity for different frequencies (Fuzessery and Pollak, 1984; Jen and Chen, 1988). A more sophisticated approach is to determine the transfer function of the ears for the entire range of behaviorally relevant frequencies discussed above. Transfer function measurements in other mammalian species (Musicant et al. 1990; Carlisle, 1990a,b; Middlebrooks, 1992) show that spectral peaks and notches are created within the ear due to multiple reflective pathways created by the shape of the external ear. These spectral peaks and notches vary strongly as a function of vertical position, and thus provide strong elevational cues. These spectral cues would likely be missed using pure-tone test frequencies spaced 10 kHz to 30 kHz apart. External-ear transfer function measurements in *Eptesicus* have revealed that the ear produces spectral peaks and notches in the ultrasonic spectrum that vary as a function of vertical location of the sound source (Figure 12.1, see also Wotton et al. 1995). The types of variation seen are similar to those reported for the external ear of the cat (Musicant et al. 1990), specifically a prominent notch in the spectrum that moves systematically with changes in the elevation of sound source. This suggests that the bat could use the detailed spectral shape of the returning echo as a cue for vertical location. Thus, we believe that spectral shape is used primarily for vertical localization while IID and ITD cues are used for horizontal localization.

Behavioral Evidence

We have quantified several cues that the bat might use for localization and have suggested possible ways these cues could be used for localization in azimuth and elevation. The question that arises is: What evidence exists, if any, for the behavioral relevance of these cues? Very few behavioral experiments have directly tested how different cues affect the bat's ability to perceive horizontal and vertical location. The most detailed behavioral studies examining the role of specific cues on localization in azimuth and elevation have been carried out in barn owls (Moiseff and Konishi, 1981; Moiseff, this volume) and humans (Blauert, 1983, Middlebrooks, 1992). In any behavioral study of localization it is necessary to have a means of controlling stimulus parameters and monitoring the localization performance of the animal. The difficulty in experiments in bats arises because it is difficult to monitor the position of the bat's head by methods that have been used in other species, which all involve attaching monitoring apparatus to the animal's head. The small size of a bat's head currently precludes such an approach.

We have constructed a system to monitor the head aim of bats during localization tasks.

The head-aim monitoring system takes advantage of the fact that the bat's sonar emission is directional. The sonar sound is beamed out of the bat's mouth and the resulting isointensity field is roughly conical in shape, with about a 30° angle of spread (Simmons, 1969; Hartley and Suthers, 1988). Thus, the sonar field can be considered analogous to an "auditory flashlight" in that it "illuminates" only a portion of space related to where the bat's head is pointed. During localization bats attempt to center their sonar beam on the object they are localizing in order to produce a maximum echo. By measuring the intensity of the sonar field using several equidistant, spatially separate microphones, it is possible to determine where the bat's head is pointed each time it produces a sonar emission.

This system is was used (in conjunction with a three-dimensional video imaging system for calibration) to determine how the vertical position of sound sources affects the bat's perceptions of the spectrum of returning echoes. Analysis of the results indicate that the bat's perception of the spectrum of the echoes is affected by the vertical location of the sound source (these results have been submitted for publication, Wotton et al. submitted). We mention this here for two main reasons. First these results provide evidence for the contention that the spectral content of the returning echo is a major cue that the bat uses for determining the vertical location of sound sources. Secondly, these experiments demonstrate that the head aim of the bat can be successfully monitored during localization experiments. Through the use of this system (or similar ones) it should be possible to investigate how a number of different parameters affect the bat's ability to localize in both the azimuth and the elevation of specific targets.

Neurophysiological Experiments

Another approach to unraveling the mechanisms of sound localization is to monitor and record how elements of the nervous system respond to specific auditory stimuli. Sounds transmitted to the cochlea are converted into action potentials by hair cells along the tonotopically organized basilar membrane. These nerve impulses are transmitted and processed by a number of nuclei in the brain of the bat. The tonotopic organization present at the cochlea is preserved through most levels of the auditory system. Through this tonotopic organization the nervous system is able to extract information about the spectral content of the sound, based on the strength of activation of different frequency bands.

Using microelectrodes it is possible to record the activity of individual neurons or small groups of neurons within specific, identifiable auditory nuclei. Our studies have been carried out mainly in the inferior colliculus (IC) of the midbrain. There are two major reasons for recording from the IC. The first is that it is easily accessible: its surface can be seen through the semitranslucent skull of the bat with the aid of a dissecting microscope. More importantly, the IC is an obligatory relay nucleus. Thus, lower centers in the brainstem transmit auditory information to higher levels of the brain via IC. In other animals the IC has been shown to be a nucleus where a number of cues about the spatial location of a sound source are first integrated (Phillips and Brugge, 1985; Pollak and Casseday, 1989).

There are two approaches to the presentation of stimuli for sound localization studies. Sounds can be presented from a speaker that is positioned at measured locations in space relative to the bat's head. The advantage of this *free-field* stimulus presentation is that it allows for the monitoring of neural responses to specific sound-source locations. However at each spatial location a number of potential cues are available (IID, ITD, spectral cues) and it is difficult to determine to which of the cues the neuron is responding. To get around this limitation it is necessary to use earphones for *dichotic* stimulus presentation. By allowing for direct control over the intensity, arrival time, and spectral content of the sound presented to each ear, dichotic stimulation provides a means for looking at neural responses to specific cues (IID, ITD, spectral cues). Dichotic stimulation is most effective if the potential cues have been quantified in a comprehensive manner, so that the cues presented during the experiment are within the range normally encountered by the animal.

Free-field studies on auditory responses of neurons in the IC have been conducted with several species of bats (Grinell, 1963; Fuzessery and Pollak, 1985; Jen and Schlegel, 1982). In these studies the rate of firing of neurons was recorded for sounds presented from a number of locations in space. Near threshold, a number of neurons in these studies had maximum responses for sounds from specific, small regions of space. In other words, the neuron exhibited tuning to that spatial location. As the intensity was increased above threshold the spatial selectivity of many of these neurons decreased, although there was a small subset of neurons in which the tuning remained relatively constant. Each of the neurons in these studies was also tuned to a specific sound frequency termed its characteristic frequency (or best frequency depending on how the frequency value was determined). These studies were carried out using pure tone stimuli presented at each neuron's characteristic frequency. This choice of stimuli dictates that the spatial tuning responses near threshold are determined by the monaural frequency-dependent directional properties of the ear on the side of the head ipsilateral to the stimulus. Due to sound shadowing effects of the head it is unlikely that for sounds near threshold the intensity of sound at the ear on the side contralateral to the stimulus would be loud enough to affect the neural response. Individual neurons are tuned to specific frequencies, thus it would be expected that each neuron would respond maximally to positions in space where the ear has the maximum amplifica-

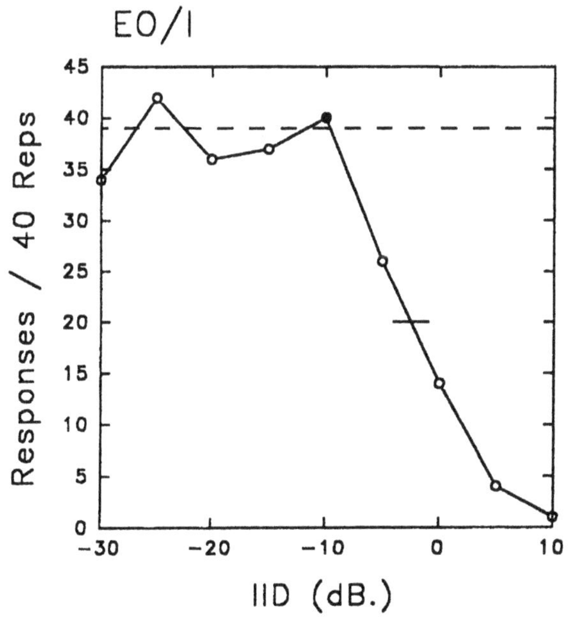

FIGURE 12.5 Example of an interaural intensity difference (IID) function or "tuning curve". The example shown in the figure is an E0/I cell, i.e. excitatory (E) response to monaural contralateral stimulation, no response (0) to monaural ipsilateral stimulation, and sound at the ipsilateral ear has inhibitory (I) effects during binaural stimulation. IID curves were generated by setting the intensity level at the contralateral ear to 10 dB. above the contralateral monaural threshold and then varying the intensity at the ipsilateral ear. The horizontal dashed line indicates the number of spikes produced by monaural contralateral stimulation at the intensity level used for IID tuning. The solid horizontal line intersecting the curve is at the point on the IID function where the number of spikes is halfway between the maximum level and the spontaneous level (50% point). The filled (solid) data point represents a conservative estimate of the starting point for the strong decline in responsiveness of the unit as a function of IID. This *inflection point* (IP) is determined using a custom recursive algorithm. The algorithm measures the difference between successive points until it finds a difference greater than the 50% level. In the example shown in the figure the difference between the filled point and the adjacent point on the right is 22 spikes. This difference is greater than the 50% level (18 spikes) so the filled point qualifies as an IP using this algorithm.

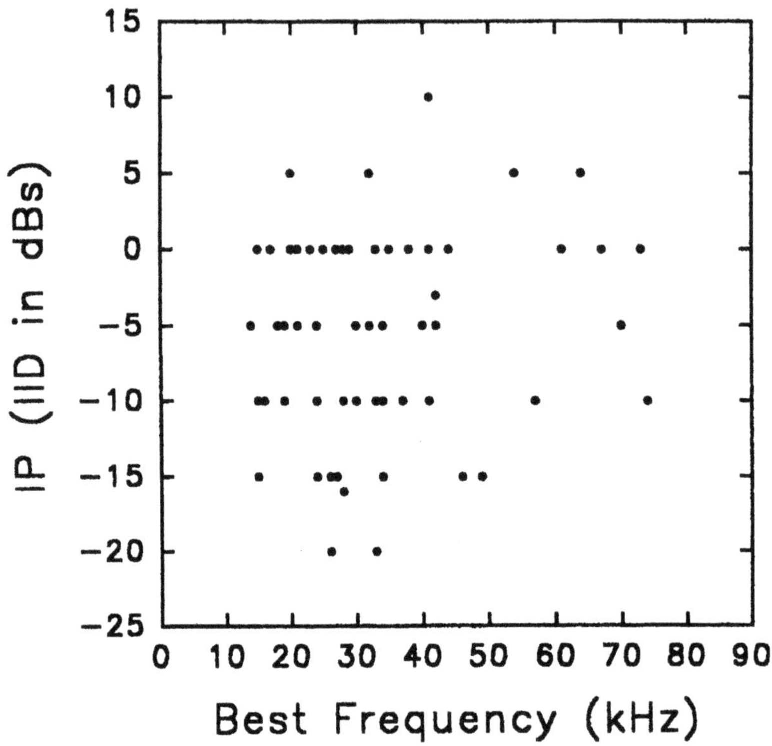

FIGURE 12.6 The plot shows the inflection point vs. the best frequency for all of the units isolated during this study for which both parameters were determined. Inflection points for each unit were determined by the method described in the caption for Figure 12.5. The best frequency for each unit was defined as the pure tone frequency that elicited the greatest response (most spikes) during 40 stimulus repetitions. All best frequency measurements were done at 10 dB above threshold. Clearly there is no relationship between the IP and the best frequency. To state this in another way: the best frequency of any given unit is in no way predictive of the likely IP of that unit.

tion for characteristic frequency of that neuron. This is indeed the case for sounds presented near threshold (monaural cues, Poon et al. 1990).

At intensities above threshold the bat is exposed to effective binaural cues, and in order to sort out the effects of these cues it is necessary to turn to dichotic studies. The type of binaural cue most dichotic studies have focused on has been interaural intensity differences. Studies carried out in *Pteronotus parnelli* have revealed a large number of cells within the 60 kHz region of the IC in which the response level is a function of IID. *Pteronotus* uses a CF-FM signal with a prominent second harmonic at about 60 kHz, and the IC of this species contains a disproportionate number of neurons with CF's in this frequency region. Some of these cells are tuned to a specific IID value, but most serve more as IID edge detectors. They are "on" (e.g. excited) when the IID level is above a certain value and "off" (e.g. inhibited) below a specific IID value. Generally, there is a relatively sharp boundary between excitation and inhibition (Wenstrup et al. 1988). The spatial receptive fields of these edge detector neurons would be expected to encompass the entire region of space in which IID values are at or above the IID cutoff value for that neuron. When these neurons were tested using a combined free-field/dichotic paradigm their spatial receptive fields roughly coincided with their IID sensitivity (Fuzessery and Pollak, 1985). We have tested IID sensitivity of neurons with a broad range of best frequencies using dichotic stimulation in the IC of *Eptesicus*. We also found that the majority of IID sensitive neurons could be classified as edge detector type neurons (Figure 12.5). With the types of tuning curves shown for the neuron in Figure 12.5 it is possible to apply an algorithm to quantitatively determine the IID at which the response of the neuron starts a continuous decline, we have termed this point the *inflection point* (Haresign et al. in prep). The inflection point value for all the neurons that exhibited inhibitory responses for some IID values is plotted against the best frequency of the neurons in Figure 12.6. The is no obvious relationship between the best frequency of the neurons and the inflection point of the IID tuning curve. This is important, because it has been suggested that tuning for auditory space might covary in a systematic way with frequency tuning in the inferior colliculus (Poon et al. 1990). If this were the case then it would be expected that some aspect of IID tuning should covary with frequency. In the dichotic studies none of the quantified features of the IID curves showed any dependence on the best frequency of the neurons (Haresign et al. in prep). Thus, while many neurons in the IC show sensitivity to IID there is no evidence for a systematic mapping of IID sensitivity across frequencies.

As mentioned in the preceding section there are a number of potential problems in relying solely on IID cues for horizontal localization. Many of these problems are diminished if the bat could make use of interaural time difference (ITD) cues. The question is: Are bats able to make use of ITDs? Attempting to answer this question leads one directly into the middle of one of the most interesting areas of contention in the field of bat sonar processing. There are a number of behavioral results from several different labs that establish that the FM bat *Eptesicus* can determine the range of a sonar target to an accuracy of less than a microsecond (Simmons, 1979; Moss and Schnitzler, 1988, Simmons, 1989), and that it also perceives the fine phase structure of returning echoes (Simmons et al. 1990a,b; Menne et al. 1989). This work has been repeated a number of times using very different methods (Simmons, 1993). Since the bat is capable of this type of performance for range discrimination there is reason believe that the same resolution could be applied binaurally to make use of ITD cues. The contention arises from the lack of evidence that the nervous system can register events with microsecond accuracy for ultrasonic stimuli. Indeed, it is generally accepted that the mammalian nervous system cannot encode cycle-by-cycle phase information at frequencies above 3 kHz.

There have been attempts to demonstrate neural sensitivity to ITDs at ultrasonic frequencies (Fuzessery and Pollak, 1985) and to ascertain the limits of time resolution of action potentials in the bat's nervous

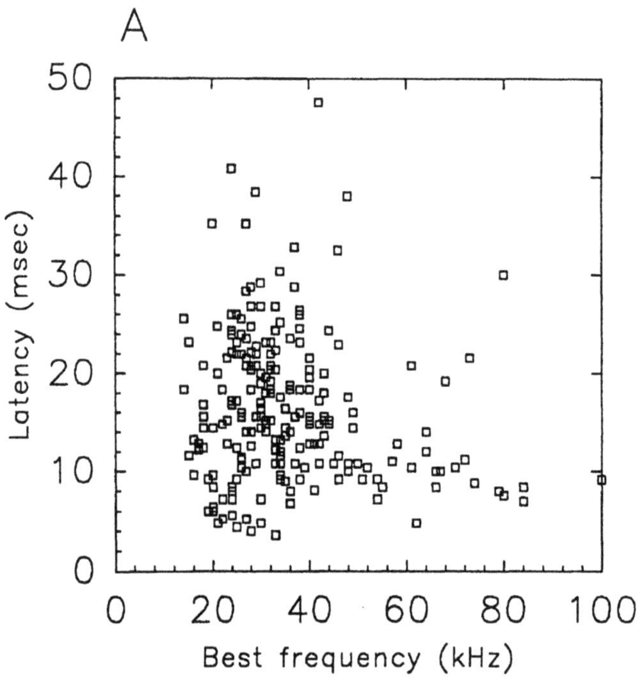

FIGURE 12.7 The response latency plotted as a function of the best frequency for units recorded in the inferior colliculus. The best frequency was determined as described in the caption of Figure 7. The latency was determined by averaging the time of occurrence of the first stimulus evoked action potential over 40 stimulus repetitions (using BF as the stimulus). All cells tended to be phasic and well time-locked to the stimulus (standard deviation in the range of 1-2 msec.) and usually there was little to no spontaneous activity. There are a wide range of latencies present among units in the IC. In addition it is clear that the best frequency of a unit has little to no predictive value in terms of the latency of the unit (there is a slight trend for units with higher frequency BFs to have shorter latencies.

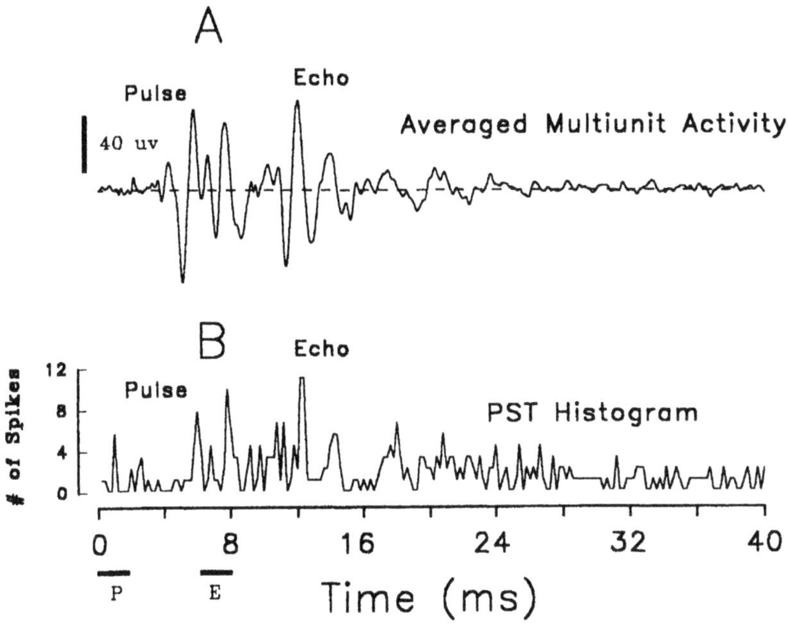

FIGURE 12.8 A) Typical averaged multiunit response to pulse-echo pair stimulus (echo delay= 6ms, see text for specific parameters of the sounds used) showing a clear pattern of neural activity. Two large peaks are present: The first at a latency of about 6 ms, corresponding to the start of the pulse-induced neural responses, and the second at a latency of about 12 ms corresponding to the start of the echo-induced neural responses. It is noteworthy that the pulse-induced activity runs into the echo induced activity and that the echo-induced activity continues for at least 10 ms beyond the initial echo induced peak. B) PST histogram obtained using the same electrode at the same recording site where the averaged multiunit response shown in 8A was obtained. The data for the PSTH was obtained using a digital window discriminator, which recorded the time of occurrence of positive crossings of a preset window threshold. The PSTH shown was constructed from 48 repetitions of the pulse-echo pair (6 ms delay). Many of the peaks in the PSTH correspond to peaks in the averaged multiunit response, indicating that the averaged multiunit response is the result of the activity of a small local population of neurons.

system (Bodenhamer et al. 1979; Bodenhamer and Pollak, 1981). These studies failed to show any sensitivity to ITDs in bats and set the lower limit of time resolution in the nervous system at approximately forty μseconds. A lower limit of forty μsec time resolution would preclude the uses of ITDs (since the maximal effective time difference experienced by the bat is ≅ 40 μsec, Figure 12.3a) and would seem to contradict the behavioral evidence showing much greater time sensitivity. A close examination of the data and interpretation of these experiments

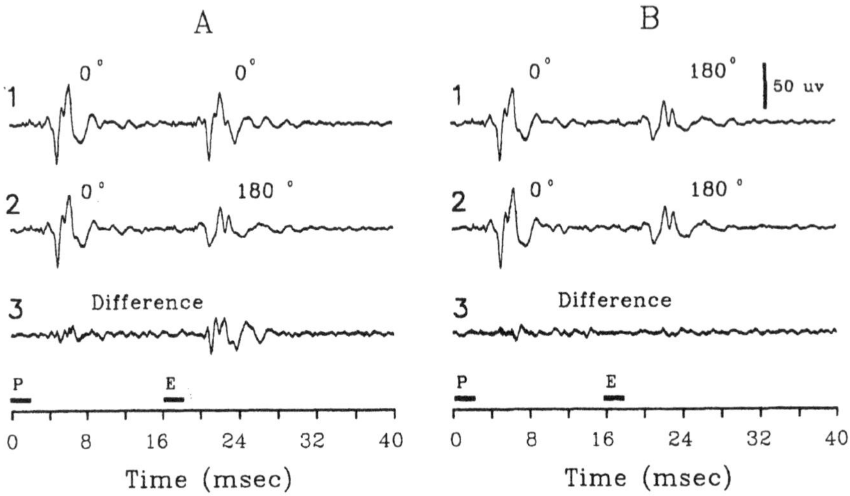

FIGURE 12.9 Differences are shown between averaged multiunit responses to phase normal and phase inverted stimuli for one recording site in the inferior colliculus. P=Pulse, E=Echo A.) In $9A_1$ The averaged multiunit response (128 reps) to a phase normal pulse (2 harmonic FM sweep, 50-25 kHz first harmonic, 100-50 kHz second harmonic, 2 ms duration, Hanning envelope) and a phase normal echo (same parameters as pulse, 15 dB less intense) is shown. In $9A_2$ the stimuli are identical, except that the phase of the echo has been inverted (multiply the digital signal by -1). There is a clear difference in the echo response between $9A_1$ and $9A_2$, while the response to the pulse remains relatively unchanged. This is further demonstrated by the difference function ($9A_3$), which is obtained by subtracting the two waveforms ($9A_1$- $9A_2$). The difference function shows a large difference between the phase normal and phase inverted echo responses. B.) Figure 9B shows the control data for the same recording site. To be sure that the changes seen between phase normal and phase inverted conditions were not due to instability in the electrode, the recording site or some other uncontrollable factor, multiple averages were obtained at different times for both the phase normal and the phase inverted conditions. A comparison of two different average multiunit responses ($9B_1$ and $9B_2$) obtained at different times for the phase inverted condition demonstrates that the response to the phase inversion is stable in time and repeatable. There is little difference between the two waveforms $9B_3$.

yields several areas through which this apparent discrepancy could be resolved. First of all, signal-to-noise constraints of extracellular recording (method employed all of the above mentioned experiments) make it theoretically impossible to record the time resolution of action potentials much better than around 20-40 μsecs. This raises the possibility that the 40 μsec arrived at in the above mentioned studies may represent a limit imposed by the recording apparatus, rather than by physiological properties of the neurons. Furthermore, even if the 40 μsec value is accepted as the limit of time resolution of a single neuron, this value should not be used as the limit of time resolution for the entire nervous system. The activity of large numbers of neurons with relatively poor time resolution could theoretically be combined to produce much higher time resolution. Modeling studies using physiologically reasonable constraints have shown that the combined activity of large groups of neurons can yield a time resolution of less than 1 μsec (Suzuki and Suga, 1990). When sensitivity to ITDs was examined at the level of single units no changes in firing rate were seen for changes in ITD (Fuzessery and Pollak, 1985). Since the presumption was that bats cannot utilize ITD information no further examination of ITD was conducted. This lack of evidence does not constitute proof that the bat cannot utilize ITD, it simply shows that if ITD sensitivity exists its neural manifestation has not yet been demonstrated. It does suggest the possibility that ITD might be encoded by something other than a rate code.

One of the interesting aspects of neural responses in the IC is the broad range of response latencies observed across the population of individual units (Figure 12.7). This latency dispersion does not occur at lower centers in the brainstem (Pollak and Casseday, 1989, Ferragamo et al. in prep). Most of the single units in the IC of *Eptesicus* respond phasically with a single spike to a sound stimulus (Ferragamo et al. in prep). The wide range of onset times that occur in the IC in response to a single sound stimulus suggests that the timing of neural firing may be as important a parameter as the rate of neural firing. We have seen evidence of large time shifts in neural response patterns to small changes in stimulus timing by recording the averaged multiunit responses of small, local neuronal ensembles.

Averaged multiunit responses were obtained using glass NaCl electrodes with a resistance in the range if 3-10 MΩ. The stimuli used in these recordings consisted dichotically presented pulse- echo pairs. The intensity and the timing of the pulse and echo at each ear could be controlled independently. Each sound consisted of a 2 msec hyperbolic FM sweep with two harmonics. The first harmonic went from 50 kHz to 25 kHz and the second harmonic went from 100 kHz to 50 kHz. A Hanning envelope was used to shape the sounds. The electrodes were advanced through the IC until a location was encountered at which the activity of several units was evident. The responses of this small group of neurons were averaged for 128 or 256 stimulus repetitions. The advantage of this method is that it reveals coordinated patterns of firing among small, local populations of neurons (approximately 5-20 neurons). The averaged multiunit responses in the IC have revealed a multipeaked, structured pattern of activity in the firing patterns of the small populations (Figure 12.8a). On the basis of single-unit data this is an unexpected result since single units in the IC tend to be phasic onset

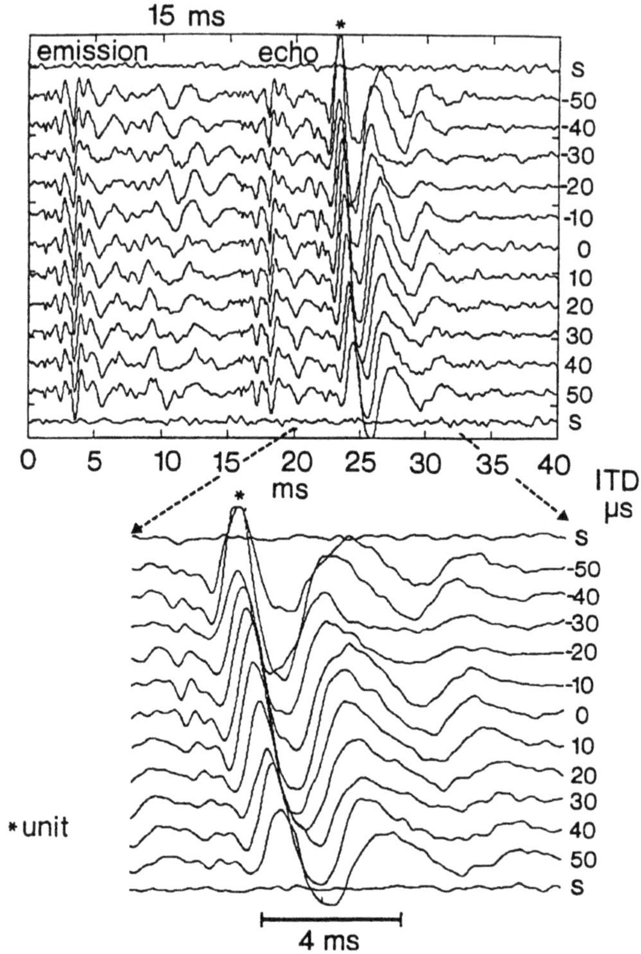

FIGURE 12.10. To simulate natural stimuli as closely as possible a dichotic stimulation system was used to present pulse echo pairs (same sound parameters described in Figure 12.9). Pulses were presented to the ears with no interaural time difference (ITD) present, while small ITDs were introduced into the echoes, with values similar to those that might naturally be encountered by the bats (Figure 12.3A). Positive values correspond to left ear leading and negative values correspond to right ear leading. A series of averaged multiunit responses is show in the upper portion of Figure 12.9 for different echo ITDs. The top-most and bottom-most traces in the upper box show the averaged spontaneous activity present at the same locations. As the ITD is changed there is a shift in the latency of several of the components of the echo response. This is shown more clearly in the time expanded section of the traces shown in the bottom portion of the figure. Looking at the case of the largest peak (possibly corresponding to the averaged activity of a single unit) it can be seen that an ITD shift of 100 microseconds (-50 msec. to 50 μsec.) results in a latency shift in the peak of approximately 1.5 milliseconds. It is interesting to note that the neural response to the ITD is manifest in a latency shift rather than in a change in the overall level (rate of firing) of the neurons.

responders, which produce a single peaked peristimulus time histogram. Hints of this have been present in multiunit recordings, where at any specific site in the IC the Post Stimulus Time (PST) histograms often show multiple peaks with a consistent latency distribution among the peaks (Figure 12.8b). Since each peak most likely represents the activity of 1 or more phasic onset neurons, the presence of multiple peaks indicates that within the local populations individual neurons have specific patterns in the timing of firing relative to other neurons in the population. The latency distribution within averaged multiunit responses is dramatically altered by changes in both stimulus phase (Figure 12.9), and small (< 50 μsec) changes in ITD (Figure 12.10) both in the IC and in the auditory cortex. These findings are the first demonstration of neural sensitivity to phase and ITDs at ultrasonic frequencies (Haresign et al. in prep). These findings also point out that the timing of neural discharges may be a more important parameter for neural processing than previous studies, which tend to emphasize firing rate, have indicated.

Summary and Conclusions

An attempt has been made in this chapter to present a brief overview of the current state of progress towards unraveling the mechanisms of sound localization in a bat that relies on the processing of ultrasonic sounds for spatial perception. The question that arises is what do these specific findings contribute towards the understanding of general mechanisms of sound localization? The external ears of bats introduce spectral peaks and notches into incident sounds. The frequencies at which these peaks and notches are imposed vary depending on the location of the sound source relative to the head of the bat. Similar patterns of peaks and notches (although shifted to lower frequencies) have been seen in several other species, including ferrets (Carlisle, 1990a,b), cats (Musicant et al. 1990), and humans (Middlebrooks, 1992). In humans, psychophysical studies have shown that the spectral cues produced by the ear are the main cues for vertical sound localization, and play a less prominent role in horizontal localization. Behavioral studies have not been conducted in cats or ferrets. Preliminary experiments with bats have shown that the vertical location of sound sources affects the bat's perception of the spectral composition of the sound. These experiments also demonstrate that *Eptesicus* can serve as a useful model animal for behavioral studies of sound localization. In the nervous system, all of the auditory nuclei are tonotopically organized and span the range of frequencies behaviorally relevant to the bat. A tonotopic organization of auditory nuclei has been seem in all mammalian species studied.

Neural sensitivity to interaural intensity differences has been demonstrated in bats, owls, and all mammalian species in which it as been examined. Thus, it seems likely that IID cues play an important role in sound localization in all of these species. The neural sensitivity to interaural time differences and phase changes at high frequencies that has been demonstrated in *Eptesicus*, has thus far not been seen in any other species. This may be primarily due to the fact that it has not been studied using the methods describe above. Experiments are needed in other species in which the timing and synchrony of neural discharges are examined in addition to discharge rate to determine if this phenomenon is a specialization unique to bats or instead represents a more general mechanism of auditory processing.

Echolocating bats are highly specialized animals, and in many species the auditory system is the primary modality which they obtain spatial information about the environment. The auditory system, however, is organized according to the same basic plan that is found in other mammals. Any differences in sound localization mechanisms between bats and other animals are likely to represent specializations of common mechanisms rather than mechanisms unique to bats. Thus bats are, and will continue to be, excellent model animals for the study of sound localization.

References

Aitken, L.M., Irvine, D.R.F., and Webster, W.R. (1984) Central neural mechanisms of hearing. In: *Handbook of Sensory Physiology.* Bethesda, MD: Am. Physiol. Soc.

Asano, F., Suzuki, Y. amd Sone, T. (1990) Role of spectral cues in median plane localization. *J. Acoust. Soc. 88:* 159-168.

Blauert. J. (1969) Sound localization in the median plane. *Acustica 22:* 205-213.

Blauert. J. (1983) *Spatial hearing: The psychophysics of human sound localization.* MIT Press, Cambridge, MA.

Bodenhamer R.D., Pollak, G.D., and Marsh, D.S. (1979) Coding of the fine frequency information by echoranging neurons in the inferior colliculus of the Mexican free-tailed bat. *Brain Res. 171:* 530-535.

Bodenhamer, R.D. and Pollak, G.D. (1981) Time and frequency domain processing in the inferior colliculus of echolocating bats. *Hearing Res. 5:* 317-335.

Carlisle, S. (1990a) The auditory periphery of the ferret. I: Directional response properties and the pattern of interaural level differences. *J. Acoust. Soc. 88:* 2180-2195.

Carlisle, S. (1990b) The auditory periphery of the ferret. II:Spectral transformations of the external ear and and their implications for sound localization. *J. Acoust. Soc. 88:* 2196-2204.

Dallos, P. (1984) Peripheral mechanisms of hearing. *Handbook of Physiology, The Nervous System, 3*: 595-637.

Ferragamo, M., Haresign, T., and Simmons, J.A. (1995) Response properties in the inferior colliculus of the ceholocating bat, *Eptesicus fuscus*: I. Frequency and latency dimensions of expanded neural spectrograms. (in prep).

Fuzessery, Z.M. and Pollak, G.D. (1984) Neural mechanisms of sound localization in a echolocating bat. *Science 225:* 725-728.

Fuzessery, Z.M. and Pollack, G.D. (1985) Derminants of sound location selectivity in bat inferior colliculus: a combined dichotic and free-field study. *J. Neurophysiol. 54:* 757-781.

Fuzessery, Z.M., Wenstrup, J.J., and Pollack, G.D. (1985) A representation of horizontal sound location in the inferior colliculus of the mustache bat (Pteronotus p. parnellii). *Hear. Res. 20:* 85-89.

Griffin, D.R. (1958) *Listening in the dark.* Yale University Press., New Haven, CT.

Grinnell, A.D. (1963) The neurophysiology of audition in bats: Directional localization and binaural interaction. *J. Physiol. Lond. 167:* 97-113.

Grinnell, A.D., and Grinnell, V.S. (1965) Neural correlates of vertical localization by echolocating bats. *J. Physiol. 181:* 830-51.

Haresign, T., Ferragamo, M.J., and Simmons, J.A. (1995) Neural sensitivity to phase in the inferior colliculus of the big brown bat, *Eptesicus fuscus.* (in prep).

Haresign, T., Ferragamo, M.J., and Simmons, J.A. (1995) Response properties of inferior colliculus neurons sensitive to IID and ITD in the big brown bat, *Eptesicus fuscus.* (in prep).

Hartley, D., and Suthers, R.A. (1989) The sound emission pattern of the echolocating bat, *Eptesicus fuscus. J. Acoust. Soc. Am. 85:* 1348-1351.

Harnischfeger, G., Neuweiler, G., and Schlegel, P. (1985) Interaural time and intensity coding in the superior olivary complex and inferior colliculus of the echolocating bat, *Molossus ater. J. Neurophysiol. 53:* 89-109.

Jen, P.H.S. and Schlegel, P.A. (1982) Auditory Physiological properties of the neurons in the inferior colliculus of the big brown bat, *Eptesicus fuscus.* 147: 351-363.

Jen, P.H.S. and Chen, D. (1988) Directionality of the sound pressure transform at the pinna of echolocating bats. *Hearing Res. 34:* 101-118.

Menne D., Kaipf, I., Wagner, I., Ostwald, and J., Schnitzler, H-U. (1989) Range estimation by echolocation in the bat *Eptesicus fuscus*: Trading of phase vs. time cues. *J. Acoust. Soc. Am. 79:* 386-397.

Middlebrooks, J.C. (1992) Narrow-band sound localization related to external ear acoustics. *J. Acoust. Soc. Am. 92:* 2607-2624.

Moiseff, A. and Konishi, M. (1981) Neuronal and behavioral sensitivity to binaural time differences in the owl. *J. Neurosci. 3:* 2553-2562.

Moss, C.F. and Schnitzler, H-U. (1989) Accuracy of target ranging in echolocating bats: Acoustic information processing. *J. Comp. Physiol. 165:* 383-393.

Musicant, A.D., Chan, J.C.K., and Hind, J.E. (1990) Direction-dependent spectral properties of cat external ear: New data and cross-species comparisions. *J. Acoust. Soc. 87:* 757-81.

Phillips, D.P., and Brugge, J.F. (1985) Progress in Neurophysiology of sound localization. *Ann. Rev. Psychol. 36:* 245-74.

Pollak, G.D., Marsh, D.S., Bodenhamer, R.D. (1977) Characteristics of phasic on neurons in the inferior colliculus of unanesthetized bats with observations relating to mechanisms for echoranging. *J. Neurophysiol. 40:* 926-942.

Pollak, G.D., and Casseday, J.H. (1989) *The Neural Basis of Echolocation in Bats.* Berlin: Springer-Verlag.

Poon, P.W.F., Sun, X., Kamada, T., and Jen, P.H.S. (1990) Frequency and space representation in the inferior colliculus of the FM bat, *Eptesicus fuscus. Exp. Brain Res. 79:* 83-91.

Simmons, J.A. (1969) Acoustic radiiation patterns for the echolocating bats *Chilonycteris rubiginosa* and *Eptesicus fuscus. J Acoust. Soc. Am. 46:* 1054-1056.

Simmons, J.A. (1973) The resolution of target range by echolocating bats. *J. Acoust. Soc. Am. 54:* 157-173.

Simmons, J.A. (1979) Perception of echo phase information in bat sonar. *Science. 204:* 1336-1338.

Simmons, J.A. (1989) A view of the world through the bat's ear: The formation of acoustic images in echolocation. *Cognition. 33:* 155-199.

Simmons, J.A., Moss, C.F., Ferragamo, M. (1990a) Convergence of temporal and spectral information into acoustic images of complex sonar targets perceived by the echolocating bat, *Eptesicus fuscus. J. Comp. Physiol. 166:* 449-470.

Simmons, J.A., Ferragamo, M., Moss, C.F., Stevenson, S.B., and Altes, R.A. (1990b) Discrimination of jittered sonar echoes by the echolocating bat, *Eptesicus fuscus*: The shape of target images in echolocation. *J. Comp. Physiol. A 167:* 589-616.

Simmons, J.A. (1993) Perception of fine echo delay and phase by the FM bat *Eptesicus fuscus. J. Comp. Physiol. A 172:* 533-547.

Simmons, J.A., Saillant, P.A., Ferragamo, M., Haresign, T., Dear, S., Fritz, J., and McMullen, T.A. (1995) *Auditory computations for biosonar target imaging in bat echolocation.* in: Auditory Computations. Hawkins, H. and McMullem, T. (Eds.) Berlin: Springer-Verlag.

Suga, N. The extent to which biosonar information is represented in the bat auditory cortex. In: G.M. Edelman, W.E. Gall and W.M. Cowan (Eds.) *Dynamic Aspects of Neocortical Function.* New York: John Wiley and Sons, pp 315-373.

Suzuki, M. and Suga, N. (1991) Acuity in ranging based upon neural responses in the FM-FM area of the mustached bat. *Neuroscience Abstracts 17:* 181.2.

Wenstrup, J.J., Fuzessery, Z.M., and Pollak, G.D. (1988) Binaural neurons in the mustache bat's inferior colliculus. II. Determinants of spatial responses among 60-kHz EI units. *J. Neurophysiol. 114:* 1384-1403.

Wotton, J.M., Haresign, T.H., and Simmons, J.A. (1995) Spatially dependent acoustical cues generated by the external ear of the big brown bat, *Eptesicus fuscus. J. Acous. Soc. of Am. 98 (3):* 1423-1425

Wotton, J.M., Haresign, T., Ferragamo, M. and Simmons, J.A. (1995) The behavioral relevance of spectral notch cues for vertical localization to the big brown bat, *Eptesicus fuscus* (submitted).

13

Sound Localization from Binaural Cues by the Barn Owl Auditory System

Andrew Moiseff, Tim Haresign, and Jianxin Wang

Abstract

The barn owl auditory system has been a productive model for the study of the neural mechanisms underlying sound localization. The two ears and their associated facial ruff, specialized feathers that form the owl's external ear, are crucial for transforming the position of a sound into interaural time differences and interaural level differences. Once separated, these binaural cues are processed along two separate neural pathways. At the level of the owl's inferior colliculus the neural information from the two pathways converge to form a neural map of auditory space. Experience plays a crucial role in allowing the nervous system to calibrate itself so that the binaural cues can be effectively interpreted. This chapter summarizes aspects of sound localization by the barn owl and presents results of a simple behavioral experiment that serves to illustrate the importance of sensory experience to produce meaningful sound localization behavior.

Introduction

Barn owls are well known for their capabilities as nocturnal predators. For many years the sensory basis of the barn owl's nocturnal hunting was not understood. Payne and Drury (Payne and Drury, 1958) demonstrated unequivocally that the barn owl could capture prey in a controlled laboratory environment under conditions of no visible light. Payne went on to show that owls relied on their hearing to detect and capture prey in total darkness (Payne 1971). Payne's work pioneered the systematic analysis of the barn owl's prey capture behavior that has since revealed much about the organization and operation of the vertebrate auditory system. No discussion about nocturnal predators would be complete without mentioning bats. Bats, which have also been widely studied, also rely on hearing for their predatory behavior. It is crucial, however, to point out that the basic strategy employed by owls to locate prey differs greatly from that employed by bats (which is discussed by Moss and Simmons (Chapter 11) and Haresign et al. (Chapter 12), this volume).

Bats actively search for prey by emitting ultrasonic pulses and processing information from the echoes that return after, for example, bouncing off insects. Bats rigidly control several parameters of their acoustical emissions. The time of the emission as well as parameters such as repetition rate, duration, and frequency characteristics of the emitted signals are under control of the central nervous system. Full 'knowledge' of the emissions allows the bats to optimize the processing of the resulting echoes. Control over when signals are produced enables bats to use temporal windows for gain control and ranging functions. Control over frequency components of the emissions enable bats to determine relative velocities and fine structure information. In contrast, the passive sound localization practiced by barn owls presents a different fundamental problem to its auditory system. Passive sound localization is characterized by the receiver (the owl) waiting silently until it detects sound made either intentionally or unintentionally by a potential prey. An example of an unintentional sound would be the noise produced when a small rodent walks over dry leaves. The owl has no control over when a prey item emits sounds, nor is there any control over the nature of the sound, i.e., its amplitude, duration, or frequency characteristics. There is also little or no information about where the sound might originate. To exploit these sounds the passive listener must be able to extract information such as the sound location whenever the brief, unpredictable sounds occur.

The study of sound localization in the barn owl has been greatly facilitated by the relative ease that sound localization capability can be assessed in the laboratory (Knudsen et al. 1979; Knudsen and Konishi 1979; Konishi 1973a; Konishi 1973b; Payne 1971). Owls localize prey acoustically, and then attempt to capture the prey. After localizing the prey, its capture consists of flying to the prey and striking it with outstretched talons. Payne measured the accuracy of owls to strike acoustic targets in the dark, as a function of the frequency components of the stimulus (Payne 1971). These elegant experiments were performed by measuring the imprints made by the talons of owls that struck a sheet of plasticine placed around a loudspeaker used to present the stimuli. These experiments proved conclusively that the localization of prey could be mediated solely by acoustic cues. Since the owls would strike at 'artificial' prey whose sounds were produced by electronic instruments that delivered sounds through loudspeakers, he was able to modify the sounds and see the effects on the owl's behavior. The results of delivering sounds comprising different frequencies indicated that frequencies between 5 kHz and 8.5 kHz were most important for sound localization. These results were pursued by Konishi, who demonstrated that wide-band noises (i.e., sounds comprising several frequencies, simultaneously) were localized more accurately than narrow-band tones (i.e., sounds containing few simultaneous frequencies) and that among pure tone stimuli, frequencies in the range of 7-8 kHz were localized most accurately (Konishi, 1973b). Further refinements in monitoring and measuring barn owl's sound localization capabilities exploited a readily observed aspect of the owl's behavior. In response to an 'interesting' sound, a perched barn owl will orient its head to face in the direction of that sound. The same head-orienting behavior can be observed when a flying owl hears an interesting sound. The practical effect of this behavior is to position the head so that any visible object or movement associated with the sound would be centered in the visual field and that subsequent sounds would be more apt to fall in more favorable areas of the acoustic axes. Such head movement is particularly important in owls because, unlike many other animals, their eyes are almost unmovable within the eye sockets, and their ears cannot move independently of the head. As a result, owls can not attend an acoustic stimulus by moving their ears (like a dog or cat) and can not perform a visual saccade to enable the eyes to foveate an image (as in humans). The movement of the head orients the owl's eyes and ears as a unit. The orientation behavior was cleverly exploited as an accurate indicator of sound localization accuracy by attaching position transducers to the head and monitoring head position in response to various auditory stimuli (Knudsen et al. 1979). In response to a brief free-field stimulus (i.e., sounds presented from a distant loudspeaker) the owl can orient its head to within 2-5° of the speaker's position. The orientation movements occur 'open-loop', or without auditory feedback. That is, the behavior is triggered by the sound and proceeds to completion even if the sound subsequently ceases. The owl determines the sound location when it first hears the sound and then moves to that location without requiring additional sound input. The delay between presentation of the sound and the initiation of the head turning behavior, also known as the behavioral latency, is approximately 60-100 msecs. Once the head begins to turn it takes no more than 100 msec to reach its desired position. Thus, the entire behavior is complete within 250 msec of the stimulus delivery. As details of the behavior were elucidated they begged the question of what acoustical cues were used by the animal to determine sound location, and what neural mechanisms were responsible for the analysis of these cues.

The acoustical cues sufficient to produce reliable sound localization are primarily interaural time differences and interaural level differences (Moiseff 1989a; Moiseff 1989b; Moiseff and Konishi 1981). To a first approximation, interaural time differences correlate with the relative azimuth of a sound source, whereas interaural level differences correlate with the relative elevation of a sound source. The exact relationship, however, is quite complicated; the two acoustical cues do not map onto orthogonal spatial axes and the importance of these cues may vary with the frequency components being analyzed (Brainard et al. 1992; Moiseff 1989b). The source of these binaural cues result from the structure and anatomical position of the owl's ears.

Both ears are spatially separate since they are located on either side of the animal's head. This results in time delays between a sound arriving at each ear that depend on the incident angle of the sound relative to the aural axis. A sound located on the medial-saggital plane reaches both ears at the same time producing an interaural time difference of 0 seconds. Maximum interaural time difference is produced by a sound 90° to the left, or right. Such a position requires that the sound travel from one ear, around the head to the other ear; this maximum interaural time delay is approximately 200 μ seconds for an adult barn owl. Specialized feathers comprising the barn owl's external ears impart directional reception for frequencies above approximately 3 kHz (Kühne and Lewis 1985; Payne 1971). The feathers produce a separate parabola-like receiver for each ear that has maximum gain along an acoustic axis and less gain as the sound location deviates from the axis. The directional axes of the two ears are displaced in elevation resulting in the left ear being more sensitive to sounds below the visual horizon and the right ear being more sensitive to sounds above the visual horizon. The interaural level difference, therefore, changes with the elevation of a sound source. Thus, the physical structure of the barn owl's external ears transforms the 2 dimensional location of a sound, relative to the head, into variations of the level and the time of arrival of a sound at each ear (Knudsen 1980; Kühne and Lewis 1985; Moiseff 1989a; Moiseff 1989b).

The behavioral significance of these binaural cues was inferred from experiments such as those that recorded the effects of plugging an owl's ear on its ability to localize sound (Knudsen and Konishi 1979). Plugging an ear resulted in predictable, systematic errors in localization. Plugging the left ear, for example, caused the owl to incorrectly locate the sound well above, and slightly to the right, of its actual position. Clearly, the owls were not using just a single ear to localize sounds, but made use of binaural information to determine location. Direct demonstration of the behavioral relevance of binaural cues was accomplished by fitting owls with custom-built stereo headphones and recording their orientation movements in response to dichotically presented stimuli (Moiseff 1989a; Moiseff and Konishi 1981). The headphones made it possible to independently control the stimuli presented to each ear. By delivering specific combinations of interaural time differences and interaural level differences it was possible to confirm that these two cues were sufficient for sound localization.

The functional importance of interaural time and level differences is associated with specializations in the peripheral auditory structures as well as in the auditory brainstem, pons, and midbrain. One of the most obvious features concerning organization of the auditory system is the segregation of neural pathways that process interaural time and interaural level cues. Auditory information organized tonotopically in nVIII (the auditory nerve) is passed to two anatomically distinct cochlear nuclei, *n. angularis* and *n. magnocellularis*. *N. angularis* is ideally suited to extract and transmit information about the intensity of a stimulus (Konishi et al. 1985; Sullivan 1985; Sullivan and Konishi 1984); *N. magnocellularis* extracts and transmits temporal information, in particular the phase of the signals component frequencies (Konishi et al. 1985; Sullivan 1985; Sullivan and Konishi 1984). The functional roles of these two cochlear nuclei were inferred by quantitative analysis of the pattern of action potentials from neurons in these areas in response to tonal stimuli. In *n. angularis*, action potential frequency varied systematically over a 10-30 dB range of stimulus intensity; stimulus intensity was encoded by action potential firing rate. *N. magnocellularis* neurons did not vary their firing frequency systematically over a wide range of stimulus intensity and thus were not good intensity coders. However, the time of arrival of *n. magnocellularis* neurons' action potentials were strongly correlated with the phase of the tonal stimulus; these neurons encoded stimulus timing information. Unlike the cochlear nuclei of most vertebrates and mammals, whose ability to encode stimulus phase is limited to stimuli with frequencies below 1.5-3 kHz, units in *n. magnocellularis* of the barn owl reliably encode phase information of stimuli as high as 8-9 kHz.

The monaural areas project, in turn, to separate nuclei that determine level differences between the two ears (nVLVp), or time differences between the two ears (*n. Laminaris*) (Carr and Konishi 1990; Manley et al. 1988; Moiseff and Konishi 1983; Takahashi and Keller 1992; Takahashi and Konishi 1988a; Takahashi and Konishi 1988b; Takahashi et al. 1989).

Both VLVp and *n. Laminaris* are organized along at least two auditory dimensions. One dimension, stimulus frequency, is evident in the tonotopic organization of both areas. The other dimension is specific to the level- or time-difference processing pathway. VLVp is systematically organized into a map of interaural level differences (Manley et al. 1988; Mogdans and Knudsen 1994a). *N. laminaris* systematically transforms interaural time differences (or more precisely, interaural phase differences) into a topological map (Carr 1993; Carr and Konishi 1990). The time-processing pathway then progresses to n. VLVa.

The separate pathways responsible for processing interaural time and level differences converge in the owl's inferior colliculus. Within the central nucleus of the inferior colliculus (ICc) neurons are found with response properties ranging from strictly monaural sensitivity, to varying degrees of interaural time or level sensitivity, as well as to both interaural time and level difference sensitivity. The ultimate transformation is evident in the external nucleus of the inferior colliculus (ICx). Neurons in ICx are characterized highly restricted spatial receptive fields that are organized in a map-like manner (Knudsen 1981; Knudsen 1984; Knudsen and Konishi 1978; Konishi 1984; Konishi 1986; Konishi and Knudsen 1982). An interesting aspect of the ICx neurons is that they respond to a broad range of frequencies (unlike neurons in the preceding stages of the auditory pathway.) By combining frequencies over a wide range the neurons can unambiguously encode discrete spatial locations; neurons that are sharply tuned to frequencies cannot do this, since the phase of a single frequency is a repeating, cyclic function (Brainard et al. 1992; Takahashi and Konishi 1986; Wagner et al. 1987).

Computational Neural Map

Knudsen and Konishi demonstrated a neural correlate of sound localization in the barn owl's midbrain auditory nucleus which is analogous to the mammalian inferior colliculus, specifically (Knudsen 1983; Knudsen and Konishi 1978). They found that the spatial position of receptive fields of auditory neurons in the external nucleus of the inferior colliculus (ICx) varied systematically with anatomical position. If they recorded the activity of an arbitrary unit in ICx they could define a distinct location from which sounds would stimulate the neuron. If they lowered their microelectrode further into ICx they found that successive neurons were selective for sounds at successively lower locations. If they shifted the electrode along the surface of ICx the receptive fields shifted accordingly. The existence of a neural map of auditory space was remarkable in that it was derived through neural computation

(Konishi 1984; Konishi 1986; Konishi and Knudsen 1982). Sound location does not exist at a single ear, rather the location is derived by comparing interaural time and interaural level differences and mapping those differences into spatial coordinates; a process that is tuned during a developmental critical period (Knudsen 1988; Knudsen 1989; Knudsen et al. 1984; Knudsen and Knudsen 1985; Knudsen and Knudsen 1986; Knudsen and Knudsen 1989a; Knudsen and Knudsen 1989b; Knudsen and Knudsen 1990). Examples of computationally derived maps of acoustical information are not unique to the barn owl. For example, the auditory cortex of the mustache bat contains many specialized areas where specific acoustical parameters are organized systematically (Suga 1982; Suga 1990). The parameters correlate with complex aspects of echolocation behavior such as determining relative target velocity (Doppler shift), target distance (echo delay) and the identity of the target (amplitude and frequency modulation).

The neurophysiological analysis of computation maps provides a powerful system to understand fundamental aspects of sensory processing. It is clear, for example, that the anatomical substrate for the necessary transformations must exist in the brain. Given these anatomical substrates, the brain must 'calibrate' itself to perform the necessary computations and have a functionally important result. In addition, calibration allows the nervous system to 'customize' its calculations for the sensory environment of each particular individual. The calibration of the sensory systems probably takes the form of supervised learning (Knudsen 1994).

In the case of the neural circuitry responsible for sound localization in the barn owl, Knudsen and colleagues demonstrated that young barn owls 'learn' the correspondence between the binaurally received sounds of an object and the object's location in 2 dimensional space (up-down, right, left). They have presented evidence localizing the anatomical sites of plasticity in the owl auditory system (Brainard and Knudsen 1993; Mogdans and Knudsen 1994b). In addition, they have demonstrated that the visual system provides the reference frame for calibrating and interpreting the auditory system's computational map (Knudsen and Knudsen 1989a). The importance of calibrating the auditory system to a spatial reference frame determined by vision is apparent when following the maturation of an owl.

What Does a Baby Barn Owl Hear?

A baby barn owl undergoes rapid growth during the first 60 days of life (Haresign and Moiseff 1988). It is interesting to look at the growth of structures that directly affect binaural sound reception. Since interaural time differences are a function of the diameter of the head, the maximum interaural time difference that an animal is subjected to, under normal conditions, is directly proportional to the head diameter. As head diameter changes, the relationship between the relative azimuth of a sound, and the resulting interaural time difference changes. Interaural level differences are produced by the differential directional characteristics of the two ears. The directional characteristics are determined by the shape, orientation, and size of the specialized feathers comprising the facial ruff. These two morphological features, head diameter and facial ruff structure, develop separately and independently during the first two months (Figure 13.1). The baby's head diameter increases dramatically from 25 mm at 10 days, to 45 mm at 30 days. The diameter change, during this time is nearly linear and varies at a rate of 1.25 mm day^{-1}. Approximately 1 week after reaching a stable head diameter, the barn owl's facial ruff feathers begin to develop their characteristic features. An indication of the growth of the ruff feathers can be obtained by multiplying the feather's length by the feather's maximum

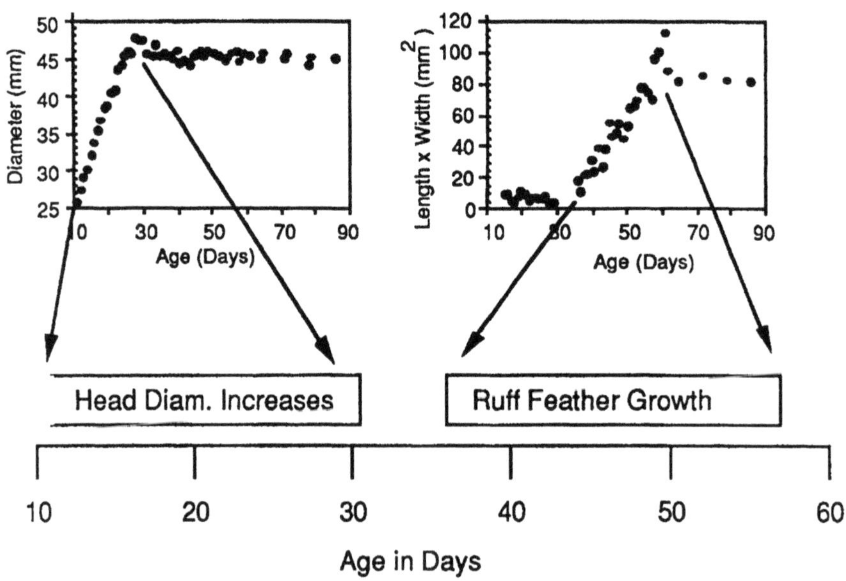

FIGURE 13.1 Head and feather growth after hatching. Top Left: Average head diameter increases during the first 30 days. Top Right: Average length x width of facial ruff feathers located above, below and behind the ear canal. This is an indication of feather area. Ruff feathers begin to grow after the head has reached its maximum diameter. For both plots, samples start at day 10, the day at which the barn owl chicks were removed from their nest. (Redrawn from Haresign and Moiseff 1988.)

width (LxW); this is proportional to the feather's area. The facial ruff develops over a period of 20 days during which time the feather's LxW increase from almost 0 to nearly 80 mm^2, a linear increase of 3.30 mm^2 day^{-1}. When the owl is 60 days old, the facial ruff feathers have reached mature size and show no further growth. Before speculating upon the effects of these morphological changes on the binaural auditory experience it was necessary to demonstrate that the owl's auditory system was functional during this time.

General brain activity is easily recorded by placing electrodes on the surface of the head and recording the output through high-gain amplifiers. This is not unlike the recording of an electrocardiogram with electrodes attached to the surface of a person's chest. Because of the complexity of the brain's activity, however, some elementary signal processing techniques are required to identify specific classes of brain activity. By repetitively stimulating the animal with specific stimuli it is possible to repeatedly evoke the brain's response to the stimuli. The responses to each of the stimuli are averaged over the many repetitions. After averaging, it is possible to detect the response evoked by the stimulus over the non-specific activity, thus deriving an 'evoked' potential. By averaging the brain's responses to selected auditory stimuli the auditory brainstem evoked potential can be obtained. Auditory brainstem evoked potentials provided us with a simple, noninvasive technique of monitoring auditory function of owls during posthatch development (Figure 13.2) (Calvo and Moiseff 1993, 1993; Haresign 1990).

One of two differential electrodes was placed on the top of the owl's head; the second electrode was placed by the right ear canal. A third, ground electrode was placed by the left ear canal. The owls were stimulated by a brief, 1 msec click presented from a speaker 1m distant. The signals were averaged by computer and we obtained consistent auditory brainstem responses (ABR). We found that the baby owl's auditory system was functional at 4 days old, which was the earliest age that we tested. This was shown by the existence of significant peaks in the ABR. However, there were several differences between-the-ABR of an 18 day old owl chick and the same bird as an adult, 115-day-old barn owl. Most notable were the lack of multiple, well differentiated peaks in the early 8 milliseconds of the response to clicks and the lack of a well defined long latency response, in the 18-day-old's ABR. The transition between the 'young' and adult forms of the ABR occurred over the first 60-70 days of age. It is unclear, at this time, what the correlates of these changes are, however, the ABR may be a useful technique to follow physiological and morphological changes. For the present time,

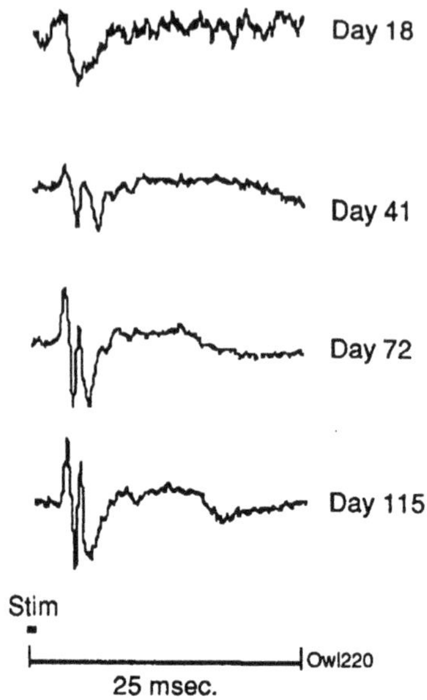

FIGURE 13.2 Auditory brainstem evoked potential in response to 1 msec click. All records were obtained from the same individual at day 18, 41, 72 and 115. Each record was at an intensity 20 dB above threshold, as detectable with the ABR. Each record is the average of 40 trials.

however, it is sufficient to limit our interpretation of the ABR results to its demonstration of the functionality of the baby owl's auditory system.

During the first 60-70 days after hatching, the growth of the head and feathers should present varying binaural cues to the owl. As the head diameter increases, the maximum interaural time differences experienced by the young owl nearly triples from an estimated 60 μsec at day 10 to 180 μsec at day 30. The development of the facial ruff from days 35-60 is accompanied by a monaural gain increase of 20 dB and the increase of the directional characteristics of both ears. As a result, the morphological substrate necessary for generating interaural level differences based on sound location also develops during this time.

One can only speculate on the importance of having the two morphological structures responsible for the two major binaural cues for sound localization develop separately and independently. It is worth further analysis to determine whether this temporally separate morphological development is correlated with separate developmental schedules for pathways in the barn owl auditory system, notably the development of the time-sensitive pathway and the anatomically separate level-sensitive pathway. Some insight into the functional implications of such a scheme may be obtained through an experiment that suggests itself directly from the growth schedule.

Since the growth of the head (varying the interaural time differences) occurs separately from the growth of the facial ruff (varying the interaural level differences) it is a simple matter to experimentally alter the growth of the facial ruff without affecting the variation in head growth. This manipulation would enable us to determine the effect of ruff deprivation on an otherwise normally developing auditory system (Haresign 1990; Knudsen et al. 1994). The ruff deprivation procedure was trivial to accomplish. During the time of feather growth (day 35+) we inspected the area that would become the facial ruff daily; all feathers in this area were cut with scissors to keep them less than 5 mm in length. At approximately 9 months age, these 'ruff deprived' owls were tested to assess the effect of this procedure on their ability to localize sound.

Behavioral Testing

As mentioned above, one manifestation of sound localization behavior is the tendency of owls to orient their heads towards the source of a sound. This biologically important behavior has proven to be very convenient to investigators studying sound localization. The position of the head can be monitored by various techniques (e.g. mirrors, light beam projectors, head coils) and it is possible to describe the head turning behavior and assess its accuracy.

A simple technique was used to measure the ability of an adult owl to localize a free-field sound source. Four loudspeakers, 1m. in front of an owl were located at the four corners of an imaginary square (Figure 13.3). The apparatus was located in a light-tight, sound attenuating, anechoic room that was solely illuminated with infrared light. The owl was presented with the task of orienting its head towards the speaker that generated a brief (<100msec) broad-band stimulus. The behavior occurred open-loop; that is, the acoustical stimulation ceased before the orientation behavior was completed. Note that the positions of the speakers require that the owl determine both the azimuth and elevation of the sound source to accurately locate them. A simple scoring system was used to characterize the azimuthal and elevational orientation. For each stimulus, we scored whether the owl discriminated the speaker position to the left or right, and up or down. Comparing the response to the speaker's true position we scored the response as *correct* (oriented left for a speaker on the left; oriented up for a high speaker, and visa versa), *incorrect* (e.g. oriented left for a speaker on the right;

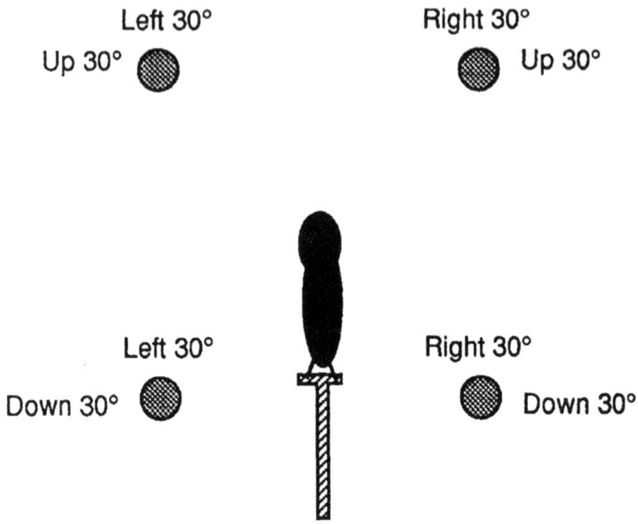

FIGURE 13.3 Diagram of behavioral orientation task. The owl was in a dark, sound and echo resistant room, 1 m. from the loudspeakers positioned as shown.

oriented up for a low speaker), or *no response*. A no response in one dimension was scored, only if the owl showed a response to the speaker's sound (correct or incorrect) in the other dimension. Each stimulus was selected randomly by computer. Although the computer recorded the location of the stimulus, this information was not made available to the person scoring the response. This insured that all scoring was done blindly.

Even this simple scoring system showed significant differences between the free-field localization capability of normal, and ruff-deprived owls (Figure 13.4). Normal owls correctly determined whether the speaker was high or low more than 75% of the time; they also correctly lateralized (left or right) the speaker more than 80% of the time. Ruff deprived owls correctly lateralized the speaker more than 79% of the time but showed a severe deficit in their ability to discriminate between a high, or low, speaker position. Thus, the lack of a facial ruff impacted the ability to determine elevation, but not the ability to determine azimuth.

These results, though significant, can be easily misinterpreted. They clearly indicate that the ruff-deprived owl has difficulty determining the elevation of a sound, but do the results reveal anything about the central nervous system? If a person places an effective earplug in one ear, he or she will have difficulty localizing sounds. The difficulty is not indicative of a reorganized central nervous system, but may merely in-

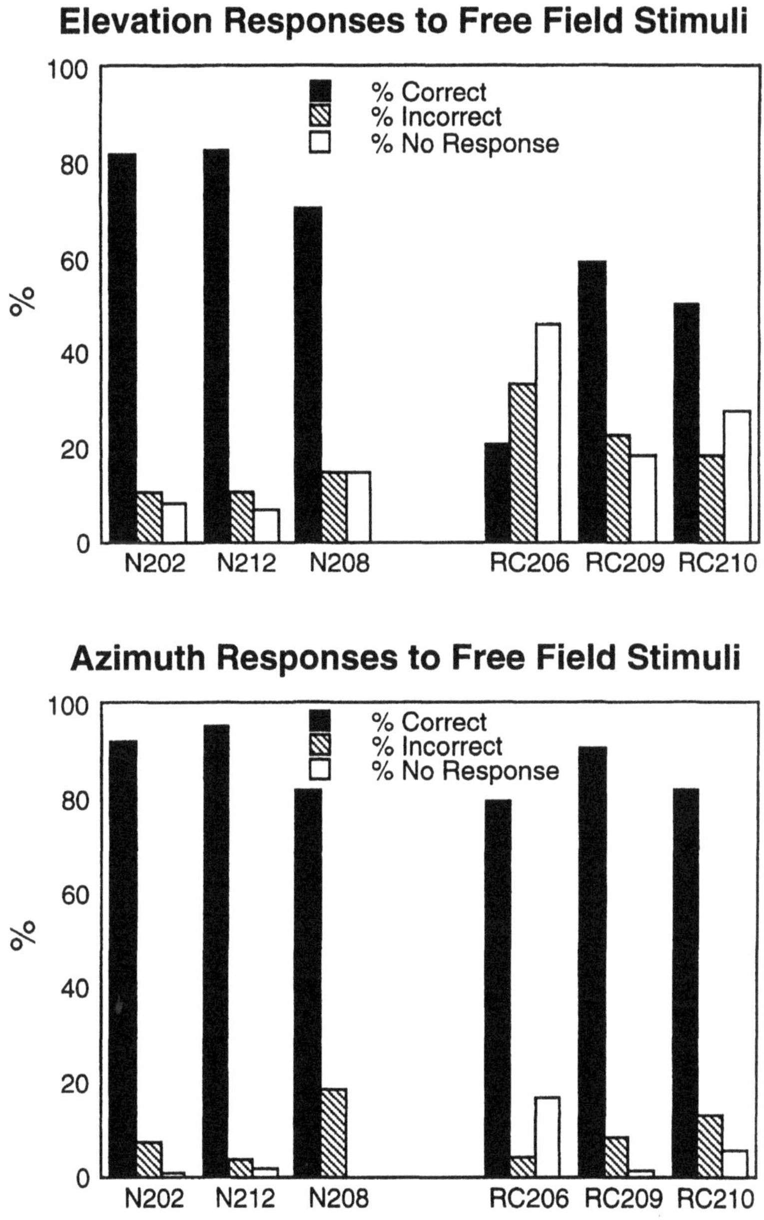

FIGURE 13.4 Accuracy of orientation behavior to free field stimulation. Normal birds (N202, N212, N208) correctly determined the relative elevation (up or down) and relative azimuth (left or right) more than 75% of the time. Birds grown without facial ruffs (RC206, RC209, RC210) correctly lateralized stimuli (left or right) more than 79% of the trials, but the were unable to determine the relative elevation (up or down) better than by chance.

dicate that the person is not receiving the binaural cues that their nervous system needs to correctly localize the sound. In the case of our owl study, we could address the problem of determining whether our results revealed changes in the organization of the auditory system or were the result of not providing the necessary cues to a normally developed auditory system. The technique we used relied on the stereo headphones mentioned earlier in this chapter.

First, miniature microphones were placed into the ear canals of a normal adult barn owl. The output of both microphones was recorded and digitized in response to stimulation by a free-field loudspeaker located in positions that included those used in the preceding experiments. We were thus able to create a digitized database of the sounds existing in the ear canal of a normal adult owl from a speaker at various free-field locations. Then, each of the ear-canal signals could be replayed through miniature earphones implanted into a ruff deprived owl's ear canals. This allowed us to present the owls with 'normal' binaural cues, and determine whether their brains were capable of interpreting this information orienting towards the direction where the speaker was originally located (Figure 13.5). Each of three ruff deprived owl's successfully determined the whether the earphone sounds were derived from the left or the right position speakers in excess of 75% of the time. Two of the ruff deprived owls showed severe deficit determining the relative elevation of the stimulus, making the appropriate up-down orientation in only about 50% of the trials. One of the ruff deprived birds was able to determine the relative elevation of the earphone stimulus over 80% of the time, though it did not show comparable ability in response to a free-field stimulus. These results suggest that ruff deprivation could, in some cases, alter the central auditory system.

Summary

The owl is an excellent system for studying the neural basis of a nontrivial behavioral task-- sound localization. Since sound localization involves sensory and motor systems, development and learning, it provides a behavioral substrate that can be used to analyze many aspects of neural function. As discussed in this chapter, the barn owl auditory system is optimized to detect and localize prey using the sounds the prey produce. Although passive sound localization is highly adaptive for owls, it is a task performed by most animals. The lessons learned from studying the barn owl auditory system are therefore not limited to an esoteric understanding of owls, but are generalizable to understanding auditory function in other animals including mammals. The infor-

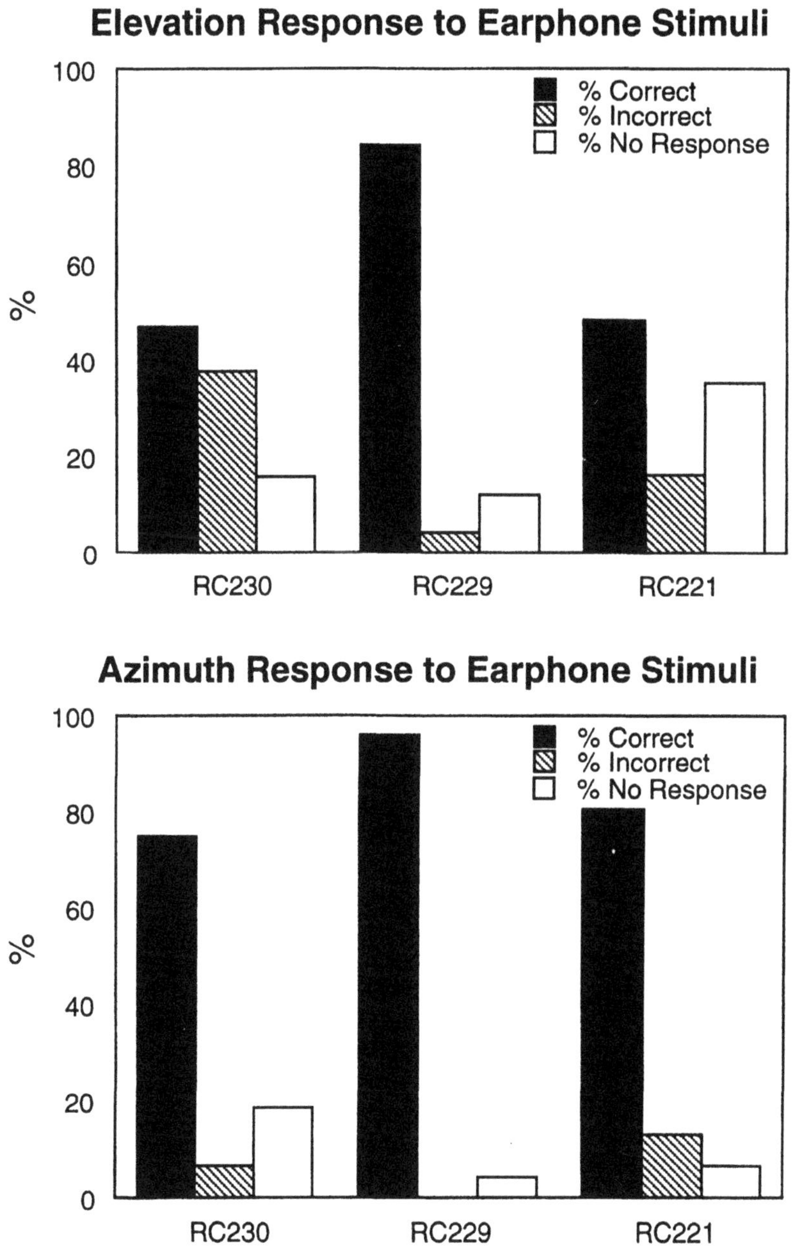

FIGURE 13.5 Response of ruff deprived birds to dichotic stimulation. The ruff deprived birds lateralized the dichotic stimulus more than 75% of the time. One bird (RC229) could accurately determine the relative elevation (up or down) but the remaining two birds were unable to do better than chance. See text for description of the dichotic stimulation technique.

mation also forms one link in our understanding of general rules for structure and organization in nervous systems. Segregation of time and intensity processing pathways is not limited to owls, nor is it restricted to the analysis of auditory signals. Electric fish, for example, rely on a similar separation of time and intensity processing pathways for the analysis of their weakly electric signals and their ability to resist signal jamming by other electric fish (Carr et al. 1986; Heiligenberg 1987; Heiligenberg 1990).

The existence and formation of computationally derived auditory maps is also non-unique to the owl, and thus of general importance. The auditory cortex of the mustache bat, for example, is organized into multiple computational maps having behavioral relevance (Suga 1990). At the present time one can only speculate about the processes responsible for the development of these complex maps. Continued studies of these various aspects central nervous system structure, development and function, have great potential to provide fundamental insights towards an understanding of sensory system organization.

Acknowledgments

This work was supported by NICD DC00277 HAR to A.M.

References

Brainard MS, Knudsen EI (1993) Experience dependent plasticity in the inferior colliculus: A site for visual calibration of the neural representation of auditory space in the barn owl. *J. Neurosci. 13*: 4589-4608

Brainard MS, Knudsen EI, Esterly SD (1992) Neural derivation of sound source location: resolution of spatial ambiguities in binaural cues. *J. Acoust. Soc. Am. 91*: 1015-1027

Calvo O, Moiseff A (1992) Neural correlates of interaural time processing in the auditory potentials of the barn owl. *Soc. Neurosci. Abstr. 18*: 149.

Calvo O, Moiseff A (1993) Monitoring of the binaural interaction in the immature barn owl. *Soc. Neurosci. Abstr. 19*: 531.

Carr CE (1993) Processing of temporal information the brain. *Ann. Rev. Neurosci. 16*: 223-243

Carr CE, Heiligenberg W, Rose GJ (1986) A Time-Comparison Circuit in the Electric Fish Midbrain. I. Behavior and Physiology. *J. Neurosci. 6*: 107-119

Carr CE, Konishi M (1990) A circuit for the detection of interaural time differences in the brain stem of the barn owl. *J. Neurosci. 10*: 3227-3246

Haresign T (1990) Ruff deprivation and intensity processing in the barn owl *Tyto alba*. Doctoral Dissertation. University of Connecticut.

Haresign T, Moiseff A (1988) Early growth and development of the Common Barn-Owl's facial ruff. *Auk 105*: 699-705

Heiligenberg W (1987) Central processing of sensory information in electric fish. *J. Comp. Physiol. A 161:* 621-631

Heiligenberg W (1990) Electrosensory systems in fish. *Synapse 6:* 196-206

Knudsen EI (1980) Sound localization in birds. *Comparative Studies of Hearing in Vertebrates.* New-York, Heidelberg, Berlin, Springer-Verlag. 289-322.

Knudsen EI (1981) The Hearing of the Barn Owl. *Scientific American 113-125:* 113-125

Knudsen EI (1983) Subdivisions of the inferior colliculus in the barn owl *Tyto alba. J. Comp. Neurol. 218:* 174-186

Knudsen EI (1984) Synthesis of a neural map of auditory space in the owl. In: *Dynamic Aspects of Neocortical Function.* Chapter 11: 375-396

Knudsen EI (1988) Early blindness results in a degraded auditory map of space in the optic tectum of the barn owl. *Proc. Nat. Acad. Sci. (USA) 85:* 6211-6214

Knudsen EI (1989) Fused binocular vision is required for development of proper eye alignment in barn owls. *Vis. Neurosci. 2:* 35-40

Knudsen EI (1994) Supervised learning in the brain. *J. Neurosci. 14:* 3985-3997

Knudsen EI, Blasdel GG, Konishi M (1979) Sound localization by the barn owl (*Tyto alba*) measured with the search coil technique. *J. Comp. Physiol. 133:* 1-11

Knudsen EI, Esterly SD, Knudsen PF (1984) Monaural occlusion alters sound localization during a sensitive period in the barn owl. *J. Neurosci. 4:* 1001-1011

Knudsen EI, Esterly SD, Olsen JF (1994) Adaptive plasticity of the auditory space map in the optic tectum of adult and baby barn owls in response to external ear modification. *J. Neurophysiol 71: 79-94*

Knudsen EI, Knudsen PF (1985) Vision Guides the Adjustment of Auditory Localization in Young Barn Owls. *Science 230:* 545-548

Knudsen EI, Knudsen PF (1986) The Sensitive Period for Auditory Localization in Barn Owls is Limited by Age, Not by Experience. *J. Neurosci. 6:* 1918-1924

Knudsen EI, Knudsen PF (1989a) Vision calibrates sound localization in developing barn owls. *J. Neurosci. 9:* 3306-3313

Knudsen EI, Knudsen PF (1989b) Visuomotor adaptation to displacing prisms by adult and baby barn owls. *J. Neurosci. 9:* 3297-3305

Knudsen EI, Knudsen PF (1990) Sensitive and critical periods for visual calibration of sound localization by barn owls. *J. Neurosci. 10:* 222-232

Knudsen EI, Konishi M (1978) A neural map of auditory space in the owl. *Science 200: 795-797*

Knudsen EI, Konishi M (1979) Mechanisms of sound localization in the barn owl (*Tyto alba*). *J. Comp. Physiol. 133:* 13-21

Konishi M (1973a) How the owl tracks its prey. *Amer. Scientist 61:* 414-424:

Konishi M (1973b) Locatable and nonlocatable acoustic signals for barn owls. *Amer. Naturalist. 107: 775-785*

Konishi M (1984) Spatial receptive fields in the auditory system. In: *Comparative Physiology of Sensory Systems.* Chapter 1.5: 103-113:

Konishi M (1986) Centrally synthesized maps of sensory space. *TINS April:* 163-168

Konishi M, Knudsen EI (1982) A Theory of Neural Auditory Space- Auditory Representation in the Owl and Its Significance. *Cortical Sensory Organization 3:* 219-229

Konishi M, Sullivan WE, Takahashi T (1985) The owl's cochlear nuclei process different sound localization. *Journal of the Acoustical Society of America 78:* 360-364:

Kühne R, Lewis B (1985) External and middle ears. *Form and Function in Birds, Vol. 3.* London, Academic Press. 227-271.

Manley GA, Köppl C, Konishi M (1988) A neural map of interaural intensity differences in the brain stem of the barn owl. *J. Neurosci. 8:* 2665-2676

Mogdans J, Knudsen EI (1994a) Representation of interaural level difference in VLVp, the first site of binaural comparison in the barn owls auditory system. *Hearing. Res. 74:* 148-164

Mogdans J, Knudsen EI (1994b) Site of auditory plasticity in the brain stem (VLVp) of the owl revealed by early monaural occlusion. *J. Neurophysiol. 72:* 2875-2891

Moiseff A (1989a) Bi-coordinate sound localization by the barn owl. *J. Comp. Physiol. A 164:* 637-644

Moiseff A (1989b) Binaural disparity cues available to the barn owl for sound localization. *J. Comp. Physiol. A 164:* 629-636

Moiseff A, Konishi M (1981) Neuronal and behavioral sensitivity to binaural time differences in the owl. *J. Neurosci. 1:* 40-48

Moiseff A, Konishi M (1983) Binaural characteristics of units in the owl's brainstem auditory pathway: precursors of restricted receptive fields. *J. Neurosci. 3:* 2553-2562

Payne R, Drury W (1958) *Tyto alba*, Part II. *Natural History 67:* 316-323

Payne RS (1971) Acoustic localization of prey by barn owls (*Tyto alba*). *J. Exp. Biol. 54:* 535-573

Suga N (1982) Functional organization of the auditory cortex. Representation beyond tonotopy in the bat. In: *Cortical sensory organization 3:* 157-218:

Suga N (1990) Cortical computational maps for auditory imaging. *Neural Networks 3:* 3-21

Sullivan WE (1985) Classification of response patterns in cochlear nucleus of barn owl: correlation with functional response properties. *J. Neurophysiol. 53:* 201-216

Sullivan WE, Konishi M (1984) Segregation of stimulus phase and intensity coding in the cochlear nucleus of the barn owl. *J. Neurosci. 4:* 1787-1799

Takahashi TT, Keller CH (1992) Comissural connections mediate inhibition for the computation of interaural level difference in the barn owl. *J. Comp. Physiol. 170:* 161-169

Takahashi TT, Konishi M (1986) Selectivity for interaural time difference in the owl's midbrain. *J. Neurosci. 6:* 3413-3422

Takahashi TT, Konishi M (1988a) Projections of nucleus angularis and nucleus laminaris to the lateral lemniscal nuclear complex of the barn owl. *J. Comp. Neurol. 274:* 212-238

Takahashi TT, Konishi M (1988b) Projections of the cochlear nuclei and nucleus laminaris to the inferior colliculus of the barn owl. *J. Comp. Neurol. 274:* 190-211

Takahashi TT, Konishi M, Wagner H (1989) Role of commissural projections in the representation of bilateral auditory space in the barn owls inferior colliculus. *J. Comp. Neurol. 281:* 545-554

Wagner H, Takahashi T, Konishi M (1987) Representation of interaural time difference in the central nucleus of the barn owl's inferior colliculus. *J. Neurosci. 7:* 3105-3116

Contributors

Johan J. Bolhuis, Department of Zoology, University of Cambridge
Susan D. Brown, Department of Psychology, University of Maryland
Rebecca C. Buxbaum, Department of Psychology, Brown University
Ruth M. Colwill, Department of Psychology, Brown University
Jeffrey Cynx, Field Research Center, Rockefeller University
Robert J. Dooling, Department of Psychology, University of Maryland
Russell D. Fernald, Department of Psychology, Stanford University
Michael Ferragamo, Department of Neuroscience, Brown University
James L. Gould, Ecology and Evolutionary Biology, Princeton University
Tim Haresign, Department of Neuroscience, Brown University
Susan D. Healy, Department of Psychology, University of Newcastle
Kazuchika Manabe, Department of Psychology, University of Maryland
Andrew Moiseff, Department of Physiology & Neurobiology, University of Connecticut
Cynthia F. Moss, Department of Psychology & Program in Neuroscience, Harvard University
Elizabeth F. Powell, Department of Psychology, University of Maryland
Sara J. Shettleworth, Department of Psychology, University of Toronto
Andrea M. Simmons, Department of Psychology, Brown University
James A. Simmons, Department of Neuroscience, Brown University
Jianxin Wang, Department of Physiology & Neurobiology, University of Connecticut
Daniel M. Weary, Department of Biology, McGill University
Janine M. Wotton, Department of Psychology, Brown University

Index

About the Book and Editors

How do bats catch insects in the dark? How do bees learn which flowers to visit? How do food-storing birds remember where their hoards are? Questions like these are addressed by neuroethology, the branch of behavioral neuroscience concerned with analyzing the neural bases of naturally occurring behaviors.

This book brings together thirteen chapters presenting findings on perceptual and cognitive processes in some of the most active areas of neuroethological research, including auditory localization by bats and owls, song perception and learning in birds, pitch processing by frogs and toads, imprinting in birds, spatial memory in birds, learning in bees and in *Aplysia,* and electroreception in fish. A variety of approaches are represented, such as field studies, psychophysical tests, electrophysiological experiments, lesion studies, comparative neuroanatomy, and studies of development.

Each chapter gives an up-to-date overview of a particular author's research and places it within the broader context of issues about animal perception and cognition. The book as a whole exemplifies how studying species and their particular specializations can inform general issues in psychology, ethology, and neuroscience.

Cynthia F. Moss is associate professor of psychology at the University of Maryland at College Park. **Sara J. Shettleworth** is professor of psychology at the University of Toronto.

For Product Safety Concerns and Information please contact our EU representative GPSR@taylorandfrancis.com
Taylor & Francis Verlag GmbH, Kaufingerstraße 24, 80331 München, Germany

www.ingramcontent.com/pod-product-compliance
Ingram Content Group UK Ltd.
Pitfield, Milton Keynes, MK11 3LW, UK
UKHW022206190926
475702UK00003B/225
9780367317003